# Ready to Haul, Ready to Fight

## U.S. Navy, Royal Australian Navy, and British Merchant Navy Cargo Ships in the Pacific in World War II

U.S. Navy cargo ships, among the most unglamorous military vessels, kept the supply lines running through the incredible expanses of the Pacific battle zones in World War II. This involved shuttling cargos of gasoline, explosives, and supplies between forward bases on an erratic, unpredictable war-time schedule. The tedious days of slow cruising were broken by an occasional enemy air raid in some atoll harbor, and the rugged work of loading and unloading cargo. Although some cargo ships exhibited the informality of tramp steamers, they got results. Cargo ships able to carry amphibious landing craft routinely steamed with other assault forces into enemy-held beachheads, and disembarked supplies and personnel under fire. A dozen or so Royal Australian Navy stores-issuing ships lived a perilous existence plying dangerous Japanese-patrolled northern Australian waters, and the coast off Papua New Guinea. In 1945, when the British Pacific Fleet joined Allied combat operations against Japan, they brought their own "fleet train." This Logistics Service Force was the most extraordinary, motley collection of shipping ever assembled in British maritime history–one that included, presumably for fleet morale, a floating brewery. One hundred sixty-eight photographs, maps and diagrams; appendices; a bibliography; and an index to full names, places, and subjects add value to this work.

To the crews of Navy and Merchant Navy cargo ships in the Pacific, the lifelines of Allied military units afloat and ashore, who supported, shared common dangers, and fought alongside cruisers, destroyers, and amphibious forces on the front lines.

# Ready to Haul, Ready to Fight

## U.S. Navy, Royal Australian Navy, and British Merchant Navy Cargo Ships in the Pacific in World War II

Cdr. David D. Bruhn, USN (Retired)

HERITAGE BOOKS
2021

**HERITAGE BOOKS**
***AN IMPRINT OF HERITAGE BOOKS, INC.***

**Books, CDs, and more—Worldwide**

For our listing of thousands of titles see our website
at
www.HeritageBooks.com

Published 2021 by
HERITAGE BOOKS, INC.
Publishing Division
5810 Ruatan Street
Berwyn Heights, Md. 20740



International Standard Book Number
Paperbound: 978-1-55613-820-1

Heritage Books by Cdr. David D. Bruhn, USN (Retired)

*Battle Stars for the "Cactus Navy":*
*America's Fishing Vessels and Yachts in World War II*

*Enemy Waters:*
*Royal Navy, Royal Canadian Navy, Royal Norwegian Navy,*
*U.S. Navy, and Other Allied Mine Forces Battling the*
*Germans and Italians in World War II*
Cdr. David D. Bruhn, USN (Retired) and Lt. Cdr. Rob Hoole, RN (Retired)

*Eyes of the Fleet:*
*The U.S. Navy's Seaplane Tenders and Patrol Aircraft in World War II*

*Gators Offshore and Upriver:*
*The U.S. Navy's Amphibious Ships and Underwater Demolition Teams,*
*and Royal Australian Navy Clearance Divers in Vietnam*

*Guns Up:*
*Naval Action in the Yellow Sea off Korea, 1950–1953*

*Home Waters:*
*Royal Navy, Royal Canadian Navy, and U.S. Navy*
*Mine Forces Battling U-Boats in World War I*
Cdr. David D. Bruhn, USN (Retired) and Lt. Cdr. Rob Hoole, RN (Retired)

*Ingram's Fourth Fleet:*
*U.S. and Royal Navy Operations Against German Runners,*
*Raiders, and Submarines in the South Atlantic in World War II*

*MacArthur and Halsey's "Pacific Island Hoppers":*
*The Forgotten Fleet of World War II*

*Nightraiders:*
*U.S. Navy, Royal Navy, Royal Australian Navy, and*
*Royal Netherlands Navy Mine Forces Battling the*
*Japanese in the Pacific in World War II*
Cdr. David D. Bruhn, USN (Retired) and Lt. Cdr. Rob Hoole, RN (Retired)

*On the Gunline:*
*U.S. Navy and Royal Australian Navy Warships off Vietnam, 1965–1973*
Cdr. David D. Bruhn, USN (Retired) and
STGCS Richard S. Mathews, USN (Retired)

*Ready to Haul, Ready to Fight:*
*U.S. Navy, Royal Australian Navy, and*
*British Merchant Navy Cargo Ships*
*in the Pacific in World War II*

*Salvation from the Sky: U.S. Navy, Royal Australian Air Force, and*
*Royal New Zealand Air Force Heroic Air-Sea Rescue in the Pacific in World War II*
Cdr. David D. Bruhn, USN (Retired) and Stephen Ekholm

*Support for the Fleet:*
*U.S. Navy and Royal Australian Navy Service*
*Force Ships That Served in Vietnam, 1965–1973*

*We Are Sinking, Send Help!:*
*The U.S. Navy's Tugs and Salvage Ships in the African,*
*European, and Mediterranean Theaters in World War II*

*Turn into the Wind*
*Volume I: US Navy and Royal Navy Light Fleet Aircraft Carriers in World War II, and Contributions of the British Pacific Fleet*

*Turn into the Wind*
*Volume II: US Navy, Royal Navy, Royal Australian Navy, and Royal Canadian Navy Light Fleet Aircraft Carriers in the Korean War and through End of Service, 1950–1982*

*Wooden Ships and Iron Men:*
*The U.S. Navy's Ocean Minesweepers, 1953–1994*

*Wooden Ships and Iron Men:*
*The U.S. Navy's Coastal and Motor Minesweepers, 1941–1953*

*Wooden Ships and Iron Men:*
*The U.S. Navy's Coastal and Inshore Minesweepers, and the Minecraft that Served in Vietnam, 1953–1976*

# Contents

## Photos and Illustrations

# Maps and Diagrams

Content Photo-1

Attack cargo ship USS *Alchiba* (AKA-6)'s mascot "Bosun' McGee" with crewmembers at the gangway at Mare Island Navy Yard, 7 August 1943.
Navy Yard Mare Island photograph #5728-4

# Foreword

In *Ready to Haul, Ready to Fight*, David Bruhn has again produced a fascinating book which throws light on a subject which, for far too long, has tended to be ignored in the annals of naval history. As David points out, 'U.S. Navy cargo ships, among the most unglamorous military vessels, kept the supply lines running through the incredible expanses of the Pacific battle zones in World War II.' Without the support offered by these 'second tier' ships, the fighting ships could never have successfully completed their role. The majority went in harm's way while supporting the fleet and combat forces ashore.

Another aspect drawn out in the book is the incredible distances involved. The landings and naval battles at Leyte Gulf in October 1944, for example, demonstrated the utility of maritime forces in power projection operations. Amphibious ships moved troops 500 nautical miles to landing beaches. Logistics ships moved vital stores, ammunition and rations, directly and indirectly, over 5,000 nautical miles to maintain land and naval forces in the area of operations. Sea-based air power provided essential air cover to the fleet and land forces in operations beyond the range of Allied land-based aircraft.

From the Australian perspective, the Pacific war saw an expansion of the infrastructure necessary to support wide ranging naval operations. Defensive minefields were laid in Australian, New Zealand and New Guinea waters. Harbour defence systems were established around Australia and forward operating bases provided logistic support to ships close to the combat areas. Shipbuilding, repair and maintenance facilities were greatly expanded, including the construction of the Captain Cook Graving Dock at Garden Island, Sydney. In addition, the RAN began requisitioning merchant vessels to supplement the fleet. These vessels, of which many were commissioned into the RAN, served as coastal patrol vessels, stores-issuing ships, amphibious landing ships and other roles in support of RAN operations. The RAN Reserve played a major role in manning the expanding fleet.

At the outbreak of war in 1939 the RAN mobilised and drew heavily on its reserve forces. When the RAN mobilised, active personnel expanded rapidly from the September 1st permanent forces total of 5,440 to

10,300. There were three types of RAN reserves at the time: RANR, RANR (Seagoing) and Royal Australian Naval Volunteer Reserve.

The RANR were the traditional reservists who undertook regular training during peacetime and included both officers and sailors. Reserve training was suspended during the War; all new entry personnel were entered through the RANR and signed an agreement for the duration of hostilities instead of the customary 12-year engagement.

RANR (S) were officers who held professional qualifications (e.g. Master's Certificate) and in peacetime were normally merchant ship officers.

The RANVR officer scheme was established to interest those who participated in maritime activities (e.g. yachtsmen) to take part in naval training. In 1943, a Special Branch of the RANVR was created and all officers appointed since the outbreak of hostilities and engaged in specialised technical and operational duties were transferred to it. These included Radar Officers, Intelligence, Metrology, Cypher Officers, Anti-Submarine and Bomb and Mine Disposal. The Royal Navy (RN) drew most of its loan personnel from Australia, from the ranks of the RANVR. In June 1944, of 500 Australians serving with the RN, more than 400 were members of the RANVR.

Perhaps the most distinguished of the RANVR officers who served in the RN was Lieutenant Commander Stanley Darling (later Captain), an Australian reserve officer commanding the Loch-class frigate, HMS *Loch Killin* who was responsible for the destruction of three U-boats and received the DSC and two bars. Another officer, Lieutenant Commander Leon Goldsworthy, GC, DSC, GM RANVR, became the RAN's most highly decorated member for his extraordinary courage and skill in rendering German mines safe.

In September 1939, the RAN Reserve Forces comprised: RANR, 245 officers, 3,900 sailors; RANR (S), 85 officers; RANVR, 245 officers. When the War ended, the total Reserve Force numbered some 2,900 officers and 29,250 sailors. This represented over 80 percent of the personnel serving in the RAN.

The Reserves were commonly known as the 'Wavy Navy' a popular title arising from the fact that the braid on the officer's uniform sleeve was applied in a waved pattern instead of the straight lines used by the regular navy. This incurred a friendly rivalry in both the RAN and RN, exemplified by a cartoon appearing in the popular British magazine *Punch*, which was to become a favourite of the Naval Reservists. Walking down a street with a lieutenant RNVR, with two wavy stripes, a young girl notices a Royal Navy commander with three straight stripes on his uniform. The RNVR officer says to the girl, 'Straight stripes? Oh, those are the fellows who run the Navy in peacetime!'

Photo Foreword-1

Reserve shoulder board (left) and regular officer (right); Surgeon Lieutenant K. M. Morris, RANVR, wearing his service jacket with Reserve lace clearly visible on his sleeve (center). (RAN)

The Navy does not exist in a vacuum, but functions as part of a wider defence organisation maintaining the security of the country. The scale and associated risk of maritime operations vary widely between peace and war. The RAN at the beginning of World War II was a prime example. When war was declared the RAN had some 5,440 men and 15 ships, with a further three under construction. By June 1945, at its peak strength, there were some 340 ships, and 39,650 serving men and women of which 80 percent were reservists. As 1943 and 1944 progressed, Australian ships were involved in the campaigns to oust the Japanese from the South West Pacific Area. In the succession of landing operations, the cruisers and destroyers carried out bombardments and provided seaward cover. The corvettes escorted merchant ships through the area. The Landing Ships Infantry HMAS *Manoora*, *Kanimbla* and *Westralia* participated in many of the amphibious operations conducted in New Guinea, the Philippines and Borneo. The cargo ships continued to keep the fleet replenished and able to fight.

David Bruhn's latest book provides a most readable outline of how the war in the Pacific was prosecuted through the eyes of the ships supporting the combat fleet. It commemorates the important role these ships contributed to ultimate victory and acknowledges the debt owed them in a time of severe adversity.

Commodore Hector Donohue AM RAN (Rtd)

# Foreword

The definitive story of the Multi-National Service Force that backstopped the naval war in the Pacific during World War II has long needed to be told. The actions and accomplishments of the various cargo and support vessels making up this Force are often overshadowed by the exploits of the combatants, or fighting ships, of the Allied navies. Having served in four (4) Service Force ships as both an enlisted sailor and a commissioned officer, *Ready to Haul, Ready to Fight*'s powerful recounting brought back memories of serving the Fleet.

As early as 1917, future Fleet Admiral Chester W. Nimitz was the U.S. Navy's leading developer of underway replenishment. When he assumed command of the naval war in the Pacific, the techniques and concepts that he developed as a young officer allowed the U.S. Fleet to operate away from port almost indefinitely. Cargo ships cobbled together from various sources by the U.S., Australia, and Great Britain formed a cohesive supply force that kept the Allied navies at sea and allowed them to take the fight to the Empire of Japan.

The ships of the Service Force could and did defend themselves against enemy attacks. They supported all the major naval engagements in the Pacific and were awarded Battle Stars, Battle Honors, and Unit Citations for their efforts. Shortly after joining the Navy, ships I served on operated with two of the top Pacific Theater Battle Star recipients. Attack Cargo Ships USS *THUBAN* (AKA-19), known affectionately as "The Thumping THUBAN," and USS *CAPRICORNUS* (AKA-57), earned 7 and 4 Battle Stars, respectively, for their World War II service. The *THUBAN* subsequently added another three in the Korean War. It is notable that into the 1960s, these two ships and others like them were still supporting and servicing naval operations globally.

Sailors work hard, no matter what type of ship they serve aboard. Knowing this firsthand, I was still struck by the volume of work performed by cargo ships in World War II, and the conditions under which, at times, this seemingly ceaseless toil took place. In particular, attack cargo ships off enemy-held islands, delivering to Allied troops that had just stormed ashore, desperately needed cargos of food, fuel, and ammunition, did not have the luxury of port facilities (piers, cranes, and stevedores) to assist in this effort. So, days of around-the-clock work were necessary, often amidst periodic enemy shore battery fire or air attack. This involved bringing cargo up out of holds, and lifting deck loads, lowering into boats and craft, sending them to the beach for unloading, return to ship, and repeat same process.

Perhaps no event in *Ready to Haul, Ready to Fight* more vividly displays the can-do and heroic actions of the officers and men of these vital ships than the submarine attack on the Cargo Ship USS *ALCHIBA* (AK-23) during the Guadalcanal Campaign. Anchored in the Solomon Islands delivering her cargo to U.S. Marines ashore engaged in the fighting, she was torpedoed by a Japanese submarine despite being screened by several destroyers. The resultant explosion "holed" her and set her aflame, and captain and crew got *ALCHIBA* under way and beached the ship to keep her from sinking. Concurrent with vigorous and determined firefighting and damage control efforts, a portion of the crew continued to deliver materiel ashore. Nine days later, still in a perilous condition, she was torpedoed a second time by a different submarine. Ultimately, the ship was saved and lived to fight again. For her dauntless and valiant efforts, *ALCHIBA* was the only Cargo Ship to be awarded the Presidential Unit Citation for heroism.

Commander Bruhn is a prolific and accomplished writer of naval-themed books depicting true historic events. His attention to the most minute details, and comprehensive research, are readily apparent to readers of his books. *Ready to Haul, Ready to Fight* continues in that vein by presenting a clear and concise portrayal of a mostly unsung group of ships and men that ultimately won the recognition and admiration of grateful nations. Once I started reading, it was impossible for me to put down. I highly recommend it, not only for aficionados of the naval war in the Pacific, but also for those individuals who would like to know more about one important and crucial maritime concept that helped lead us to victory in World War II.

Cdr. Lee M. Foley, USN (Retired)

# Foreword

Historically, success in battles, campaigns and even wars has relied as much, if not more, on efficient and effective logistics as any other single factor. The RFA (Royal Fleet Auxiliary), which has supplied ships of the Royal Navy at sea with ammunition, stores and fuel since its establishment in 1905, is tangible proof of this truism.

As an organisation, the RFA is still run on merchant navy lines although its personnel come under the Naval Discipline Act during hostilities. Such an arrangement has afforded its ships, unarmed until relatively recently, useful flexibility in accessing ports and other areas normally forbidden to warships. On the downside, the Royal Navy has had to tolerate the occasional strike and work-to-rule in peacetime.

The quasi-civilian status of the RFA has led to certain other incongruities. For example, when a frigate in which I was serving was engaged in receiving stores and fuel during a RAS (Replenishment at Sea) in distant waters during the 1970s, members of our all-male ship's company were often discombobulated by the sight of bikini-clad wives of RFA officers sunning themselves around a makeshift swimming pool on the upper deck of the tanker or stores ship steaming close abeam at 15 knots or so. In the meantime, we toiled under the same relentless sun to fill our tanks with fuel and strike below palletised ammunition, food, naval & victualling stores and spare parts. Perhaps of greatest importance on a personal level were the crates of tinned beer which had a disconcerting habit of disappearing somewhere along the human chain between the dump area on the upper deck and their assigned stowages in the bowels of the vessel.

The most demanding historical challenge for the RFA, and other British and Dominion supply vessels and auxiliaries, was the provision of logistical support for the abnormally far-flung BPF (British Pacific Fleet) in the latter stages of the Second World War. The BPF was no small concern and comprised 6 fleet carriers, 4 light carriers, 2 aircraft maintenance carriers and 9 escort carriers, with a total of more than 750 aircraft, 4 battleships, 11 cruisers, 35 destroyers, 14 frigates, 44 smaller warships, 31 submarines, and 54 large vessels in the 'Fleet Train.'

The unglamorous but no less important 'Fleet Train' comprised the motley collection of vessels supporting the BPF. In his unique fashion, David Bruhn has brought its essential activities to life and woven a rich tapestry of stories ranging from fleet to individual unit and personal level.

Until I read David's typically well researched work, I hadn't heard of the Canadian-built Fort and Park cargo ships, nor realised that they and the more familiar American-built Liberty and Victory ships were all based on the design of the British 'North Sands' type freighter. I also learned more about MOWT (British Ministry of War Transport) vessels which complemented those of the RFA.

Many British merchant ships were taken up by the Royal Australian Navy, including those that escaped Singapore before it fell to the Japanese, and commissioned as auxiliaries including 'stores-issuing ships.' Despite their lack of armament compared to their US Navy equivalents, these ships bravely plied the dangerous waters off northern Australia and Japanese-occupied Papua New Guinea, ever vulnerable to attack by enemy aircraft, submarines and surface craft. While doing their part for the war effort, they suffered grievous casualties from enemy action and this book helps act as a tribute to their amazing efforts as it does for all those allies who stood with them to achieve ultimate victory in the Pacific theatre and thus, with the help of a couple of atomic bombs, bring an end to the Second World War.

Rob Hoole
Vice Chairman & Webmaster
RN Minewarfare & Clearance Diving Officers' Association
www.mcdoa.org.uk

# Foreword

David Bruhn sets out in this, his latest book, to retrace the march of naval actions that saw the Allies drive the Japanese invaders back to their homeland from the verges of the vast Pacific that they had rapidly attacked and occupied in 1941 and 1942. David has made these vicarious voyages before employing a format rich in vignettes and illustrations. Each has been from the perspective of specific types of ships with emphasis on those that have performed heroic actions and/or earned battle stars for their campaign ribbons.

The subject this time is cargo ships that brought troops, arms and supplies to the various landing beaches on the islands that had to be secured to enable the Allies to continue their march across the Pacific. Although other types of ships are mentioned such as repair ships and fleet oilers and large troop transports, the focus with respect to US Navy ships is limited to two specific types of cargo vessels. Otherwise, considering the vast numbers and types of vessels employed by US Forces, in various combat operations, their stories could not be contained in a single volume.

*Ready to Haul, Ready to Fight* also covers the valuable contributions of supply ships of the Royal Australian Navy, and British merchantmen. The latter vessels were a part of the "Fleet Train" of the British Pacific Fleet that joined the Pacific war in 1945. Because of fewer such ships, David was able to expand his coverage, somewhat, of Allied ships beyond only cargo ships. In particular he has included stories of three former passenger liners that the RAN requisitioned and employed in various roles including troop transports.

The US Navy ships featured are Cargo Ships (AKs) and Attack Cargo Ships (AKAs). Initially, all Navy cargo vessels were designated AK. In early 1943, "Navy brass," wanting to employ some of its cargo ships (AK) and transport ships (AP) in direct support of island assaults, modified, then redesignated these vessels assault cargo ships (AKA) and assault transports (APA). The most significant change to the ex-AKs was equipping them to carry amphibious landing craft.

Canada participated in the Pacific war along with its huge contribution to the one in Europe. Even before Pearl Harbor she sent troops for the fruitless defence of Hong Kong and then in 1942 for the defence of Alaska and the Pacific Northwest. It is amazing that the Japanese while penetrating south to the very doorstep of Australia and New Zealand were also able to attack the North American coast with their submarines, shelling light houses, attacking shipping and throwing fear into local inhabitants. It was a remarkable feat given the vastness of the Pacific. They learned well from the lesson of Commodore Perry and, as a result of their experience in the Sino-Pacific, Russo-Japanese and First World wars, developed into a first class "Blue-water Navy."

Photo Foreword-2

Corvette HMCS *Edmundson* (K106) alongside the sinking SS *Fort Camosun* at the entrance of Juan de Fuca Strait – vessel was salvaged because cargo of lumber and timber helped her stay afloat after being torpedoed by a Japanese submarine, 20 June 1942.
Photograph source unknown – acquired from *For Posterity Sake*

Following the death and destruction wrought by Japan on ill-prepared Allied forces during the early months of the war, US and Australian soldiers, sailors, airmen, and Marines picked themselves up, and slowly began to drive the Japanese aggressors back toward their home islands. This progress came at a great cost, in terms of lives and (speaking only regarding the Allied navies) of ships and aircraft lost. The most intense battles were thousands of miles from sources of supply, and victories at sea, in the air, and on land, could not have been achieved without supply ships – hundreds of them, ships such as described in this book.

Existing older ships were first employed but their numbers were limited and steadily being diminished by German North Atlantic submarine wolf packs. With these losses, and even more on expanding battle fronts in many theaters of war, North American shipyards and their personnel came to the fore. Records compiled by naval architect Tim Colton give proof of the huge contribution of these yards in supplying the war efforts in the Pacific and in Europe (whilst keeping the British people from starving). He lists a total of 5,858 vessels as "Ships, Boats and Large Barges Built for the U.S. Maritime Commission," and 436 "Merchant Ships" constructed in Canada.

It is gratifying to me that 255 of the large cargo ships (referred to as "10,000-tonners") were launched from seven yards in my home province of British Columbia. These were of three related types. North Sands which employed three Scotch (fire tube) type coal fired boilers, Victory type with two oil-fired water-tube boilers and the "Canadian Type" a modified Victory ship which could burn both coal and fuel oil.[1]

Although only our cruiser HMCS *Uganda* participated in actions with the British Pacific Fleet, our naval aviators distinguished themselves with the Fleet Air Arm aboard the Royal Navy carriers. Additionally, our shipyards produced many of the "Fleet Train" vessels which supported the BPF. These included eight "Fort" merchant vessels employed as "Stores-Issuing" ships, and at least two RN commissioned repair ships HMS *Farmborough Head* and HMS *Beery Head*—all based on Victory ship hulls with oil-fueled boilers. All of these vessels were built and modified in British Columbian yards. Of special importance to fleet sailors, West Coast Shipbuilders Ltd. of Vancouver converted the former RN minelayer HMS *Menestheus* to an amenities ship. Her service was short-lived, but greatly esteemed; equipped with her own brewery, she was affectionally known as "The British Brew Barge of WWII."[2]

Canada's North Sands and Victory type ships were similar to their cousins the US Liberty ships. The principal difference in construction was that they were, for the most part, of rivetted construction, whereas the US ships were welded. Riveting was generally performed by four-men crews working eight-hour shifts, round-the-clock, 7 days a week. The rivet-heater used a portable forge located close by to the riveting to raise the rivets to a yellow heat and then he threw them with tongs to the rivet-passer who caught them in a sheet-metal funnel like-contraption called a "bucket" and ran them to the nearby work, and inserted them with tongs into the waiting holes where the riveter, with his holder-on (or sometimes called "bucker") pressing the rivet heads in place, and quickly flattened the tail ends with his pneumatic gun. With 383,000 rivets required per hull, multiple hulls under construction and multiple crews at work, the dock mates had to be alert, lest they get hit and burned by one in a swarm of airborne rivets.[3]

Photo Foreword-3

Riveters (man on right may be Elio Lus), Date: possibly 1943.
NVMA #27-795 Versatile Pacific Shipyards Inc. Fonds,
courtesy Archives of North Vancouver

George H. S. Duddy, P.Eng Ret.

# Acknowledgements

Brilliant, and prolific maritime and aviation artist Richard DeRosset has vividly captured on canvas in the cover art for this book, one of the many occasions during the Pacific war in which the unsung crew of a cargo ship found themselves fighting for their lives. This painting pays tribute to the Service Force, and should enjoin pride in former sailors who have trod the deck of an oiler, ammunition, or cargo ship, delivering product to the perceived more glamorous carriers, cruisers, destroyers, and submarines. Richard himself is a former sailor, with much blue water experience. His time at sea included tours as a Navy "bluejacket" aboard the destroyer USS *Epperson* and amphibious transport USS *Paul Revere*; crewmember of the fishing vessel *Petrel*—which caught fire and sank seventy-five miles off the southern California coast—and master of the small merchant tanker MV *Pacific Trojan.*

Photo Acknowedgement-1

Richard DeRosset with his painting *Brutal Ambush off Wonsan*, which depicts the savage attack on the USS *Pueblo* (AGER-2) by an overwhelming force of North Korean torpedo boats, MIG-21 fighter aircraft, and sub-chasers, on 23 January 1968. Courtesy of Richard DeRosset

I am particularly indebted to four distinguished individuals, citizens of Australia, the United States, Britain, and Canada, respectively, who graciously lent time to this book, and penned forewords offering their own unique perspective born of hard-won, considerable naval and/or maritime experience. Each has authored published articles, books, or both, and, as such, are "men of letters" as well as "men of action."

Commodore Hector Donohue, AM RAN (Retired), has provided much assistance with previous books, and has done so again for *Ready to Haul, Ready to Fight.* His foreword and postscript add much richness to the book, and provide perspective of a Royal Australian Navy flag officer. Donohue began his career in the RAN, in 1955, as a seaman officer and subsequently sub-specialized as a clearance diver and torpedo and anti-submarine officer. His service in the RAN included command of the destroyer escort HMAS *Yarra* and the guided missile frigate HMAS *Darwin.* Ashore, he held a number of senior positions in Defence policy and Force development prior to retirement in mid-1991.

Photo Acknowledgements-2

Commodore Hector Donohue, AM RAN (Rtd.)
Courtesy of Hector Donohue

Commander Lee M. Foley, USN (Retired), served thirty-two years on active duty, as both an enlisted man and an officer, between 1961 and 1993. He served at sea aboard twelve ships, including ones in the amphibious, service force, mine force, and salvage force, as well as an aircraft carrier, and a destroyer. After advancing from seaman recruit to master chief petty officer, and up through the warrant officer ranks, and limited duty officer ranks to lieutenant commander, the Navy ordered him to command of the ocean minesweeper USS *Excel* (MSO-439). His last sea-going service was as the executive officer of USS *Kansas City* (AOR-3). Following this assignment, he served as Personal Liaison of the director of the Defense Nuclear Agency to commander in chief, Pacific, in Hawaii; and head of a directorate at the Defense Nuclear Agency's Field Command in Albuquerque, New Mexico.

Photo Acknowledgements-3

Courtesy of Lee Foley

Rob Hoole co-authored with me the *Home Waters/Nightraiders/Enemy Waters* trilogy of mine warfare books and, for this book, reviewed the Royal Navy-related material, and graciously penned a foreword. Rob is a former Royal Navy mine clearance diving officer who, like Commodore Donohue, joined the Service as a seaman officer. He first qualified as a Ships' Diving Officer, then as a

Minewarfare & Clearance Diving Officer, and later commanded the *Hunt*-class mine countermeasures vessel HMS *Berkeley*, now HS *Kallisto* in the Greek Navy.

Rob recently spearheaded successful efforts to establish a monument at Gunwharf Quays in Portsmouth, UK, as a tribute to the tens of thousands of service personnel who passed through the gates of the training establishment HMS Vernon. These personnel included the mine warfare and diving specialists, mine designers, minefield planners, bomb & mine disposal personnel and the crews of the minelayers, minesweepers, and minehunters who were trained or based there. The monument also honours all those involved – past, present and future – in naval and military mine warfare, diving and EOD (Explosive Ordnance Disposal).

Photo Acknowledgement-4

Rob on board a *Ton*-class minehunter off Gibraltar in 1984 and, more recently, in front of the Vernon Mine Warfare & Division Monument at Gunwharf Quays (formerly HMS Vernon) in Portsmouth.
Courtesy of Rob Hoole

Canadian George Duddy has served as critical reviewer and content editor for a number of my books. In addition to having the keen eye and analytical mind of a professional engineer, he also has much knowledge of maritime subjects. In his retirement, he has developed into an expert on the maritime history of western Canada and the Arctic, but his lineage stretches across the Atlantic to Britain. His father, as a schoolboy, took photographs of surrendered German battleships in the Firth of Forth near the end of World War I; his great grandfather was a pioneering Leith steamship owner and his great great grandfather, as Master in both sail and steam, ended his career as marine superintendent for the Leith, Hull & Hamburg Steam Packet Co. One of his other relatives, Midshipman Percival George, tragically fell to his death from the mast of a Royal Navy ship during the age of sail. His dirk is on display in a museum in South Africa.

Photo Acknowledgements-5

George Duddy in Glacier Bay aboard the MS *Volendam* during an Alaskan cruise in 2019. A contributor and U.S. Military & Naval Vessel Correspondent for Nauticapedia.ca Project, he is sporting an organization ballcap.
Courtesy of George Duddy

Finally, a crisp salute to my editor Lynn Marie Tosello, who pores lovingly over manuscripts, while her pen works its will in making the written word more concise and more understandable. She has learned much about the Navy, but is still a gentle lady, who has yet to be heard to utter, "bear hand," "turn to," or "get hot" to anyone.

# Preface

*Linked inseparably with combat is naval logistic support, the support which makes available to the fleet such essentials as ammunition, fuel, food, repair services--in short, all the necessities, at the proper time and place and in adequate amounts.*

*In 1940 the Base Force Train [later called the Service Force] included a total of 51 craft of all types, among them 1 floating drydock of destroyer capacity. By 1945 the total was 315 vessels, every one of them needed.*

—Rear Adm. Worrall Reed Carter, USN, in his seminal book on U.S. Navy fleet logistics, *Beans, Bullets and Black Oil*, 1953.[1]

Photo Preface-1

Attack cargo ships USS *Aquarius* (AKA-16) at left, and *Titania* (AKA-13) at right, at Pavuvu, Russell Islands, 28 April 1944. U.S. Marine Corps photograph #USMC 86265

Photo Preface-2

One of the attack cargo ship USS *Rankin*'s (AKA-103) cargo holds, opened. The upper level is the main deck, with cargo-handling winches visible. The lower level is the deck onto which cargo is combat loaded. Housed in the "tween decks" area are the mess decks where the crew and, if embarked, troops, ate their meals.
U.S. Navy photograph

Readers of a certain age will likely be well familiar with the comedy-drama film *Mr. Roberts*, staring Henry Fonda as Mister Roberts, James Cagney as Captain Morton, William Powell as Doc, and Jack Lemmon as Ensign Pulver. The 1955 film was nominated for three Academy Awards, with Jack Lemmon winning the award for Best Supporting Actor. The action takes place aboard the fictitious U.S. Navy cargo ship *Reluctant* in a "backwater" area of the Pacific, which services combatant ships on their way to the ever-expanding battlefront. The executive officer/cargo officer, Lt. (jg) Douglas A. "Doug" Roberts, feels that the war, in its waning days, is leaving him behind. He yearns to be aboard a destroyer, repeatedly requests a transfer, and finally gets his wish.

Some U.S. Navy and Royal Australian Navy cargo ships, and those of the British Merchant Marine supporting the British Pacific Fleet, all the subject of this book, did serve in backwater areas. Most did not, and went in harm's way while supporting the fleet, and combat forces ashore. Naval commanders well understood the advantages of supply logistics afloat and near areas of fleet operations. Having cargo ships, oilers, ammunitions, and other such support close at hand was often the only option, particularly early in the war in the Pacific.

In the spring of 1942, Japanese forces were pushing south and southeast from newly gained positions in the Philippines, Netherlands East Indies, and the northern coast of New Guinea; and older established bases in the Caroline and Marshall Islands. The battles of Midway and Coral Sea were yet to be fought, and Australia and New Zealand were threatened. The only American naval establishment south of the Equator was at Tutuila, America Samoa. The British and New Zealand forces had some facilities at Suva in the Fijis. It was not known then, how near the enemy it would be safe to establish major supply bases, yet it was highly desirable to provide logistics support to the U.S. Navy's limited South Pacific Force as close to areas of potential operations as possible.[2]

## BREADTH OF *READY TO HAUL, READY TO FIGHT*

Before progressing further into this preface, it's important to clarify for readers exactly what the book encompasses, and what it doesn't. When war broke out in the Pacific, on 7 December 1941, United States and Royal Australian naval forces were miniscule in comparison to the threat posed by Japan. As overburdened shipyards already fully engaged in supplying the European war began building desperately needed combatant ships, both navies acquired and commissioned civilian merchant vessels to meet fleet auxiliary ship requirements. The U.S. Navy took up existing freighters for use as cargo ships and, as the size of the sea service expanded greatly, acquired hulls laid down by the Maritime Commission as merchant vessels, for completion as USN cargo ships. (The identity of these type hulls, and brief explanation of their differences, may be found in Appendix A.)

Needing increasingly more cargo ships as the war expanded, U.S. shipyards began mass producing the famous Liberty ship. Americans are rightly proud of the contributions of the Liberties and their successors, Victory ships, in World War II. These type vessels served both as commissioned ships and as merchantmen, with Navy Armed Guard units aboard the latter to man naval guns fitted in the civilian ships to provide a means of self-defense.

The U.S. Navy sent nearly 300 cargo vessels to sea in World War II in all theaters. Summary information about ship numbers may be found in the following table, and identities of the ships in Appendix B. Fourteen cargo ships (AK) later converted to attack cargo ships (AKA) are counted once in the table, and the small numbers (7 in total) of net cargo ships (AKN) and aircraft ferry cargo ships (AKV) are omitted.

General stores-issue ships (AKS) are included in the table, because they are referenced in this book although there is little material devoted to them. There is no mention of the AKNs and AKVs.

**U.S. Navy Cargo Vessels (AK, AKA, AKS) in World War II**

| Ship Class | AK | AKA | AKS |
|---|---|---|---|
| Miscellaneous cargo ships (14 of 47 AKs converted to AKAs in 1943) | 33 | 14 | |
| *Carter*-class cargo ships | 66 | | |
| *Alamosa*-class attack cargo ships | 48 | | |
| *Andromeda*-class attack cargo ships | | 30 | |
| *Artemis*-class attack cargo ships | | 32 | |
| *Tolland*-class attack cargo ships | | 32 | |
| General stores-issue ships | | | 16 |
| Total (271 cargo vessels) | 147 | 108 | 16 |

Canadians are proud of the many Park and Fort ships their yards churned out during the war for use by Canada and Britain, respectively. (It was North American shipyards that won the war of attrition against the German submarine wolf packs replacing ships faster than they could be sunk and ultimately prevented the British from starving.) The ships retained by Canada were named after parks, and those constructed for use by Britain, forts.

The Liberty and Park/Fort ships were closely related, "kissing cousins" so to speak, owing to the design of each type being based on working drawings for the British "North Sands" vessel, a product of the J. L. Thompson & Sons North Sands shipyard in Sunderland, England. (Greater information about this may be found in Appendix C.)

The freighters and ships built on other types of merchant vessel hulls, formerly known as AKs (cargo ships) pressed into duty as commissioned USN cargo ships, were faster than the more pedestrian Liberty ships. In early 1943, AKs fitted with additional guns (in some cases), and in receipt of other modifications, including the ability to carry amphibious landing craft, were reclassified AKAs (attack cargo ships). Attack cargo ships assigned to divisions of transport ships carrying assault troops, participated in amphibious landings and, thus, generally were in harm's way more that AKs, which also participated but more often landed their cargos after beach heads were established.

The U.S. Navy owed much to the efforts of America's shipyards, and was fortunate to have merchant vessels, and hulls under construction, that could be taken up for use as cargo ships. The Royal Australian Navy (RAN) also had a need for cargo ships, but very limited resources available to acquire them. Although a large country,

Australia had then only a small population. In this respect, Australia and Canada were very similar. The RAN found cargo ships, which it termed "stores-issuing ships," wherever it could; most were former British merchant ships. Some of these ships were from among the lucky few able to escape Singapore, before it fell to Japanese forces.

It's important to note that the RAN, having only a modest number of ships in comparison with the much larger USN, employed some of them in varying roles during the war, in an effort to meet its most immediate needs. Thus, as examples, following their acquisition, some of the stores-issuing ships were initially classified and used as minesweepers, or as harbor entrance control vessels before their ultimate role. Because cargo ships often worked closely with transports (meaning they carried troops) in amphibious operations, a portion of the book is devoted to the RAN's three infantry landing ships (LSIs). Former passenger ships, they first served as armed merchant cruisers before their modification for service in amphibious assault roles.

British merchant ships played an important role in the Pacific war, first as acquisitions for the RAN, and later as units of the Fleet Train (logistics force) in support of the British Pacific Fleet, which joined in the final operations against Japan in 1945. General Douglas MacArthur opposed the Royal Navy under his Southwest Pacific command unless it was in the form of a task force attached to the U.S. Seventh Fleet. Admiral Ernest King did not want the Royal Navy in the Central Pacific at any price, allegedly because of logistics.[3]

Ships of the Royal Navy were "short-legged," accustomed to putting into a base every two or three weeks for replenishment and upkeep. American Navy ships were long-legged, able to "keep the sea" for months at a time, supported by a mobile Service Force. At the time when the entry of the British was under consideration, this force was stretched very thin. King feared that the RN fleet would be a drain on the USN supply system and would hamper their attack forces.

The Royal Navy's solution to this problem was to bring a Fleet Train to the Pacific to support the British Pacific Fleet. This logistics service force was comprised of Royal Fleet Auxiliary (RFA) units and civilian merchant ships cobbled together from Britain and other Allied countries. Among the force were some "Fort" stores-issuing ships, an important Canadian contribution to the war effort.

## U.S. NAVY'S MOST EXTOLLED CARGO SHIP

On 21 November 1942, the cargo ship USS *Alchiba* (AK-23) left Noumea, New Caledonia, with a cargo of aviation gasoline, bombs, ammunition, and provisions for American forces in the Guadalcanal-Tulagi area. She was a part of Task Unit Baker, which also consisted of the transport *Barnett* (AP-11), and the destroyers *Lardner* (DD-487), *Lamson* (DD-367), and *Hughes* (DD-410).[4]

Map Preface-1

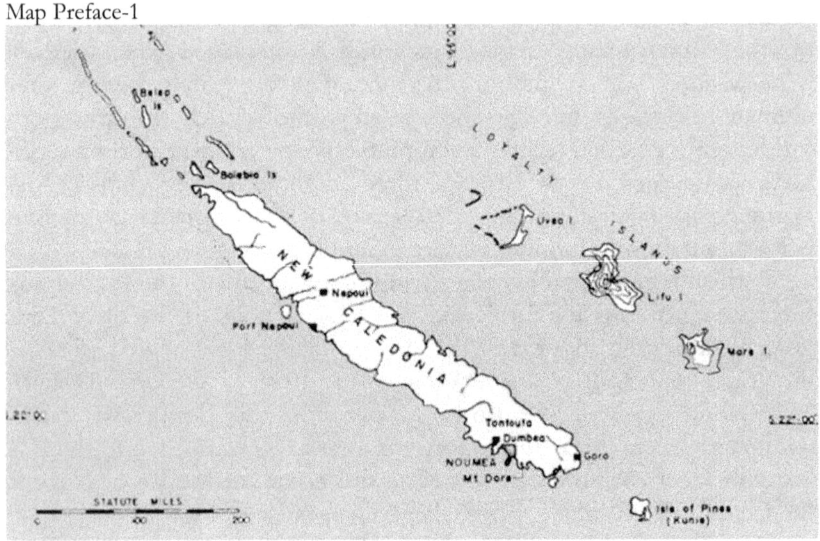


The French Colony of New Caledonia (comprising dozens of islands, located to the north-northeast of Brisbane, Australia, across the Coral Sea)
*Building the Navy's Bases in World War II: History of the Bureau of Yards and Docks and the Civil Engineer Corps, 1940-1946, Volume II*

Noumea, at the southern end of the island of New Caledonia, had recently been developed as the main fleet base in the South Pacific, and served as a staging area for the development of other advance bases such as Guadalcanal. It also served as headquarters for Vice Adm. Robert L. Ghormley, USN, commander, South Pacific Force.[5]

On the morning of 28 November, *Alchiba* was at anchor off Lunga Point, on the northern coast of Guadalcanal in the Solomon Islands. At 0616, without warning a torpedo fired by the Japanese submarine *I-16*, hit her port side at approximately frame 125. It seemed inconceivable such an attack could have succeeded. She and *Barnett*, which lay at anchor nearby, were being screened (protected) by the destroyers *Lamson*, *Lardner*, *Hughes*, *McKean*, and *Manley*.[6]

Photo Preface-3

Attack cargo ship USS *Alchiba* (AK-23) afire off Lunga Point, Guadalcanal. She was torpedoed, on 28 November 1942, by the Japanese submarine *I-16*. U.S. Marine Corps photograph #USMC 66457

The explosion from the warhead of the torpedo opened *Alchiba*'s hull to the sea, and set her aflame. She quickly took a 17-degree list, with No. 2 hold a mass of flames, and fire spreading to No. 1 hold. Acting quickly, her commanding officer, Comdr. John S. Freeman, beached the ship to prevent her total loss. Details about the heroic actions of her crew following this combat damage, and after she was again torpedoed, on 7 December 1942, are provided in Chapter One.

The fact that the cargo ship was saved under these circumstances was the result of superlative leadership by her captain and resolute efforts of his crew. Navy "top brass," which initially believed *Alchiba* had been lost, recommended her for award of the Presidential Unit Citation (PUC). She was the only cargo ship in the war to receive a PUC, the highest unit award for heroism, considered to be equivalent to receipt of the Navy Cross by an individual.[7]

Presidential Unit Citation (PUC) pennant. Awarded in the name of the U.S. president for extraordinary heroism in action against an armed enemy. Pennant is yellow with broad stripes of "Old Glory" blue and red along the upper and lower edges.

## RAN SUPPORT OF MACARTHUR'S ALLIED FORCES

Map Preface-2

Northern Australia and surrounding areas
https://www.ibiblio.org/hyperwar/USN/Building_Bases/maps/bases2-p278.jpg

While the U.S. Navy was supporting U.S. Marine and later Army ground forces fighting on Guadalcanal, prior to a push up through the Solomon Islands toward the Japanese stronghold at Rabaul in New Britain, the Royal Australian Navy (RAN) was fully engaged in the Southwest Pacific. Over the course of the war, the RAN acquired thirteen stores ships, mostly merchant vessels formerly under British charter, and pressed them into more hazardous duties. Danger was not new for two of these ships (Chinese steamers *Ping Wo* and *Whang Pu*) which barely escaped Singapore before its fall to the Japanese.

The activities of the stores-issuing ships involved delivery of critical supplies and cargo to outposts in Northern Australia and points north. These included Cairns, Darwin, Thursday Island in the Torres Strait, and Port Moresby, Papua. Later, as MacArthur's Allied forces moved forward to Milne Bay, and then began fighting their way up the New Guinea coast, the operations of these stores ships were extended to those dangerous waters. Passage in these New Guinea waters exposed them to the likely chance of attack by Japanese float planes, operating from captured bases in the Netherlands East Indies, and enemy bastions in and around New Guinea.

Of the thirteen RAN stores ships identified below, two were sunk by Japanese aircraft—HMAS *Maroubra* and HMAS *Patricia Cam.* A third ship, HMAS *Matefele*, disappeared at sea while en route to Milne Bay, and was never seen again. (Additional information about the characteristics of all thirteen ships may be found in Appendix D.)

| Ship | RAN Service | Ship | RAN Service |
|---|---|---|---|
| HMAS *Adele* | 24 Oct 39-<br>7 May 43 | HMAS *Mombah* | Mar 44-48 |
| HMAS *Baralba* | 31 May 42-<br>11 Feb 43 | HMAS *Patricia Cam* | 9 Feb 42-<br>22 Jan 43 |
| HMAS *Falie* | 17 Jul 40-<br>2 Aug 46 | HMAS *Ping Wo* | 22 May 42-<br>26 Jun 46 |
| HMAS *Gerard* | 1 Jul 41-<br>8 Apr 46 | HMAS *Poyang* | 6 May 43-<br>6 Mar 46 |
| HMAS *Maroubra* | 20 Mar 42-<br>10 May 43 | HMAS *Wang Pu* | 1 Oct 43-<br>22 Apr 46 |
| HMAS *Matefele* | 1 Jan 43-<br>24 Jun 44 | HMAS *Yunnan* | 20 Sep 44-<br>31 Jan 46 |
| HMAS *Merkur* | 12 Dec 41-1949 | | |

HMAS *Matafele* and HMAS *Patricia Cam* earned Battle Honours NEW GUINEA 1942-44 and DARWIN 1942-44, respectively. (This subject is taken up at the end of the preface.)

## ALLIED CHINESE SHIPS THAT SERVED THE ROYAL NAVY AS VSIS SHIPS

In addition to the four Chinese steamers—HMAS *Ping Wo*, HMAS *Poyang*, HMAS *Wang Pu*, HMAS *Yunnan*—commissioned into the RAN and mentioned in the previous section, there were two others that served as RFA (Royal Navy Fleet Auxiliary) ships. Before the war, SS *Changte* and SS *Taiping* (4,500-ton sister merchant ship) plied between Melbourne and Tokyo for the Australian Oriental Line, carrying general cargo and frozen meat. *Changte* was requisitioned by the Royal Navy as a VSIS (victualing stores-issuing ship), on 27 August 1939, and designated RFA *Changte* (Y1-9). She was returned to her owners in 1946. *Taiping* was similarly acquired in 1941 for the same type duties, and returned to her owners in 1947.[8]

Photo Preface-4

Plaque commemorates those who served aboard Allied Chinese Ships during World War II, at Garden Island Naval Base, Garden Island, NSW, Australia. Courtesy of Peter Williams

## HALSEY'S MOVEMENT UP THE SOLOMON ISLANDS

*Before Guadalcanal the enemy advanced at his pleasure—after Guadalcanal he retreated at ours.*

—Adm. William F. Halsey Jr., USN, commander,
South Pacific Force and South Pacific Area.

On 18 October 1942, Admiral Chester Nimitz, commander in chief, Pacific Fleet, appointed Vice Adm. William F. Halsey as commander, South Pacific Force and South Pacific Area, relieving Vice Adm. Robert L. Ghormley. After forces under Halsey's command defeated the Japanese in the Battle of Santa Cruz, on 26 October, and the following month routed them in the battles at Guadalcanal, Halsey was promoted to full admiral. In February 1943, the Japanese evacuated their remaining troops from Guadalcanal.[9]

In subsequent movement of his forces up the Solomons, Halsey became the first practitioner of the bypass strategy. After a bloody campaign on New Georgia, he executed the first "leapfrog" when he bypassed Kolombangara for Vella Lavella. It was a major contribution to American strategy, and General MacArthur took note and utilized this tactic in New Guinea.[10]

Map Preface-3

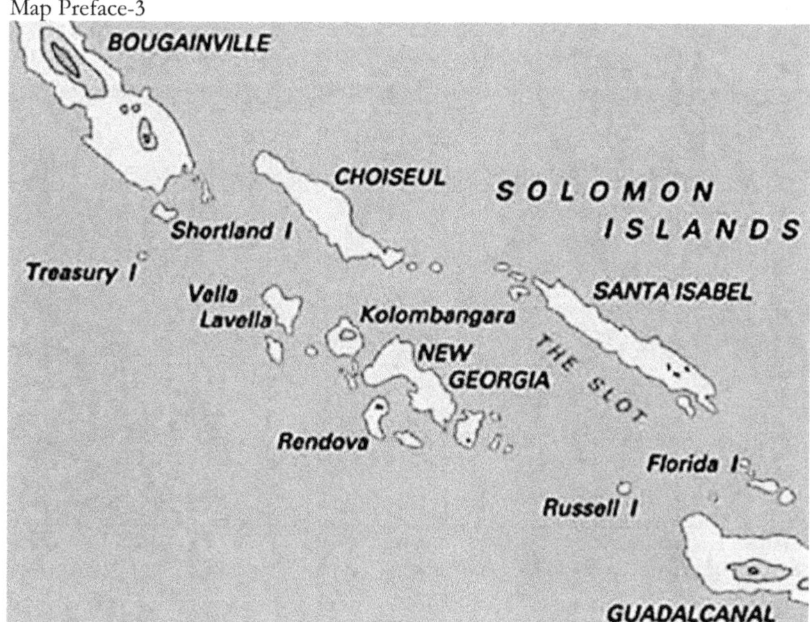


Central Solomon Islands

Halsey remained at the helm of the South Pacific theater until early 1944, when he took command of the U.S. Third Fleet, occasionally coordinating with MacArthur's adjoining Southwest Pacific theater.[11]

## USN AND RAN SUPPLY SHIPS JOIN IN PHILIPPINES

In autumn 1944, after MacArthur's Southwest Pacific Force, and Halsey's U.S. Third Fleet (the former South Pacific Force) breached the Bismarck Archipelago, they joined forces for assault landings at Leyte, marking the return of MacArthur to the Philippines. The naval component of MacArthur's force was the U.S. Seventh Fleet, commanded by Vice Adm. Thomas C. Kinkaid, USN.

The Bismarks, stretching between the northwest coast of New Guinea and the Green Islands (northern part of the Solomon Islands), were the site of powerful Japanese air and naval bases at Rabaul on New Britain Island and at Kavieng on New Ireland. These bases were repeatedly attacked, but not occupied by Allied forces.

Map Preface-4

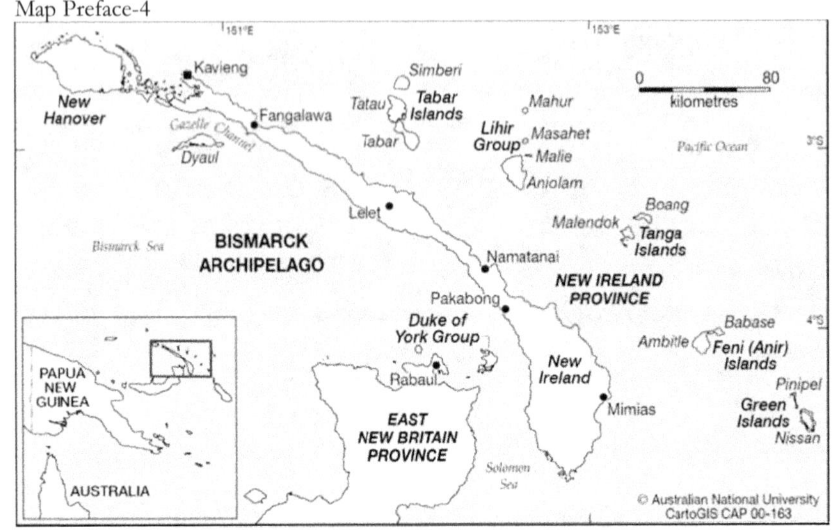


Bismarck Archipelago
Courtesy of Australian National University

Forming part of Task Group 77.7 (the Leyte Gulf Service Group of the Seventh Fleet) were the ammunition ships HMAS *Yunnan* and *Poyang*, the provision ship HMAS *Merkur*, and the oiler HMAS *Bishopdale*. The service group was a component of the larger Leyte Gulf Force of some 550 ships, consisting of battleships, cruisers, escort carriers, destroyers, destroyer escorts, attack transports, cargo ships, landing craft, survey vessels, minecraft, and supply ships. In total, thirteen RAN ships were a part of the Leyte Gulf Force:

- Task Force 74: Heavy cruisers HMAS *Australia* and *Shropshire*, and destroyers *Arunta* and *Warramunga*
- Supply ships HMAS *Bishopdale*, *Poyang*, *Yunnan*, and *Merkur*
- Landing ships HMAS *Manoora*, *Westralia*, and *Kanimbla*
- The frigate HMAS *Gascoyne* and motor launch *HDML 1074*, both of which were part of the minesweeping and hydrographic group[12]

As the Philippine Islands Campaign continued into 1945, with support from the U.S. Seventh Fleet, the Third and Fifth Fleets continued their march across the Pacific toward Japan.

Map Preface-5

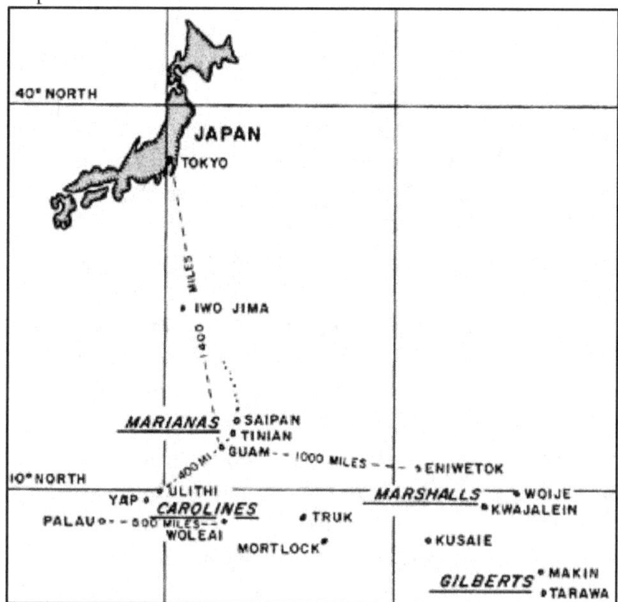


Japanese home islands
George C. Dyer, *The Amphibians Came to Conquer: The Story of Admiral Richmond Kelley Turner*

## BRITISH PACIFIC FLEET ARRIVES IN THEATER

*American warships being "dry," the daily British rum ration was a source of great envy. On one occasion the battleship King George V requested an American destroyer to come alongside. Facing one of the perennial British Fleet Train shortages, the British warship offered to exchange a bottle of whisky for certain key radar components. The signalman on the American destroyer achieved a certain brand of immortality with his immediate response, "Hell, for a bottle of whisky you can have the whole damn ship!"*

—When the Royal Navy returned in force to the Pacific to help finish the war with Japan, there was much to learn of Pacific warfare, 12,000 miles from home. Moreover, His Majesty's Ships suffered limitations in comparison to those of the U.S. Navy, such as their design for shorter-range operations and fitting of far fewer anti-aircraft guns. However, they were the envy of USN sailors in at least one respect.[13]

With the Allied invasion in Europe well advanced in late 1944 and aircraft carriers no longer required there in great numbers, the Royal Navy offered to assist the U.S. forces by supplying a fleet of carriers and other vessels to work in battling the Japanese in the Pacific. Initial U.S. opposition to this offer was overcome when it was established that the British force would be largely self-sufficient and would not put additional pressure on an already stretched America logistics system. The British Pacific Fleet (BPF), ultimately mobilized to aid in the potential invasion of Japan, was the most powerful strike force ever assembled by the Royal Navy. It included 6 fleet carriers and their squadrons, 4 light carriers, 2 maintenance carriers, 9 escort carriers, 4 battleships and dozens of cruisers, destroyers and lesser combat ships, as well as a huge train of supply and maintenance ships, oilers, and assorted auxiliaries.[14]

The Fleet Train supporting the BPF was a remarkable example of the national British genius for what is known as "muddling through." A masterpiece of improvisation, it was formed from what ships were available, manned by such personnel as were available, and sent out to the Pacific as they became available in various states of capability, efficiency, and morale. An American remarked, after coming alongside one of the British Fleet train vessels (an old tramp steamer), that "the work of getting the lines across was being carried out by a Geordie mate (in a waistcoat and bowler hat) assisted by three consumptive Chinamen."[15]

The BPF Fleet Train was the most extraordinary motley collection of shipping ever assembled in British maritime history. It included Norwegian masters and Chinese deckhands, Dutch mates and Lascar firemen, Royal Navy captains and Papuan winchmen. Commonwealth Navy personnel abounded, with officers and men from Australia, New Zealand, South Africa, India and Canada. Ships ranged from brand new to 30 years old: There was a Panamanian collier, a Dutch hospital ship, a Panamanian tanker, and Norwegian and Belgian ammunition ships. There were also floating repair ships, floating docks and, presumably for fleet morale, a floating brewery.[16]

Providing other fleet requirements were ships and netlayers, salvage tugs, water distilling ships, aircraft ferry ships, aircraft maintenance ships, and armament stores, air naval stores, and victualling storage and supply ships, with personnel of different nationality, different charter parties, and articles of agreement. The problems of administration were enormous. U.S. Navy men, with their modern Service Force ships, each one commissioned and under naval discipline, looked upon the British Fleet Train (Task Force 117) with frank amazement.[17]

Only the stores-issuing ships are identified in the table. While many Fleet Train units, particularly the oilers, were a part of the Royal Fleet Auxiliary (RFA), all of the stores ships were civilian merchant vessels.

**British Fleet Train Stores-Issuing Ships (August 1945)**

| **Stores-Issuing Ships Victualling)** | **Air Stores-Issuing Ships** |
|---|---|
| British SS *Fort Alabama* (B577) | British SS *Fort Corville* (B531) |
| British SS *Fort Constantine* (B578) | British SS *Fort Langley* (B532) |
| ing British SS *Fort Dunvegan* (B579) | **Mine-Issuing Ship** |
| British SS *Fort Edmonton* (B580) | British SS *Prome* (B432) |
| British SS *City of Dieppe* (B558) | **Armament-Stores Carriers** |
| **Stores-Issuing Ships (Naval)** | Danish MS *Gudrun Maersk* (B538) |
| Norwegian MS *Bosphorus* (B557) | British MV *Kistna* (B542) |
| British SS *Glenartney* (B584) | British MV *Kola* (B543) |
| British SS *Fort Providence* (B582) | **Armament Stores-Issuing Ships** |
| British SS *Fort Wrangell* (B583) | Australian MV *Corinda* (B536) |
| British MV *Hickory Burn* | British SS *Darvel* (B537) |
| British MV *Hickory Dale* | Norwegian MS *Hermelin* (B539) |
| British MV *Hickory Glen* | British SS *Heron* (B540) |
| British MV *Hickory Stream* | British SS *Kheti* (B541) |
| Dutch SS *Jaarstroom* (B562) | British MV *Pacheco* (B544) |
| (unknown country) SS *Marudu* (B563) | Belgian SS *Prince de Liege* (B545) |
| Norwegian MS *San Andres* (B564) | Belgian SS *Prinses Maria-Pia* (B546) |
| (unknown country) SS *Slesvig* (B565) | Danish MS *Robert Maersk* (B547) |
| | Danish MS *Thyra S.* (B548)[18] |

## ROYAL FLEET AUXILIARY

Importantly, Royal Navy discipline existed aboard Royal Fleet Auxiliary (RFA) units of the Fleet Train—many of the oilers, one water carrier, and one distilling ship. The Fort ships listed above, became members of the RFA after the war.[19]

**Royal Fleet Auxiliary established in 1905**

**RFA personnel wear Merchant Navy rank insignia with naval uniforms**

**Personnel are part of the naval service and are under naval discipline**

**Ships carry the prefix RFA, and fly the Blue Ensign defaced with an upright gold killick anchor**

## BPF SUPPORTED FROM ADMIRALTY ISLANDS

Original plans to base the Fleet Train at Sydney proved untenable, it being 3,500 miles to where British Pacific Fleet carrier forces would be conducting combat operations against the Empire of Japan. From 26

March to 20 April 1945, while supporting the invasion of Okinawa, the British Pacific Force was responsible for neutralizing Japanese air bases in the Sakishima Islands and on Formosa (Taiwan), which were a constant threat to the Allies from the southwest. Gunfire and air attack were used against potential Kamikaze staging airfields that might otherwise be used to support attacks against U.S. Navy ships at Okinawa. The Sakishima Island Group, southwest of Okinawa, are a part of the same Ryukyu Island chain as is Okinawa.[20]

Map Preface-6

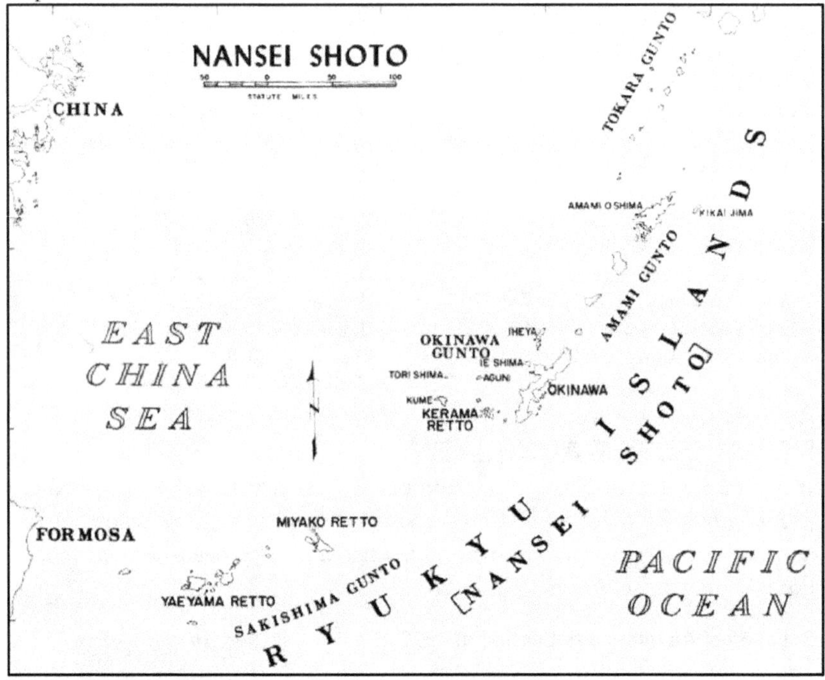


Ryukyu Islands

With American assistance, the defeat of a small Japanese garrison and a great deal of hard work, a base was constructed on Manus in the Admiralty Islands. Though nominally British, 37,000 Americans were stationed there as well, making the British Fleet as much a "visitor" as an owner. Apparently, Seeadler Harbor was not greatly beloved by at least one British sailor—perhaps because of its isolation and lack of amenities ashore, as Sydney would have offered—who described the island as "Scapa Flow with bloody palm trees."[21]

# BATTLE HONOURS AND BATTLE STARS

Photo Preface-5

Battle Honours board of the Royal Australian Navy infantry landing ship HMAS *Manoora*, depicting the ship's battle honours and badge.
Australian War Memorial photograph 118843

The ships of the Royal Navy and the other Commonwealth navies, with successful war service, earn "Battle Honours." Because His/Her Majesty's warships (whether RN, or units of other Commonwealth nations) do not carry Army regimental colours, battle honours are instead displayed on a battle honour board. Traditionally teak, this solid wooden board is mounted on the ship's superstructure, carved with the ship's badge and scrolls naming the ship and the associated honours. The board is either completely unpainted, or with the lettering painted gold. To pay tribute to past ships of the same name, their honours are displayed as well. Battle Honours (which date back to the year 1588, when 'ARMADA 1588' was authorized for the first honor ever) are awarded for six types of action:

- Fleet or Squadron Actions
- Single-ship or Boat Service Actions
- Major Bombardments
- Combined Operations
- Campaign Awards
- Area Awards[22]

Photo Preface-6

Ribbons board on the bridge wing of the battleship USS *New Jersey* (BB-62). Her Asiatic-Pacific campaign ribbon, with one silver star and four bronze stars (denoting nine battle stars), is the one located in the center, third row from the top. Courtesy of John Werda

The United States had a similar method of battle honours for naval vessels and their crews based on campaign ribbons and "Battle Stars." Naval personnel serving in the Pacific in World War II warranted sporting on their uniform blouses an unadorned Asiatic-Pacific campaign ribbon. Those whose ships earned one or more battle stars were authorized to affix representative stars to their campaign ribbon. Ships similarly proudly garnished their Asiatic-Pacific campaign ribbon, displayed with other type ribbons earned, on their deckhouses.

The *Navy and Marine Corps Awards Manual*, 1953, specified that U.S. Navy ships and units had to meet one of the following criteria to be considered to have participated in combat operations (and thereby earn a battle star):

- Engaged the enemy
- Participated in ground action
- Engaged in aerial flights over enemy territory

- Took part in shore bombardment, minesweeping, or amphibious assault
- Engaged in or launched commando-type raids or other operations behind enemy lines
- Engaged in redeployment under enemy fire
- Engaged in blockade of Korean waters (Korean War)
- Operated as part of carrier task groups from which offensive air strikes were launched
- Were part of mobile logistic support forces in combat areas[23]

## USS *LIBRA*'S COMBAT SERVICE IN THE SOLOMONS, CENTRAL PACIFIC, PHILIPPINES, AND JAPANESE HOME WATERS

Photo Preface-7

USS *Libra* (AK-53) working cargo in the port of Wellington, New Zealand, in July 1942 while preparing for the upcoming Guadalcanal campaign, in September 1942. National Archives photograph

While entertaining, the movie *Mr. Roberts* suggests to viewers that duty aboard a U.S. Navy cargo ship in the Pacific was tedious, but benign, with the only excitement being "high jinks" by the crew. This tranquil portrayal is belied by the fact that 119 cargo ships—67 attack cargo ships (AKA), 49 cargo ships (AK), and 3 stores-issuing ships (AKS)—

collectively earned 254 battle stars in the Pacific Theater in World War II. Top honors went to the cargo ship USS *Libra* (AK-53), which was later redesignated an attack cargo ship (AKA-12).

Connection of the geographic areas in which *Libra* earned her nine battle stars would yield a general track of the movement of Allied naval forces across the Pacific into the heart of the Japanese empire:

- Guadalcanal-Tulagi landings
- Capture and defense of Guadalcanal
- Guadalcanal – Third strike
- Consolidation of Southern Solomons
- New Georgia Group operation: New Georgia-Rendova-Vangunu occupation
- Treasury-Bougainville operation: Occupation and defense of Cape Torokina
- Marianas operation: Capture and occupation of Guam
- Luzon operation: Lingayen Gulf landing
- Iwo Jima operation: Assault and occupation of Iwo Jima

Twenty-one AKs/AKAs (including *Libra*) were awarded four or more battle stars for duty in the Pacific Theater. Some ships also earned additional stars for service in the American and/or European Theaters. The three AKSs (stores-issuing ships) each earned three or fewer stars in the Pacific. (Appendix E provides summary information for all the AKs, AKAs, and AKSs awarded battle stars for Pacific combat operations.)

**Top Twenty-one Pacific Theater AK/AKA Battle Star Recipients**

| Ship | Battle Stars | Ship | Battle Stars |
|---|---|---|---|
| *Libra* (AKA-12) | ★★★★★★★★★ | *Auriga* (AK-98) | ★★★★★ |
| *Aquarius* (AKA-16) | ★★★★★★★★ | *Electra* (AKA-4) | ★★★★★ |
| *Alcyone* (AKA-7) | ★★★★★★★ | *Fomalhaut* (AKA-5) | ★★★★★ |
| *Thuban* (AKA-19) | ★★★★★★★ | *Hercules* (AK-41) | ★★★★★ |
| *Virgo* (AKA-20) | ★★★★★★★ | *Mercury* (AK-42) | ★★★★★ |
| *Centaurus* (AKA-17) | ★★★★★★ | *Arned* (AKA-56) | ★★★★ |
| *Jupiter* (AK-43) | ★★★★★★ | *Bellatrix* (AKA-3) | ★★★★ |
| *Titania* (AKA-13) | ★★★★★★ | *Betelgeuse* (AKA-11) | ★★★★ |
| *Alhena* (AKA-9) | ★★★★★ | *Capricornus* (AKA-57) | ★★★★ |
| *Almaack* (AKA-10) | ★★★★★ | *Chara* (AKA-58) | ★★★★ |
| *Alshain* (AKA-55) | ★★★★★ | | |

## THREE AKAS AWARDED NINETEEN TOTAL NUCS

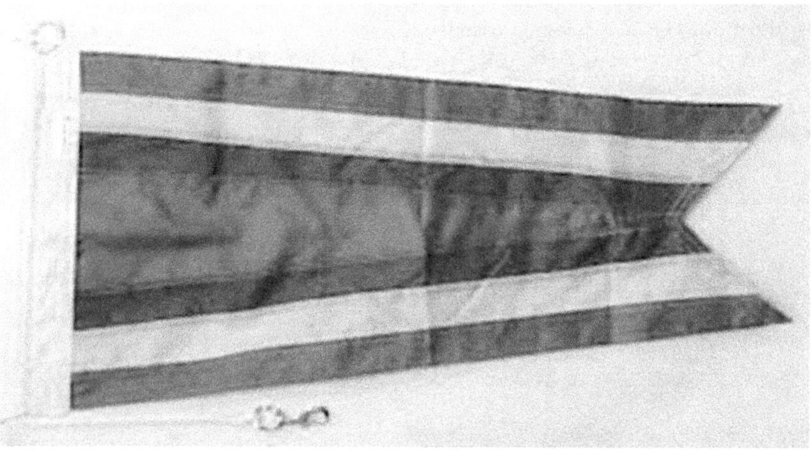

Navy Unit Commendation (NUC) - Authorized 1944. Awarded for outstanding heroism in action against the enemy or extremely meritorious service in support of military operations. Pennant is hunter green with bands of royal blue, Spanish yellow, and scarlet along the upper and lower edges (blue at the edge).

Three attack cargo ships from the above group—*Libra*, *Alcyone*, and *Titania*—collectively received an astounding nineteen Meritorious Unit Commendations (MUCs) during the war. The first was earned for the landings in North Africa, America's initial entry into the war in Europe. The final one was for landings on Borneo—an operation that cost the lives of many Australian soldiers, and which remains controversial to this today, regarding whether questionable gains were warranted so late in the war.

### MUC Recipients

#### EUROPE

North African Landing (Africa)
USS *Titania* (AKA-13) 8-14 November 1942

Sicilian Occupation (Italy)
USS *Alcyone* (AKA-7) 10 July 1943

#### PACIFIC

Guadalcanal – Tulagi Landings (Solomon Islands)
USS *Libra* (AKA-12) 7-9 August 1942

Capture and Defense of Guadalcanal (Solomon Islands)
USS *Libra* (AKA-12) 11 November 1942

Battle of Rendova (Solomon Islands)
USS *Libra* (AKA-12) 30 June 1943

Battle of Bougainville (Solomon Islands)
USS *Libra* (AKA-12) 1-8 November 1943
USS *Titania* (AKA-13) 1-8 November 1943

Gilbert Islands (Central Pacific)
USS *Alcyone* (AKA-7) 20 November 1943

Kwajalein Atoll (Central Pacific)
USS *Alcyone* (AKA-7) 31 January-6 February 1944

Marianas Operation (Central Pacific)
USS *Alcyone* (AKA-7) 15 June-22 July 1944
USS *Libra* (AKA-12) 21-25 July 1944
USS *Titania* (AKA-13) 21-26 July 1944

Leyte Landings (Philippines)
USS *Alcyone* (AKA-7) 20 October-18 November 1944
USS *Titania* (AKA-13) 20 October-13 November 1944

Lingayen Gulf Landings (Philippines)
USS *Alcyone* (AKA-7) 9-13 January 1945
USS *Libra* (AKA-12) 11 January 1945
USS *Titania* (AKA-13) 8-12 January 1945

Iwo Jima Occupation (Japan)
USS *Libra* (AKA-12) 19 February-6 March 1945

Borneo Operations (Netherlands East Indies)
USS *Titania* (AKA-13) 27 April-5 May 1945

## ODE TO WAR SERVICE OF STORES-ISSUING SHIPS

George Ruxton, South Australia's secretary of the RAN Allied Chinese Ships' Association, noted "the record of service for these ships ranks high among the units that were formed to meet emergencies in the Pacific war zone" about the old, "clapped out" former Chinese steamers, pressed into wartime duties when their contributions were desperately needed. An engineer aboard HMAS *Po Yang*, Stoker Butler, paid tribute to her, as well as to his shipmates, in verse:

So proudly we hailed her
Though she's battered and worn,
And there's no doubt about it,
She came in for scorn
By men who were thoughtless
Of a job so well done,
While she carried ammo
To defeat Jap and the Hun[24]

## CARGO SHIPS FIGHT WITH BOOMS

> *It is said that the wars are fought with guns and that Navy ships fight with guns, but the AK-110 fought this war with cargo booms in a manner in which the American people and the American Navy may be well proud.*
>
> —From the war history of the cargo ship USS *Alkes* (AK-11)

The above sentiment metaphorically describes, fighting the war with cargo booms, in reference to the contributions of cargo ships to the Pacific war. However, by happenstance, as shown in the sketch, USS *Mercy* (AK-42) literally employed a boom, in addition to her guns, in combat action to down an enemy aircraft. (Chapter 16 offers a more detailed account of this incident.)

USS *Mercury*, Ship's History, 17 April 1946

With this introduction to the stalwart cargo ships of the U.S. Navy, Royal Australian, and British and other Allied merchant navies, it is time to stand out of port, and (vicariously) ply enemy waters and dangerous coasts with the crews of these ships.

Photo Preface-8

Cover art by Richard DeRosset depicting gun crews aboard the cargo ship USS *Betelgeuse* (AK-28) shooting down two Japanese torpedo planes. Her attackers were part of a group of twenty-one Mitsubishi, type 96, heavy bombers attacking shipping off Guadalcanal, Solomon Islands, on 12 November 1942.

# 1

# *Alchiba* Torpedoed Twice

ALCHIBA *undoubtedly would have sunk after the first torpedo struck had she not been quickly beached. Fully laden and open for the discharge of cargo she was very vulnerable to both flooding and fire. The remarkably persistent and skillful efforts of her entire crew not only saved her but also most of her cargo, sorely needed at that time. The history of* ALCHIBA, *from the time she was torpedoed on 28 November 1942 until she was placed back in service on 7 August 1943, is marked by the inflexible determination of her personnel. This factor is the key to her ultimate survival and return to service.*

—From report titled, "USS *Alchiba* (AKA6) Torpedo Damage Solomon Islands 28 November and 7 December 1942."
(When damaged, her hull number was AK-23.)[1]

Photo 1-1

USS *Alchiba* aground and on fire off Lunga Point, Guadalcanal, late November 1942.
U.S. Marine Corps photograph #USMC 52796

On the morning of 28 November 1942, the cargo ship USS *Alchiba* (AK-23) was anchored off Guadalcanal in the Solomon Islands, making preparations for the discharge of cargo. She had left Noumea, New Caledonia, a week earlier, on the 21st, in company with the attack transport USS *Barnett* (APA-5), and destroyers USS *Lardner* (DD-487), *Lamson* (DD-367), and *Hughes* (DD-410). The French Colony of New Caledonia (comprising dozens of islands) lay to the north-northeast of Brisbane, Australia, across the Coral Sea.[2]

In addition to her cargo of aviation gasoline, bombs, ammunition, and provisions for American forces in the Guadalcanal-Tulagi area, *Alchiba* also towed a flat-top barge loaded with Marston mats. This type of perforated steel matting was typically used for the rapid construction of temporary runways and landing strips.[3]

Photo 1-2

Landing craft clustered offshore, during landing operations at Lunga Point, Guadalcanal, in November 1942.
National Archives photograph #80-G-30521

At 0616 on 28 November, *Alchiba* was struck by a torpedo (port side, abreast No. 2 hold) fired by the Japanese midget submarine *Ha-10*, carried to the area on the back of a mother submarine, *I-16*. The approach of the torpedo was not observed, nor was the submarine that had fired it. The detonation of the warhead was accompanied by a column of orange-colored flame and dark brown smoke, with its destructive effects immediate, as described in a report of torpedo damage to the cargo ship:

Gasoline from ruptured drums stowed in No. 1 deep tanks and oil from ruptured fuel oil tanks in the double bottom were blown up through the "tween deck spaces of No. 2 hold. The vapors ignited almost simultaneously with the explosion, resulting in a severe fire which spread into No. 1 hold through fragment holes in the bulkhead between Nos. 1 and 2 holds. Small arms ammunition on upper and lower 'tween decks of No. 2 hold soon began to explode.[4]

Diagram 1-1

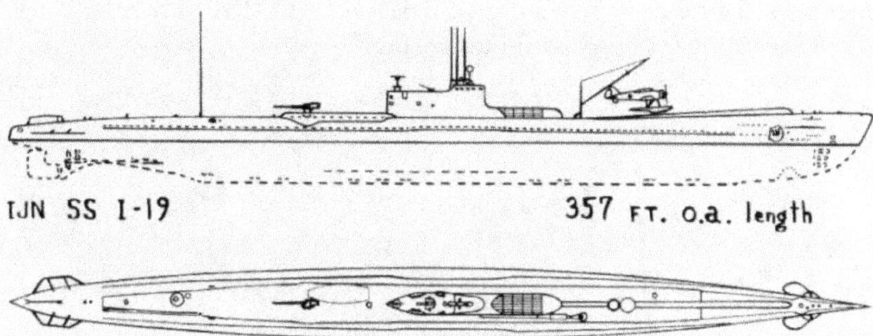


Japanese submarine *I-19* of the *I-15*-class (same type as the *I-16*).
Naval History and Heritage Command photograph #NH 111756

Photo 1-3

Japanese Type-*A* midget submarine aground on an eastern Oahu beach, following attempts to enter Pearl Harbor during the 7 December 1941 Japanese attack.
Naval History and Heritage Command photograph #NH 91331

As a result of the hole blasted in *Alchiba*'s hull, and associated ruptured bulkheads, Nos. 1, 2, and 3 holds flooded, causing a port list which increased rapidly. With his ship afire and listing badly, Comdr. James S. Freeman beached *Alchiba* ten minutes after the torpedo hit to prevent her total loss. Her crew then fought the fires aboard for 4 grueling days and nights, in spite of exploding cargo and other hazards, before bringing them under control, on 2 December. Fires in the 'tween deck spaces were fought by *Alchiba* sailors, assisted by additional hose lines passed over by the fleet tug USS *Bobolink* (AT-131), a former World War I vintage *Lapwing*-class minesweeper.[5]

Photo 1-4

Fighting fires in USS *Alchiba*'s (AK-23) forward holds, with the assistance of the fleet tug USS *Bobolink* (AT-131), circa late November 1942.
U.S. Marine Corps photograph #USMC 52795

## *ALCHIBA* TORPEDOED A SECOND TIME

Unloading of *Alchiba*'s cargo continued during the period that fires were being fought on board and, as soon as they were under control, salvage operations commenced. Primary salvage work was being carried out, on 7 December, when the cargo ship was torpedoed a second time. At 0800 that morning, a torpedo fired by another midget submarine *Ha-38* launched from Japanese submarine *I-24*, struck *Alchiba* on her port side at about frame 50. Detonation of the warhead caused considerable structural damage and flooding of the engine room and No. 4 hold. As water rushed in, the ship began to settle aft and rise forward.[6]

Unlike the first torpedo attack, on 28 November, there was brief warning of this one. At 0759, what appeared to be the conning tower of a midget submarine was sighted close aboard the port quarter of the *Alchiba*. Almost immediately, a torpedo struck her, and a second one was seen to pass close under her stern. Much damage resulted:

> The blast from this torpedo appeared to be much heavier than the [earlier] one forward. It was accompanied by a "tearing" sound, like tearing cloth in a gigantic manner. No smoke or flame was noticed. Water and debris were thrown into the air to at least masthead height. Explosive fumes were noted in the crew's quarters immediately after the water settled. No fires resulted from this hit.[7]

As reported in the damage report, the detonation of this torpedo blasted an irregular opening in her shell plating, which extended about 40 feet between frames 42 and 59, and about 25 feet vertically across E, F, G, and H strakes. Damage to *Alchiba*'s hull extended both upward and downward from the opening:

> Above the opening the indentation of the shell extended up through the "K" strake to the shelter deck edge. Below the opening, "D" and "C" strakes at the turn of the bilge were deflected inward and split. Distortion of "A" and "B" strakes on both sides of the keel occurred between frames 43 and 58. The flat and vertical keel plates were buckled considerably but held intact.[8]

Other pages of the report, detail much additional harm done to topside areas, and to bulkheads, tanks, machinery and equipment inside the ship. Fortunately, as she settled by the stern, *Alchiba* did not slide off the sand and gravel ledge located a little off the beach on which she came to rest.[9]

The disheartening effect of the second hit, after so much effort by *Alchiba*'s crew to control the damage from the first hit, can well be imagined. Nonetheless, the unwavering determination of personnel to save their ship remained. Salvage efforts, begun shortly before she was torpedoed a second time, continued. As salvage tugs became available, they bent their efforts toward the primary task of restoring buoyancy. Temporary shoring and patching of bulkheads, and pumping of Nos. 1 and 3 holds continued for two weeks.[10]

Finally, on 27 December, the fleet tug *Navajo* (AT-64) and submarine rescue ship *Ortolan* (ASR-5) pulled *Alchiba* free of the bottom, and towed her across Sealark Channel to Tulagi.[11]

Map 1-2

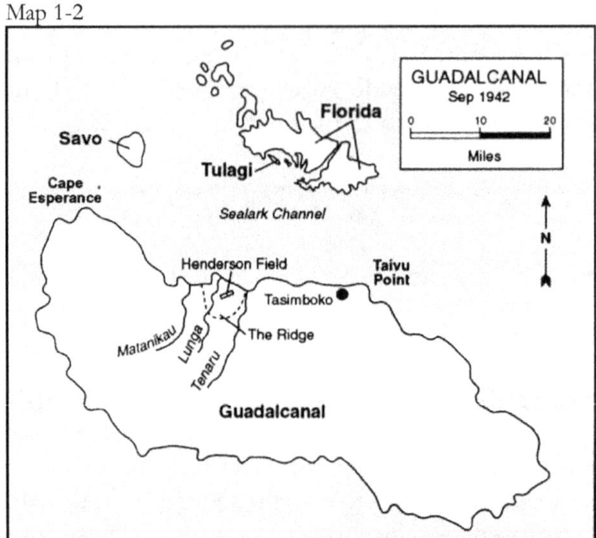


Guadalcanal, Savo, Tulagi, and Florida islands

Map 1-3

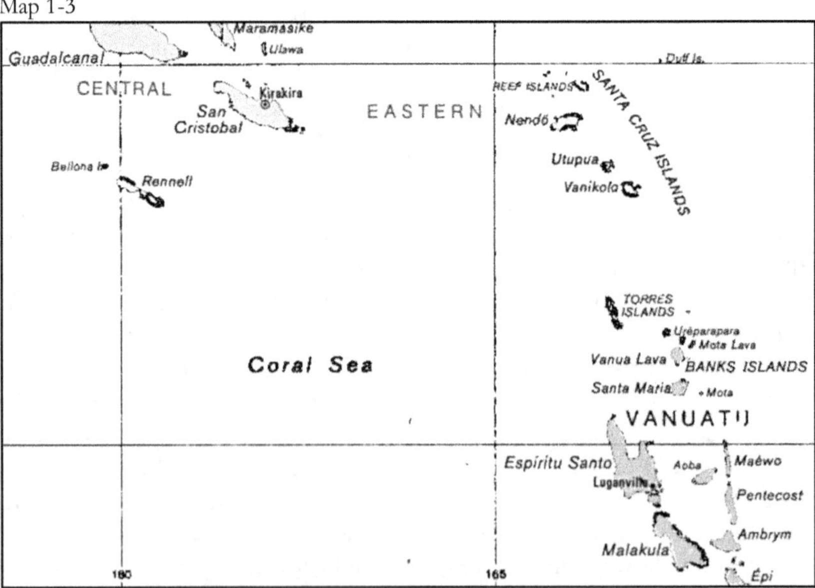


Southern Solomon, Santa Cruz, and Vanuatu islands

At Tulagi, the remainder of *Alchiba*'s cargo, with the exception of a number of drums of aviation gasoline in port deep tank No. 1, was discharged. Repair work aboard the ship consisted largely of replacing sufficient structural strength to allow her to be towed to a more capable repair base. On 19 January 1943, *Alchiba* was taken in tow by *Navajo*, assisted by *Ortolan* on her port quarter, to Espiritu Santo in the Vanuatu Islands, which lay to the southeast. While at "Santo," on 1 February, *Alchiba*'s designation was changed from AK-23 to AKA-6.[12]

The attack cargo ship got under way on her own power, on 6 May 1943, and made the Pacific crossing to Navy Yard Mare Island in the San Francisco Bay. Following her arrival, on 3 June, repairs and alterations were begun that would lead to her return to service, on 7 August 1943. In addition to receipt of battle stars for the Guadalcanal-Tulagi landings (7-9 August 1942) and Capture and defense of Guadalcanal (15 October-28 November 1942), *Alchiba* would earn one more during the war. From 1-13 November 1943, she took part in the occupation/defense of Cape Torokina as part of the capture of Treasury and Bougainville islands.[13]

## LAURELS FOR SHIP AND CREW

Example of the type of award that USS *Alchiba* received. Presidential Unit Citation Pennant awarded to Task Group 22.3 for its successful capture of German submarine *U-505*. The above PUC pennant was flown from the escort carrier USS *Guadalcanal* (CVE-60) and presented to the Museum of Science and Industry in Chicago by Admiral Daniel V. Gallery. It is heavily soiled from the smokestacks of the ship and frayed from whipping in the wind.

If ever there was a crew that lived up to Captain James Lawrence's dying words, "Don't give up the ship" (in reference to his 38-gun wooden-hulled, three-masted heavy frigate USS *Chesapeake* during the War of 1812), it was that of the unlikely cargo vessel USS *Alchiba* (AK-23). The officers and men serving under Comdr. James S. Freeman, proved that American sailors would exhibit the greatest of bravery, regardless of whether they were on the newest warship or a former merchant ship. *Alchiba* became the only cargo ship to be awarded a Presidential Unit Citation (August – November 1942 – South Pacific). The PUC is the highest award a military unit may receive for heroism, and is considered the equivalent of the Navy Cross for an individual. Comdr. James S. Freeman, the ship's commanding officer and a future rear admiral, was awarded the Navy Cross for heroism, and her executive officer, Comdr. Howard R. Shaw, the Silver Star.[14]

Freeman's medal citation follows:

> The President of the United States of America takes pleasure in presenting the Navy Cross to Commander James Shepherd Freeman, United States Navy, for extraordinary heroism and distinguished service in the line of his profession as Commanding Officer of the Cargo Ship U.S.S. *ALCHIBA* (AK-23), during operations in the Solomon Islands during the period 7 August 1942 through 28 November 1942. Commander Freeman successfully maneuvered his cargo ship through enemy-infested waters and, in spite of attacks by high altitude bombers and aerial torpedoes, landed supplies and equipment for our forces during the initial operations of occupying the Solomon Islands. On the third subsequent trip to this area his ship was attacked and hit by an enemy submarine. Commander Freeman beached his ship, despite numerous gasoline and ammunition explosions, in order that the cargo and vessel might be saved. Commander Freeman's inspiring leadership and the valiant devotion to duty of his command contributed in large measure to the outstanding success of these vital missions and reflect great credit upon the United States Naval Service.[15]

**2**

# U.S. Pacific Fleet Service Force

On 1 February 1941, the existing United States Fleet was organized into three separate fleets: Atlantic, Pacific, and Asiatic. Four months later, on 1 June, three task forces were created within the Pacific Fleet. Number One consisted of the Battle Force; Two was primarily aircraft carriers as a Reconnaissance and Raiding Force; and Three, a Scouting Force to support Expeditionary and Amphibious Operations.[1]

In July 1941, the Base Force (later to become Service Force, Pacific Fleet) was being reorganized to accomplish both peace and war missions. The drumbeats of war were sounding in both the Atlantic and in the far Pacific and, while the size of the entire United States Fleet was then extremely small, Navy leadership was endeavoring to place itself on an efficient and ready for war footing, as available resources and circumstances permitted. In this new organization, there were to be four Train Squadrons:

- Squadron Two: Harbor services
- Squadron Four: Transportations of personnel, landing force equipment, etc. (later to grow into the Amphibious Force)
- Squadron Six: Offensive and defensive mining and general services
- Squadron Eight: Transportation of bulk cargo[2]

On 7 December 1941, when the Japanese attacked Pearl Harbor, Train Squadron Eight, which came into being in July 1941, consisted of a mere eighteen ships. They were the stores issue ships *Antares* and *Castor*; ammunition ship *Pyro*; provision ships *Bridge*, *Arctic*, *Boreas*, and *Aldebaran*; and eleven fleet tankers—*Kanawha*, *Cuyama*, *Brazos*, *Neches*, *Ramapo*, *Sepulga*, *Tippecanoe*, *Neosho*, *Platte*, *Sabine*, and *Kaskaskia*.[3]

Responsibility for these ships when on the west coast of the United States, fell to the subordinate Service Force command at San Francisco, commissioned, in June 1941, as the West Coast representative of Base Force. On 7 December, there were only five units of Train Squadron

Eight at Pearl Harbor—the stores issue ships *Antares* and *Castor*, ammunition ship *Pyro*, and tankers *Neosho* and *Ramapo*.[4]

The cargo ship *Vega* (AK-17) was at nearby Honolulu Harbor. Commissioned on 21 December 1921, she was assigned to the U.S. Naval Transportation Service. *Vega* had arrived there the previous afternoon, and moored to berth 31A. Shortly after 0800 the following morning, 7 December, explosions were heard in the direction of Pearl Harbor. After receipt of a message from commander in chief, Pacific Fleet, advising all ships present of a Japanese air raid on Pearl Harbor, General Quarters was sounded immediately and 3-inch and 5-inch ammunition ready boxes were filled to capacity. Stevedores engaged in discharging cargo assisted in the transfer of ammunition.[5]

Photo 2-1

Cargo ship USS *Vega* (AK-17) at the Brooklyn Navy Yard, date unknown.
Naval History and Heritage Command photograph #L45-295.05.01

About 0830, Japanese bombers were seen circling the city, with the insignia on the planes plainly visible through binoculars. The initial formation consisted of six planes flying in a V-formation, headed in the direction of the docks. *Vega* opened fire with her 3" anti-aircraft guns. This formation of planes circled back over the city without, it appeared, having dropped any bombs.[6]

Another formation of planes approached overhead at about 0930 and began circling, evidently in preparation for a dive-bombing attack on the waterfront docks and oil tanks. *Vega*'s guns resumed firing, apparently joined by shore batteries at the harbor entrance. The planes released several bombs: one appeared to have dropped on the Honolulu Gas Company property; a second fell astern of the cargo ship, about 30 yards away in the channel; and a third struck Sand Island approximately 300 yards distant from *Vega*. Concurrently a burst of machine gun fire struck the water off the port bow of the ship.[7]

*Vega*'s action report describes the effect of her gunfire on the enemy aircraft and bomb-induced casualties to personnel on the pier, and highlighted that sailors attired in stark white uniforms, presented easy visual targets to the enemy:

> A total of fourteen rounds of 3"-23 caliber were fired. One burst was seen to rock one of the planes badly and appeared to bring about a marked change of course of the formation.
>
> At approximately 1020, a small bomb or shrapnel fragment landed on the pavement on the starboard quarter of the ship, distant about 200 yards, injuring a number of civilian workmen.
>
> It was noted that the undress white uniforms of the ship's company stood out in sharp contrast to the dark gray paint of the ship, thus making conspicuous targets for machine gun fire.[8]

## *ANTARES* SIGHTS JAPANESE MINI SUBMARINE

At 0630 on the morning of the attack, the stores issue ship *Antares* (AKS-3) arrived from Canton and Palmyra islands off the entrance to Pearl Harbor, with a 500-ton steel barge in tow. While awaiting a tug to which it was to transfer the tow before entering port, the supply vessel sighted what appeared to be a small submarine with upper conning tower awash and periscope partly raised, about 1,500 yards on her starboard quarter. The destroyer USS *Ward* (DD-139) which was the inshore patrol ship in the vicinity was notified and proceeded to investigate. At 0633 a PBY-5 Catalina patrol plane from Squadron VP-14 circled and dropped two smoke pots near the object.[9]

At 0645 the *Ward* commenced an attack, firing one salvo each from her Number 1 and 3 guns, followed by depth charges, on the submarine. The engagement lasted only a minute or so. The first shot passed directly over the sub's conning tower, the second, from No. 3 gun located on the galley-deckhouse starboard side, struck the submarine at the junction of hull and conning tower. As the sub heeled over and sank, the *Ward* ran over its last position and dropped a pattern of four depth charges set to explode at a depth of one-hundred feet. The resulting pressure wave from the explosions lifted the submarine, causing its bow to surface, surrounded by concentric circles of foamed water and falling spray, before it disappeared on its final dive to the bottom. A large quantity of oil bubbling to the surface gave evidence of its demise.[10]

At 0715, the rescue ocean tug *Keosanoua* (ATR-125) arrived to receive the tow. Forty-three minutes later, explosions were observed in Pearl Harbor and Japanese planes were seen delivering an attack. *Antares* came under machine gun attack from a Japanese plane at 0800, and was hit in a few places topside. Several bomb and numerous shell fragments continually fell in close proximity, and she was shaken by bomb bursts. Having no armament, no offensive actions were possible.[11]

After finally passing the tow, *Antares* maneuvered (zigzagging and turning to a position between the Pearl Harbor restricted area and Honolulu entrance) inshore of combatant ships. *Antares*' commanding officer, recognizing that his ship and personnel were in great jeopardy, requested permission to enter Honolulu Harbor. The Harbor Control Officer granted this request at 1054, and at 1146 *Antares* was moored at Berth 5A, having suffered no personnel or material casualties. Her crew had reacted to the emergency in an exemplary manner, with those disconnecting the tow and others on exposed stations calm and steady.[12]

## *CASTOR* AT MERRY POINT, PEARL HARBOR

> *The majority of the crew and all officers except the Commanding Officer and Executive Officer are reservists. Their steadiness and fortitude in their first action, their fine discipline and battle spirit were superb.*
>
> *Number 3 (after) 3"/23 caliber gun in action against enemy torpedo plane, at range of about 500–700 yards altitude about 500 feet bearing due aft heading across toward Ford Island observed tail assembly and about two feet of fuselage shot away.... Cannot definitely state damage caused by* Castor, *due to number of guns in action. Plane was observed to lose altitude and appeared to ground either on Ford Island or beyond.*
>
> —Commanding Officer, USS *Castor*, report on December 7th Raid.[13]

A second stores issue ship, *Castor* (AKS-1), was at Pearl Harbor. At the first warning of danger, her executive officer, Lt. Comdr. Harold Burton

Herty, USN, ordered General Quarters set at 0755, and informed the commanding officer, Comdr. Harold James Wright, USN, that the fleet was being bombed.[14]

Photo 2-2

Aerial view of the Pearl Harbor Submarine Base, with part of the supply depot beyond and the fuel farm at right, on 13 October 1941. USS *Castor* (AKS-1) is at bottom left, berthed at Merry Point, with the old derelict minelayer *Baltimore* (C-3) forward of her. National Archives photograph #80-G-451125

*Castor*'s guns were quickly brought into action. At 0800, her 3"/23 caliber AA battery and .30 caliber machine guns were firing at torpedo planes flying low and close aboard, and against dive bombers. Ready ammunition boxes were kept full, fuzes set, and gunfire maintained as targets presented themselves. Topside personnel not part of gun crews, were armed with Springfield rifles.[15]

At 0930, crewmembers were detailed to handle the tanker *Neosho*'s lines, berthing astern of *Castor*. *Neosho* had gotten under way from Berth F-4 without orders in order to clear the way for the battleship USS *Maryland* in the event the latter desired to move. During this evolution, enemy planes strafed *Castor* and *Neosho* but, fortunately, no personnel casualties or damage to the ships resulted.[16]

During action that day, several machine gun bullets hit *Castor*'s No. 3 anti-aircraft AA gun, which scored the recoil cylinder nut and barrel, but failed to knock the gun out of commission. In his report, Comdr. Harold James Wright commended actions by the gunnery officer, Ensign J. E. Kendall, D-M, USNR, and Seaman First Class R. A. Schwerdtfeger, USN, which resulted in the removal of a lighter loaded with 450 aerial depth charges from the ship's side. The lighter and its dangerous load of explosives, was moved across the slip by ship's personnel and ship's boat.[17]

## LOGISTICS SUPPORT SCARCE IN EARLY MONTHS

It was fortunate that the cargo ship *Vega*, and stores issue ships *Antares* and *Castro* emerged unscathed from the attack on Pearl Harbor. These type ships being in such short supply, as it were, during the first six months of the war (which included delaying action in the Philippines and the Battle of the Java Sea in the Netherlands East Indies), naval forces in the Southwest Pacific were pretty much on their own. They received little support from either the continental United States or from the Hawaiian area, which was having supply problems of its own. Preparations for the defense against a possible second attack were in progress, amid concerns that another attack might be supported by transports and landing craft.[18]

## INITIAL PACIFIC FLEET OFFENSIVE OPERATIONS

Pacific Fleet naval activities were limited to local patrols and hit-and-run raids on some of the Japanese-held islands, particularly in the Marshall and Gilbert groups. These raids were supported from Pearl Harbor, to which the combatant ships returned. Except for minimal fueling-at-sea by fleet tankers, there was no logistics support afloat nor provided west of the Hawaiian Islands. The major task of the then very small Service Force, and of Squadron Eight in particular, was to furnish logistics support to the Fleet at Pearl.[19]

These efforts reached their peak in June 1942, in support of the Battle of Midway, 1,100 miles west of Pearl Harbor. Fuel and food were delivered in heavier issues than ever before, and this effort was not equaled until more than a year later, when operations in the Central Pacific commenced.[20]

In the meantime, some changes took place. The name of Base Force was changed to Service Force, in April 1942. Train Squadron Eight then became Service Squadron Eight, and the title of its commander, Capt. Charles H. Maddox, was changed to ComServRon 8. Maddox remained in command until succeeded, in March 1943, by Captain (later Commodore) Augustine H. Gray. Gray would serve in this role until the squadron was decommissioned after the end of the war. By July 1942, *Antares* had gone to the South Pacific and been replaced by *Castor* as flagship for Squadron Eight.[21]

## LOGISTICS IN THE SOUTH PACIFIC

> *From the summer of 1942 to the summer of 1943, the South Pacific was the most active Pacific theater of operations. Supply lines, for the most part, ran directly from the West Coast of the mainland to bases in Samoa, the Fijis, New Zealand, and New Caledonia, and soon to the New Hebrides Islands and to the Solomons. Supplies also passed through this area en route Australia and the Southwest Pacific area, ruled off on charts as being under General MacArthur.*
>
> —Commander Service Force, U.S. Pacific Fleet, History of Service Force – forwarding of, 31 January 1946.[22]

Map 2-1

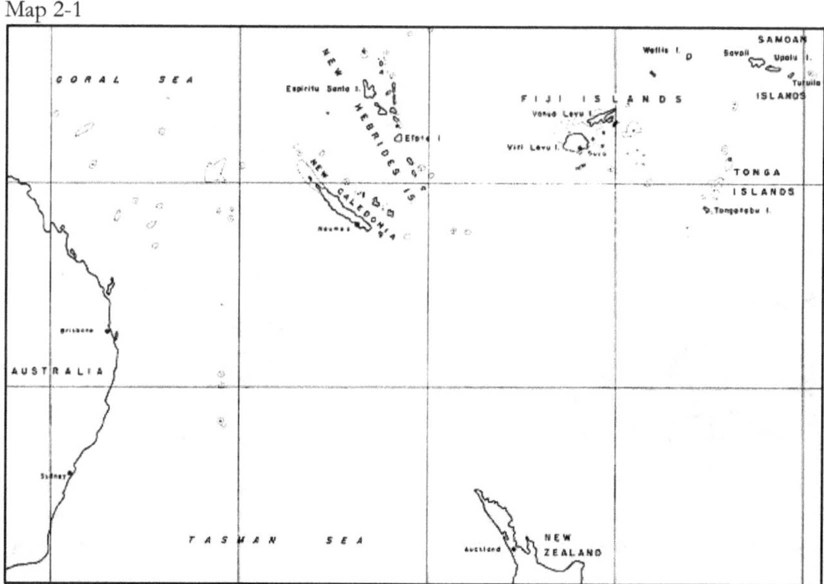


Western South Seas Islands

In the spring of 1942, Japanese forces were pushing south and southeast from areas captured in the Philippines, Netherlands East Indies, and the northern coast of New Guinea, and from older positions in the Caroline and Marshall islands. It was obvious that an attempt was being made to cut all direct supply lines and direct routes from the United States and the two "down under" British Commonwealth countries. Australia and

New Zealand were threatened, and keeping open the line of logistics with these two great bastions of the South Pacific was a strategic imperative. The northern route, west-southwest from Hawaii, was already controlled by Japanese positions. Only the southern route was available for use.[23]

From Borabora, the midway point of voyages from the U.S. West Coast or Panama to Sydney, shipping destined for Australia would have to sail through or close to a number of the South Pacific island groups. First would be the Cook Islands, then the Samoa, Tonga, and Fiji groups, and finally, the New Hebrides group and New Caledonia, forming the eastern rim of the Coral Sea. If the Allies could hold this indispensable shipping route, future offensive operations would be possible. Striking power could be built up and, when it was great enough, action could be taken to wrest the initiative from the enemy.[24]

Photo 2-3

Borabora, Society Islands of French Polynesia.
https://www.ibiblio.org/hyperwar/USN/Building_Bases/bases-24.html

The only American naval establishment south of the equator was at Tutuila, American Samoa. British and New Zealand forces had some facilities at Suva in the Fijian Islands. It was unclear whether the French civil administration at Noumea, New Caledonia, would support the Allies desire to develop a strategically located base there. Adding to this uncertainty, it was not known how near the enemy it would be safe to establish major supply bases, but it was desirable to provide logistics support to the limited South Pacific Force as near the area of potential operations as possible.[25]

# Landings at Guadalcanal and Tulagi

*The situation at 2400 on August 8th [1942] may be summarized as follows: The Marines held Tulagi, Gavutu, and Tanambogo, although several nests of enemy snipers remained on the last island. We had put 7,500 men ashore in that area at the cost of 248 casualties. Japanese casualties, however, were virtually 100 percent, or about 1,500. At Guadalcanal we had about 11,000 men ashore, with a fairly firm hold on the northern shore from Kukum to Koli Point. The airfield we had captured was ready for almost immediate use by fighters and dive bombers. With few tools and no barbed wire, our troops were busily digging in to defend their valuable beachhead. The enemy forces on Guadalcanal had all withdrawn to the bush.*

—*Combat Narratives, Solomon Islands Campaign: I, The Landing in the Solomons 7-8 August 1942* (Washington DC: Office of Naval Intelligence, 1943).[1]

Map 3-1

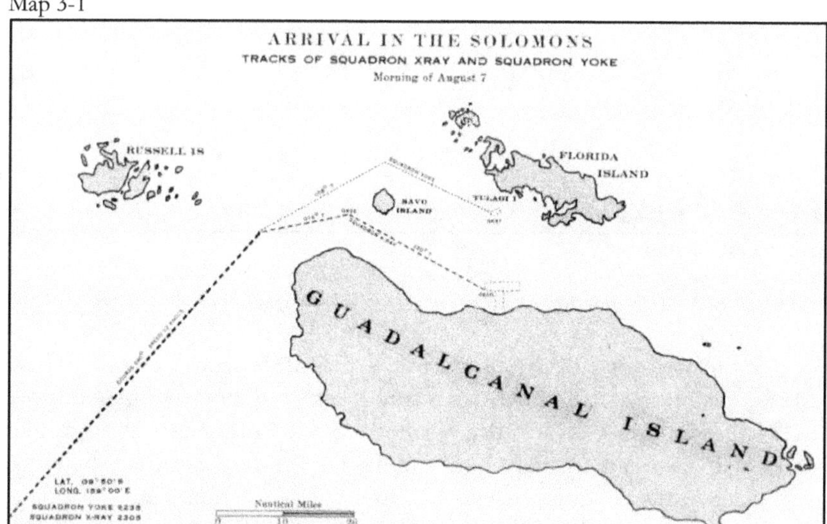


Guadalcanal and adjacent islands
*Combat Narratives, Solomon Islands Campaign: I, The Landing in the Solomons 7-8 August 1942* (Washington DC: Office of Naval Intelligence, 1943)

East of New Guinea, a barrier of islands extends from the Bismarck Archipelago (of which New Britain and New Ireland are a part) in a NW-SE direction to New Caledonia off the northeast coast of Australia. As the potential site of a series of air and sea bases, these islands offered the Japanese, in the spring of 1942, the possibility of severing Australia and New Zealand's lifeline to America's west coast. Conversely, the islands provided a means by which Allied naval forces might advance northwestward to the Japanese bases in the Carolines, bypassing the Gilbert and Marshall islands, and possibly move westward into the southern Philippines.[2]

Map 3-2

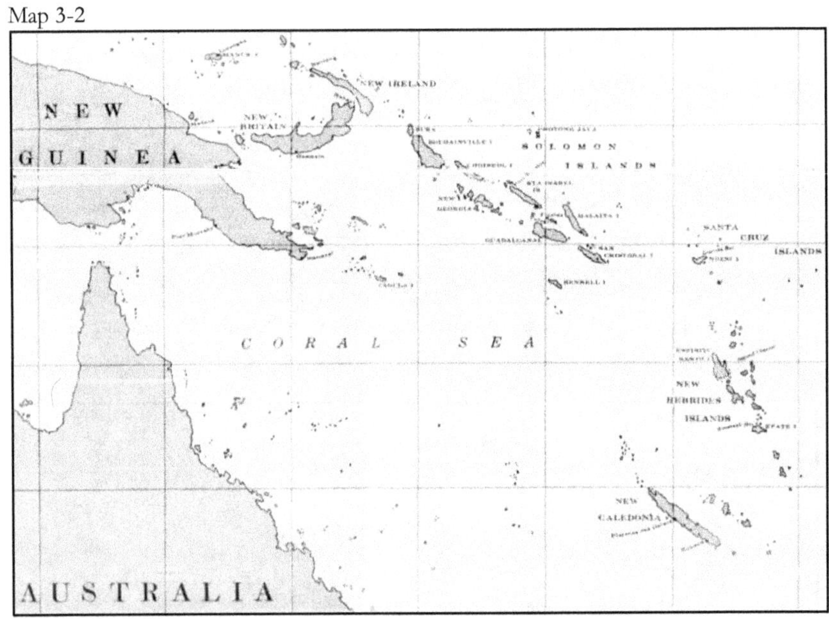


Southwest Pacific
*Combat Narratives, Solomon Islands Campaign: I, The Landing in the Solomons 7-8 August 1942* (Washington DC: Office of Naval Intelligence, 1943)

Japan was the first to attempt to exploit this strategic route when, as early as January 1942, her naval forces invaded and established bases in New Britain, and in the northern Solomons, both then under Australian mandate. In April, the enemy moved farther down the chain to Tulagi. This small island, nestled in a bay at Florida Island opposite Guadalcanal, was a British possession. Seizure of Tulagi was part of a first step to also take Port Moresby in Papua in order to secure air mastery of the Coral Sea, before pursuing much more ambitious aims to knock the United States Navy out of the war.[3]

Noted naval historian Samuel Eliot Morison described the Imperial Japanese Navy's grand design, and intended follow-on actions, had not the U.S. Navy prevailed in the Battle of Midway:

> The Combined Fleet would cross the Pacific to "annihilate" the United States Pacific Fleet, and at the same time capture Midway Island and the Western Aleutians; then set up a "ribbon defense" anchored at Attu, Midway, Wake, and the Marshalls and Gilberts. This effort, had it not been thwarted by the Battle of Midway, would have been followed by invasion of New Caledonia, the Fijis and Samoa, and Australia.[4]

The Battles of Coral Sea and Midway were fought 3-8 May, and 3-4 June, respectively, but a decision by the Combined Chiefs of Staff to make Tulagi the Allies' principal objective in the Solomons had been reached earlier, in April. Tulagi had been the seat of the commissioner of the British Solomon Islands. In addition to his residence, there were several public buildings including a hospital, prison, and radio station, as well as a golf course and cricket grounds. Nearby Gavutu Island served as the headquarters of Lever Pacific Plantations, Ltd., which had extensive coconut groves throughout the islands. It was also the site of the company's offices, stores, and machine shops for the repair and upkeep of its small fleet of schooners and motorboats.[5]

Japanese forces had overrun Tulagi and its environs, in early April, and soon began making use of the various facilities available. On 5 July, a considerable force landed on Guadalcanal. A few days later, Allied reconnaissance planes observed that efforts were underway to build a landing field on the north coast of the island, not far from Lunga Point. The operation of enemy land-based aircraft from Guadalcanal would immediately threaten the Allies' control of the New Hebrides and New Caledonia area, and the necessity of regaining that island became even more of an imperative.[6]

## SOUTH PACIFIC AREA COMMAND ESTABLISHED

In order to support the Allies' first major offensive in the Pacific, a new command was created under Vice Adm. Robert L. Ghormley, USN. On 19 June 1942, he became commander, South Pacific Area and South Pacific Force, and served in that command, until 20 October 1942, with headquarters at Auckland, New Zealand. Subsequently, Ghormley was recognized as the commander of all Allied land, sea, and air forces in the South Pacific area, with the exception of land forces specifically assigned to the defense of New Zealand.[7]

On 16 July, Ghormley issued an operation plan for forces under his command to successively seize, occupy, and defend Tulagi and adjacent positions, and the Santa Cruz Islands, for the purpose of denying these areas to the enemy and in preparation for future offensive action. The date (D-Day) for launching an attack on Tulagi-Guadalcanal was tentatively set for 7 August. The plan in the main, provided for three major task forces, two under the command of Vice Adm. Frank Jack Fletcher. His subordinate commanders Rear Admirals Leigh Noyes and Richmond K. Turner commanded Task Force NEGAT (aircraft carriers) and Task Force TARE (amphibious force), respectively.[8]

The third force, Task Force MIKE, under Rear Adm. John S. McCain, was to supply aerial scouting and advance bombing of enemy facilities ashore by land-based aircraft and seaplanes. McCain was directly responsible to commander, South Pacific Force.[9]

## SIX CARGO SHIPS A PART OF TASK FORCE TARE

Comprising part of Task Force Tare were thirteen amphibious transports, and six cargo ships. The latter, recently converted merchant vessels, carried tanks, armaments, and other war materiel and supplies in support of the assault forces aboard the transports.

**Cargo Ships Involved in Guadalcanal-Tulagi landings (Awarded a Battle Star for period 7-9 August 1952)**

| Cargo Ship/ Former vessel | Comm/ Decom | Commanding Officer |
|---|---|---|
| *Alchiba* (AK-23)<br>MS *Mormacdove* | 15 Jun 41<br>14 Jan 46 | Comdr. James S. Freeman, USN |
| *Alhena* (AK-26)<br>SS *Robin Kettering* | 15 Jun 41<br>22 May 46 | Comdr. Charles B. Hunt, USN |
| *Bellatrix* (AK-20)<br>SS *Raven* | 17 Feb 42<br>1 Apr 46 | Comdr. William F. Dietrich, USN |
| *Betelgeuse* (AK-28)<br>MV *Mormaclark* | 14 Jun 41<br>15 Mar 46 | Comdr. Harry D. Power, USN |
| *Fomalhaut* (AK-22)<br>SS *Cape Lookout* | 2 Mar 42<br>25 Jun 46 | Comdr. John D. Alvis, USN |
| *Libra* (AK-53)<br>SS *Jean Lykes* | 13 May 42<br>12 Apr 48 | Comdr. William B. Fletcher Jr., USN[10] |

## MOVEMENT TO THE OBJECTIVE AREA / LANDINGS

The bulk of the ships that would be involved in the assault landings at Guadalcanal and Tulagi assembled at Wellington, New Zealand, and the rest at San Diego and Pearl Harbor. Once joined, the expeditionary force totaled approximately eighty ships. The weather, on 6 August, for the final approach to the Solomons was ideal—an overcast sky and a mist rendered enemy reconnaissance hopeless. In late afternoon, the force changed to an "approach disposition" for entering the area of operations, with Transport Group YOKE, destined for Tulagi in the lead, and Transport Group XRAY, which would conduct the landing at Guadalcanal, six miles astern.[11]

**Guadalcanal: Transport Group XRAY (Task Group 62.1)**

| Transports | Cargo Ships |
|---|---|
| USS *American Legion* (AP-35) | USS *Alchiba* (AK-23) |
| USS *Barnett* (AP-11) | USS *Alhena* (AK-26) |
| USS *Cresent City* (AP-40) | USS *Bellatrix* (AK-20) |
| USS *Fuller* (AP-14) | USS *Betelgeuse* (AK-28) |
| USS *George F. Elliott* (AP-13) | USS *Fomalhaut* (AK-22) |
| USS *Hunter Liggett* (AP-27) | USS *Libra* (AK-53) |
| USS *McCawley* (AP-10) flagship | |
| USS *President Adams* (AP-38) | |
| USS *President Hayes* (AP-39) | |

**Tulagi Area: Transport Group YOKE (Task Group 62.2)**

| Transports | High Speed Transports |
|---|---|
| USS *Heywood* (AP-12) | USS *Colhoun* (APD-2) |
| USS *Neville* (AP-16) | USS *Gregory* (APD-3) |
| USS *President Jackson* (AP-37) | USS *Little* (APD-4) |
| USS *Zeilin* (AP-9) | USS *McKean* (APD-5)[12] |

The aircraft carriers of Task Force NEGAT arrived southwest of Guadalcanal, about 75 miles from Tulagi, at midnight. Early the next morning, Groups YOKE and XRAY separated at 0300 on 7 August to proceed to their respective assault beaches. Sunrise was at 0633. With the break of dawn came the signal "Land the Landing Force"—at 0637 for YOKE off Tulagi—followed thirteen minutes later by the same order for XRAY off Lunga Roads, Guadalcanal (0650).[13]

Although a smaller number of vessels and troops was involved in the Tulagi area, the amphibious operations there would be considerably more complex than on Guadalcanal. On the latter, much larger island, it was merely a matter of pouring first troops and then supplies onto a single beach. The Tulagi region required several landing areas, as detailed below, necessitating a more elaborate schedule for operating landing craft and conducting fire support.

- 0740: Landing at village of Haleta, Florida Island
- 0800: Landing on Blue Beach, southwest shore of Tulagi
- 0845: Landing at Halavo Bay, Florida Island
- 1200: Landing at Gavuto Island[14]

## COMBAT ASHORE

The first troops landed on Red Beach at Guadalcanal at 0913. In contrast to the fighting on Tulagi and Gavuto, the occupation of Guadalcanal proceeded smoothly. The Allies had expected there to be greater resistance on Guadalcanal and had, accordingly, concentrated a majority of the landing forces there. Additionally, Japanese forces pulled back into the hills of Guadalcanal, enabling establishment of a beachhead along the northern shore. On Tulagi, and other small islands off much larger Florida Island, the enemy was trapped and, refusing to surrender, fought almost to the last man.[15]

## ATTACKS BY ENEMY BOMBERS / DIVE BOMBERS

In late morning, a warning was received at 1045 from a coast watcher, that a flight of twin-engine bombers had passed Bougainville (located farther up the Solomons) on a southerly course, evidently headed for the Guadalcanal area. At once, the order was given, "Repeal air attack." At 1320, twenty to twenty-five bombers were sighted coming over Savo Island at 10,000 to 14,000 feet. Aboard the attack transport USS *President Adams* (AP-38), it was noted that they resembled DC-3 Douglas transport aircraft, and were a bright silver color, exactly like commercial airlines. The planes were sighted by ships off Florida Island, but were out of range, and apparently only interested in the large number of transports off Guadalcanal.[16]

Twenty-seven land attack planes of the Fourth Air Group and eighteen Zero fighters of the Taima Air Group had been dispatched. When the attack group struck the ships at Guadalcanal at 1325, the planes encountered opposition from carrier-based fighters and anti-aircraft fire from the ships. As the heavy bombers approached, the screening ships and transports opened fire. One bomber staggered, then crashed into the sea. Another crashed in flames, and a third descended, last seen gliding downward over the hills of Guadalcanal trailing smoke. The remaining planes, attempting a bombing pattern, exhibited poor marksmanship. Within ten minutes, the enemy raiders disappeared behind the island's mountain peaks, having failed to hit or damage any ships.[17]

Three enemy attack planes and two fighters were shot down, two attack planes made emergency landings, and nineteen attack planes and two fighters were damaged. The Japanese in turn shot down nine Allied fighters and dropped all their bombs. No hits were made on the ships, since the bombs fell well clear of the surface targets, landing between the transports and their screening cruisers.[18]

A little over an hour later, seven to ten Japanese single-engine dive bombers suddenly attacked the ships. They apparently approached from the direction of Tulagi, flying high above Transport Group YOKE off Florida Island, and then dove steeply down on Group XRAY off Guadalcanal. Three near misses and one bomb hit on USS *Mugford* (DD-389) off Lunga Point couldn't prevent the destroyer from downing two of her attackers, but she suffered eight killed, 17 wounded, and 10 missing. Although considerable damage was done to the ship and a fire started, repairs were carried out in a few hours.[19]

The dive bombers were one of two groups that together, totaled sixteen Type-99 (land-based) carrier bombers of the Japanese Second Air Group, launched to attack the Allied forces. The flight commander had divided his force into two attack groups, one of which attacked the Allied fighters thereby allowing the other to make its attack unopposed by naval air. The unopposed attack group made one hit on the *Mugford*, with two of the attacking planes crashing close aboard the destroyer. The opposed attack group suffered the loss of six bombers and damage to three which made emergency landings.[20]

## ACTION ON 8 AUGUST

Photo 3-1

Cargo ship USS *Betelgeuse* (AK-28), circa 1942, displaying her camouflage paint scheme. National Archives photograph #80-G-31906

At 1038 the following morning, 8 August, a warning was received from an Australian coast watcher on Bougainville of 40 enemy twin-engine bombers proceeding southeast. Admiral Turner immediately ordered all ships to move out of the transport areas, and assume a cruising disposition. When the enemy planes arrived almost precisely at noon, the transport groups and screening ships were in formation and maneuvering at top speed. Task Force 62 (South Pacific Amphibious

Force) was attacked by about twenty-three torpedo planes, and by four dive bombers, escorted by fighters, which approached around the southeast end of Florida Island. The aircraft were flying at an extremely low altitude, possibly only 50 feet above the water.[21]

The enemy attack group consisted of approximately twenty-three land-attack planes of both the Misawa Air Group and the Fourth Air Group, escorted by fifteen fighters of the Tainan Air Group. The cost to the Japanese of the ensuing engagement would be nineteen planes, shot down or missing. The planes did not drop any bombs or torpedoes in the Tulagi area. As before, their major targets were the ships of Transport Group XRAY off Guadalcanal. At 1200, the formation of which the cargo ship *Betelgeuse* was a part, was attacked by a flight of twin-engine torpedo planes. Several of the aircraft were soon burning and falling into the sea, two shot down by *Betelgeuse*.[22]

The first plane approached the *Betelgeuse* at a height of 15 feet above the water, crossing from her starboard quarter to port, then down her port side. As the plane came abreast the bridge, a ragged hole was visible in the after part of the fuselage, resulting from a direct hit by No. 4 gun. The round had passed through the plane without exploding. The plane was then hit by several 20mm projectiles from No. 2, 4, and 6 machine guns. After turning to the left, away from the ship, it crashed in flames about 3,000 yards from *Betelgeuse*, near a destroyer.[23]

Three minutes later, a plane approached the *Betelgeuse* from her starboard bow, crossing to port. No. 2 gun opened fire at an estimated range of 500 yards. The third or fourth salvo exploded either directly over the plane, or in front of it at a range of about 2,500 yards, causing it to burst into flames and crash.[24]

In between the two aircraft attacks on *Betelgeuse*, a flaming enemy plane impacted the boat deck of the nearby transport *George F. Elliott* (AP-13). The anti-aircraft fire of ships in the formation was very accurate, and many planes had burst into flames and plunged into the water, when one sped toward the *Elliott*'s starboard beam, in level flight, about 30 feet above the water. As her starboard 20mm machine guns were making direct hits, the aircraft swerved slightly and crashed into the transport amidships. Immediately, the starboard areas of the boat and bridge deck were in flames, which soon spread to the port amidship section.[25]

The crash of the plane had ruptured the ship's fire main and, as there was no water available with which to fight the fire, the flames were soon out of control. The destroyer *Hull* (DD-350) came alongside and hose lines were led on board the *Elliott* in an effort to get the fire under control. *Hull* also attempted, with assistance from a second destroyer,

*Dewey* (DD-349), to tow the *Elliott* to the beach, but was unsuccessful. Despite valiant attempts by *Elliott*'s crew, the fire continued to burn out of control, and the transport was finally abandoned about an hour after dark.[26]

Photo 3-2

Left-center: Transport USS *George F. Elliott* (AP-13) afire after hit by an enemy aircraft. The other two smoke plumes mark the locations of aircraft that crashed into the water. US Navy photograph #NH 69114

Upon orders from commander, Task Force 62, after her crew had been taken aboard destroyers, the unsalvageable *George F. Elliott* was to be sunk by a destroyer. At 1730, the *Dewey* (DD-349) fired three torpedoes into her, and she settled in shoal water off Florida Island and continued to burn, illuminating the overcast after dark. Following a night action, described below, the amphibious forces left the area the following day. The transport *Hunter Liggett*, with the commanding officer and survivors of the *Elliott* now aboard her, was among the first to leave. The hulk of the *Elliott* was still burning when last seen by her commanding officer, Capt. Watson Osgood Bailey, USN. Casualties identified at that time were: 1 officer and 1 crewman dead; 2 officers and 6 men believed dead; 1 man missing; 15 Marines missing, 10 of whom were believed to have reached Red Beach.[27]

## BATTLE OF SAVO ISLAND, 9 AUGUST

The Imperial Japanese Navy, in response to Allied amphibious landings in the Tulagi area and on Guadalcanal, undertook a night surface attack on the ships screening the Allied landing forces. The Japanese task force, consisting of seven cruisers and one destroyer, sailed from Rabaul, New Britain, and Kavieng, New Ireland. These Japanese naval and air bases were located in the Bismarck Archipelago, immediately northwest of the Solomon Island chain.[28]

Proceeding down New Georgia Sound (also known as "the Slot") under the cover of darkness, the Japanese force caught the Allied surface forces unaware and routed them in the worst defeat in a single fleet action suffered by the United States Navy. The Amphibious Force was composed of eight separate task groups. Three of these, not previously discussed—the Screening Group, and Fire Support Groups LOVE and MIKE—were involved in the Battle of Savo Island.[29]

**Screening Group (Task Group 62.6)**

| **Heavy Cruisers** | **Destroyer Squadron 4** |
|---|---|
| HMAS *Australia* (D84) | USS *Helm* (DD-388) |
| HMAS *Canberra* (D33) sunk by gunfire and torpedoes | USS *Henley* (DD-391) |
| USS *Chicago* (CA-29) | USS *Jarvis* (DD-393) sunk by torpedoes |
| **Light Cruiser** | USS *Mugford* (DD-389) |
| HMAS *Hobart* (D63) | USS *Patterson* (DD-392) |
| **Destroyer Squadron 4** | USS *Ralph* (DD-390) |
| USS *Bagley* (DD-386) | USS *Selfridge* (DD-357) |
| USS *Blue* (DD-387) | USS *Talbot* (DD-114) |

| **Fire Support Group Love (TG 62.3)** | **Fire Support Group Mike (TG 62.4)** |
|---|---|
| USS *Astoria* (CA-34) sunk by gunfire | USS *San Juan* (CL-54) |
| USS *Quincy* (CA-39) sunk by gunfire | USS *Buchanan* (DD-484) |
| USS *Vincennes* (CA-44) sunk by gunfire | USS *Monssen* (DD-436) |
| USS *Dewey* (DD-349) | |
| USS *Ellet* (DD-398) | |
| USS *Hull* (DD-350) | |
| USS *Wilson* (DD-408)[30] | |

The Allied surface forces lost one Australian and three American cruisers. Allied dead totaled 1,023; there were 709 personnel wounded. In a brief twenty-minute action involving a night gunfire and torpedo attack, the heavy cruisers USS *Quincy* and *Vincennes* were so badly holed that they sank within one hour. USS *Astoria* burned throughout the night, and sank the following day at noon. In a separate action, USS *Jarvis* was sunk by torpedo planes 130 miles southeast of Tulagi, with the loss of all hands. Vice Adm. Mikawa Gunichi's task force suffered only light damage. Had he elected to also destroy the transports, the Guadalcanal campaign—and possibly the next phase of the Pacific war—might have evolved differently.[31]

**4**

# MacArthur's Papua Campaign

*The campaign on New Guinea is all but forgotten except by those who served there. Battles with names like Tarawa, Saipan, and Iwo Jima overshadow it. Yet Allied operations in New Guinea were essential to the U.S. Navy's drive across the Central Pacific and to the U.S. Army's liberation of the Philippine Islands from Japanese occupation. The remorseless Allied advance along the northern New Guinea coastline toward the Philippines forced the Japanese to divert precious ships, planes, and men who might otherwise have reinforced their crumbling Central Pacific front.*

—From *New Guinea: The U.S. Army Campaigns of World War II.*[1]

On 22 February 1942, on the eve of the fall of the Philippines to Japanese forces, President Franklin D. Roosevelt ordered Gen. Douglas MacArthur to proceed from the Philippines to Australia and establish a new headquarters. American and Filipino forces were then trying desperately to hold the Bataan Peninsula, which formed the northern boundary of the entrance to Manila Bay, and nearby Corregidor Island that divided the opening into two channels. Lt. John D. Bulkeley's Motor Torpedo Boat Squadron 3—*PT-32*, *PT-34*, *PT-35*, and *PT-41*—transported MacArthur as well as his family and members of his staff, and Rear Adm. Francis W. Rockwell and his staff, to Mindanao on 11 March, from whence B-17s flew them to Australia.[2]

Six weeks later, following a rearguard campaign, Lt. Gen. Jonathon M. Wainwright surrendered Corregidor and Manila Bay forts and all the armed forces in the Philippines to the Japanese. A bitter pill to swallow, and intended to prevent further effusion of blood, this action followed an epic struggle by Filipino and American soldiers to withstand the constant and grueling fire of a superior enemy for more than three months. Besieged on land and blockaded by sea, cut off from all sources of help in other parts of the Philippines and in America, the intrepid fighters had endured all that they could.[3]

On 17 March, two weeks after General MacArthur's arrival in Australia, the Joint Chiefs of Staff ordered the Pacific divided into two commands, the Pacific Ocean Areas under Adm. Chester W. Nimitz and the Southwest Pacific Area under MacArthur. The Australian government nominated MacArthur for the post of Supreme Commander in a proposed new Allied command, the Southwest Pacific Area (SWPA); within a month the unified command was established under him.[4]

Photo 4-1

General MacArthur, accompanied by his family and friends, arriving at Spencer Street Station in Melbourne.
Australian War Memorial photograph 043114

The Joint Chiefs directed him to hold the key military regions of Australia as bases for a future offensive, and to check the Japanese southward advance by destroying enemy shipping, aircraft, and bases in the Netherlands East Indies, New Guinea, and the Solomon Islands. Pending completion of formal arrangements, steps were taken to integrate the United States and Australian forces. The Australian Army was reorganized and regrouped, with increased emphasis given to training new recruits. The combined Australian and United States air forces were placed under the command of Lt. Gen. George H. Brett, USAAF, while Vice Adm. Herbert F. Leary, USN, assumed command of the naval forces.[5]

Upon constitution of the SWPA, on 18 April 1942, MacArthur initially established his general headquarters at Melbourne, but transferred it three months later to Brisbane. The immediate problem confronting him was the defense of Australia, however, the forces available were wholly inadequate to cover such an extensive coastline against an expected invasion. Five thousand Japanese troops of the elite jungle-trained South Seas Detachment had stormed ashore at Rabaul in the Australian Territory of New Guinea, on 23 January 1942, and quickly overwhelmed the small Australian garrison. The Japanese then began to develop Rabaul into a major base for further military operations in the Southwest Pacific area.[6]

MacArthur felt strongly that passive defense was strategically unsound, and decided to move the bulk of his forces forward more than a thousand miles to Port Moresby and force the Japanese to fight on his terms across the barrier of the Owen Stanley Range. These great mountains, nearly 14,000 feet high with large peaks and deep gorges, formed a natural barrier which ran the entire length of eastern New Guinea. Passage along the Kokoda Trail, a one-man-wide foot trail through dense jungle was the only feasible means of crossing the range.[7]

Photo 4-2

Panoramic view of Port Moresby Harbour and township, August 1942. Australian War Memorial photograph 034029

These geographical challenges were compounded by problems of climate and health. Penetrating, energy-sapping heat was accompanied by high humidity resulting from frequent torrential rains totaling as much as 300 inches per year. At the lower altitudes on the westward side of the mountains, there were swollen streams, reeking nipa palm woodlands and mangrove swamps, and thick mud. Millions of insects abounded, and clouds of mosquitoes, flies, leeches, chiggers, ants, fleas, and other parasites pestered man night and day. Disease was an unrelenting foe; malaria took a heavy toll, dengue fever was common, and deadly blackwater fever, though not so prevalent, was no less an adversary. Bacillary and amoebic dysentery were possibilities; tropical ulcers, easily formed from the slightest scratch, were hard to heal; and scrub typhus, ringworm, hookworm, and yaws all awaited careless soldiers.[8]

Despite these adverse conditions, Japanese forces under Gen. Tomitaro Horii had advanced by late summer to within thirty-two kilometers of Port Moresby when ordered by Imperial General Headquarters to retract. Horii had suffered heavy casualties fighting

Australian forces, his troops were exhausted, and he had outpaced his supply lines. The reason for the retreat was on Guadalcanal, where an effort by Japanese forces to recapture Henderson Field, on 14 September, had been soundly defeated by the opposing U.S. Marines with a high number of casualties inflicted on the Japanese. After news of the defeat reached Japan, the top generals decided they could not support concurrent battle fronts on both New Guinea and Guadalcanal, and directed Horii to withdraw.[9]

Having abandoned the effort to reach Port Moresby over land, the only feasible approach by which the enemy could invade Australia was via a sea route through Milne Bay at the eastern tip of New Guinea. While Horii had been mounting his southward attack over the mountains, other Japanese forces had captured the towns of Lae and Salamaua on the north coast of New Guinea, and established bases. Although temporarily halted in their advance toward Australia, the Japanese still controlled all the chains of islands across its northern sea approaches, with the single exception of southeastern New Guinea. The enemy also held the initiative as to the point of attack.[10]

Holding command of the sea enabled the Japanese to concentrate on particular objectives and overwhelm the defenders with superior forces. While this was occurring, the Allies continued to strengthen Port Moresby as a land and air base. Located on the southeastern coast of the Papuan Peninsula, protected on the north by the Owen Stanley Range and flanked on the east by Milne Bay, it was well placed strategically except for its vulnerability to amphibious assault—which the Battle of the Coral Sea, fought in early May 1942, had prevented.[11]

## THE PAPUA CAMPAIGN

Two attempts by Japanese forces to capture Port Moresby—an initial one by sea, followed by an effort on land described above—were unsuccessful. A planned amphibious assault was thwarted by the Battle of the Coral Sea, which turned back the invasion fleet. An advancement on land as previously indicated was stopped by desperate fighting by Australian forces on the Kokoda Trail, across the Owen Stanley Mountains.[12]

In autumn 1942, the Allies prepared for an offensive involving attacks on the Japanese strongholds along the northeast coast of Papua. MacArthur allocated American and Australian troops to this operation, new airfields were built to support the campaign, and supplies were moved around by sea. The offensive began in mid-November. Gona fell, on 9 December, and Buna was captured, on 14 December, although fighting in the area lasted until 2 January 1943. Sananada, the final Japanese position, fell on 22 January, ending the Papua campaign.[13]

Map 4-1

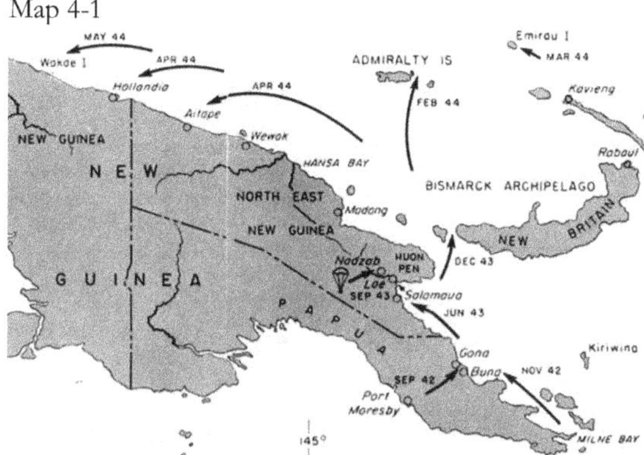


MacArthur and Halsey's Advancement. After defending Port Moresby, Papua, from a Japanese invasion, and establishing a base at Milne Bay, MacArthur's Australian and American forces advanced northwestward up the east coast of Papua and New Guinea, along his road back to the Philippines. Halsey's forces moved northwestward up the Solomons to Green Island (on the northwest edge of the Solomon Island chain, not shown) and Emirau Island in the Bismarck Archipelago (upper right of map). (http://www.history.army.mil/books/AMH/Map23-43.jpg)

Photo 4-3

Kokoda Trail. Wounded being carried by native bearers. Australian War Memorial photograph 013642

Despite having been ejected from Papua, and sustaining heavy losses in the continuing struggle for Guadalcanal, in January 1943, Japan still held the preponderant air, naval, and ground strength in the Southwest Pacific and retained the strategic initiative in New Guinea.[14]

## VESSEL SUPPORT OF COMBAT FORCES

In 1942, MacArthur and Halsey both needed small vessels to operate in shallow, often uncharted or poorly charted waters bespread with hazardous reefs in support of combat forces ashore. In spring of 1943, prior to the construction and arrival in the theater of steel landing craft, they utilized whatever type small craft they could obtain; MacArthur acquiring his from Australia and New Zealand, and Halsey from the United States. (My book, *MacArthur and Halsey's "Pacific Island Hoppers": The Forgotten Fleet of World War II*, is devoted to this subject.)[15]

Throughout the first half of 1942, a fleet of Australian-based Army cargo ships regularly sailed from ports on the east coast of Australia to deliver equipment, and supplies to support the troops at Port Moresby. That port was the main northern terminal up until the Japanese advance across the Owen Stanley Mountains was stopped. MacArthur had ordered, in June, the secret construction of an air, land, and naval base at Gili Gili, a village on Milne Bay, about 250 miles to the east-southeast of Port Moresby. When combat moved, in November 1942, to the Japanese-held Buna-Gona area on the northeastern coast of Papua, Milne Bay became the main terminal.[16]

The cargo ships, and small vessels that took their cargos farther northwesterly along the Papua coast, were part of what was known as the "Catboat Flotilla," operated by the U.S. Army Small Ships Section, whose administrative offices were in Sydney, Australia. Some Royal Australian Navy stores-issuing ships were also part of the resupply of Allied troops in New Guinea. For these ships, such operations could be dangerous, as was the escape of some of their sisters from Singapore before commencement of new duties as RAN cargo ships. This subject is taken up in the next chapter.[17]

# 5

# RAN Chinese Steamers, and HMAS *Baralaba*

Photo 5-1

Stores carrier HMAS *Ping Wo* (an old Chinese river steamer) while operating as a tender to HMAS Assault, the Amphibious Operations Training Establishment at Port Stephens. Although fitted for a 12-pounder gun forward, it is not mounted. Her only armament is four single Vickers .303 machine guns; one each on port and starboard sides, behind the deckhouse on the forward superstructure, and two on top of the after superstructure, 28 September 1942.
Australian War Memorial photograph 301176

On 3 September 1939, in response to Adolf Hitler's invasion of Poland, Britain and France declared war on Germany. Other members of the British Commonwealth followed suit separately, all within one week of each other; these countries were Canada, Australia, New Zealand, India, and South Africa.

With the onset of World War II, the Royal Australian Navy (RAN) began requisitioning merchant vessels to supplement her fleet and release her warships for operational duties around the world. These vessels, some commissioned into the RAN, served as coastal patrol vessels, stores-issuing ships, amphibious landing ships, and other activities where there was the greatest need within the RAN commitments. Among them were four Chinese coastal steamers, His

Majesty's Australian ships (HMAS) *Ping Wo*, *Poyang*, *Whang Pu*, and *Yunnan*. These ships were owned by Chinese subsidiaries of British shipping companies, and were all of a similar size, between 2,600 and 3,300 tons.[1]

## ESCAPE FROM SINGAPORE

Photo 5-2

An unidentified soldier walks along the beach at Keppel Harbour, on 16 February 1942, the day after the fall of Singapore.
Australian War Memorial photograph P02569.119

In 1942, *Ping Wo* and *Whang Pu* were undergoing refit in Singapore following their requisition by the Royal Navy. When it became apparent that the island would soon fall to the Japanese, they joined a large number of Allied warships and merchant vessels evacuating the port up until the final surrender of the British Crown Colony, on 15 February.[2]

As the two steamers made their way to the Australian west coast, *Ping Wo* endured the more eventful passage as she participated in the longest continual tow in Australian naval history. She was a 3,105-ton, flat-bottomed Yangtze riverboat built in 1922, and afterward operated by the Indo-China Steam Navigation Co. Ltd. *Ping Wo* was requisitioned by the British Admiralty, in December 1941, commissioned into the Royal Navy, and later served the Royal Australian Navy. Facilitating her escape from Singapore, two officers and European crewmembers from the badly damaged 10,253-ton merchant ship *Talthybius* (which had been bombed by the Japanese in Singapore Harbour and abandoned), as well as some RANVR and RAN sailors, formed a portion of her crew.[3]

Captain Kent, under orders from the Royal Navy Port Authority, had his crew load coal and food supplies aboard *Ping Wo*. After taking on board nearly 200 European and Eurasian refugees, the old steamer left Singapore on the night of 11 February 1942, bound for Java in the Netherlands East Indies. Following a number of bombing attacks by enemy aircraft, *Ping Wo* arrived in Java. There, she transferred the evacuees to other ships, refueled, and loaded more supplies.[4]

Information about other ships that fled Singapore is provided later in the book, in material associated with merchant vessels that were part of the Fleet Train supporting the British Pacific Fleet.

Photo 5-3

HMAS *Vendetta* in tow of *Ping Wo*, in March 1942. Her tow by the steamer from Singapore to Melbourne marked the longest by any Australian warship in RAN history. Australian War Memorial photograph 128104

Ready for sea, *Ping Wo* received orders to tow HMAS *Vendetta* to Australia. The destroyer had been undergoing repairs and a refit at the naval dock in Singapore, with her engines removed. To avoid her capture by the Japanese, the Royal Navy destroyer HMS *Stronghold* had towed her to Palembang, Netherlands East Indies. The Royal Navy tug HMT *St. Just*, and the Royal Australian Navy sloop HMAS *Yarra* under

Lt. Comdr. R. W. Rankin, then took over the tow to Tangjong Priok, Java, where *Ping Wo* had arrived.[5]

*Ping Wo* set off from Tangjong Priok with HMAS *Vendetta* in tow in early evening on 24 February, headed for Fremantle, Australia. She was accompanied by the British freighters *Giang Ann* and *Darvel*, and under escort by Australian sloop HMAS *Yarra* and the British destroyer HMS *Electra.* The small convoy arrived at their destination at Fremantle in early morning darkness on 4 March.[6]

*Ping Wo* was carrying a precious cargo of 10,635 ounces of gold bullion worth approximately £85,080 sterling, contained in twenty-one boxes. The gold, which belonged to the Bank of England, was from the Straits Settlements Bank of Singapore. It was quickly picked up by Bank officials in Fremantle, and transferred by train, under escort, to Melbourne for assay. On 3 May, the shipment was moved by rail to Broken Hill Gaol (jail in American) in New South Wales, where all gold bullion in Australia was stored for the remainder of World War II.[7]

An interested side note to this story is that, on 4 April 1946, the Australian Prime Minister, Joseph Benedict Chifley, revealed that the sole prisoner in the Broken Hill Gaol had been turned out of his comfortable quarters, and sent to Long Bay, to make way for storage of the large quantities of gold. During the war, foreign-owned gold reserves were brought to Australia because its owners, including the Bank of England and the Dutch authorities and banks, were reluctant to incur greater risk of losses to enemy attack by shipping it via the Pacific, Indian, and Atlantic oceans back to Europe. The place of the former prisoner had been taken by gold ingots worth millions of pounds, including gold salvaged from the *Niagara*, sunk off the New Zealand coast.[8]

The sinking of the trans-Pacific liner RMS *Niagara* off Northland's Bream Head, on 19 June 1940, by German mines, had shocked the New Zealand public and shattered any illusions that distance would protect their islands from enemy attack. On the night of 13 June, the *Orion*, a German raider disguised as a merchant ship, had slipped undetected into New Zealand waters and laid 228 contact mines in the approaches to the Hauraki Gulf. Six days later, the 13,415-ton *Niagara*, which had left Auckland on her regular run to Suva and Vancouver, struck two mines and sank quickly by the bow.[9]

Fortunately, all 349 passengers and crew abandoned safely in eighteen lifeboats; the only casualty being the ship's cat, Aussie. Lost was the *Niagara*'s secret cargo of small-arms ammunition and gold ingots worth £2.5 million (equivalent to nearly $250 million today). In late 1941, an epic salvage effort successfully recovered almost all of the gold from the ship wreckage, which lay at a depth of 60 fathoms.[10]

## HILFSKREUZER (WOLVES IN SHEEP'S CLOTHING)

> *The crews carried are large, at least 300 and probably 400 [officers and men]. They are mainly young regular naval ratings, but some are reservists from the merchant navy who have had peace-time experience of the waters in which the raider is to operate. This is particularly true of the officers, who are in a position to advise the captain of the types and habits of shipping likely to be encountered.*
>
> *Importance is attached to the nuisance value of minelaying. It is not thought that many mines are ever laid at one time, as the chances of the raider being detected would be too great. It is possible that the ferocity of raiders' attacks on ships are due to their fear of a victim scoring one lucky hit on their stock of mines.*
>
> —Raider Supplement to the Weekly Intelligence Report No. 64, of 30 May 1941, issued by the Naval Intelligence Division, Navy Staff, Admiralty.[11]

Photo 5-4

German raider *Orion*.
Raider Supplement to the Weekly Intelligence Report No. 64, of 30 May 1941, issued by the Naval Intelligence Division, Navy Staff, Admiralty

The 485-foot raider HK *Orion* was a former merchant vessel, the SS *Kurmark*. In early 1940, in support of its goal of starving Britain of the goods and materials it needed to survive and stay in the war, Germany had introduced the use of Hilfskreuzer (auxiliary cruisers) as commerce raiders. These armed ships—nondescript, converted German freighters, disguised as "clapped out" merchantmen of other nations—went to sea and fought as true warships. The 300-plus man crews of the Hilfskreuzer were led by high-ranking officers of the rank of captain or commander, all of whom, excepting one, had seen active service before the war. For different reasons, they were considered unsuitable for command of a regular Navy warship, and either too senior in rank or too old for U-Boat command. Generally, these men were former training ship commanders.[12]

*Orion* was heavily armed with six 5.9-inch (15 cm) guns taken from the battleship *Schleswig-Holstein*, one 3-inch (7.5cm) gun, two 1.5-inch (3.7cm) guns, four 20mm (2cm) anti-aircraft guns, six 21-inch (53.3cm) torpedo tubes, and 228 EMC moored-contact mines. The raider also carried one Arado Ar 196 A-1 seaplane to scout for prey at sea. The guns fitted on the seemingly harmless ex-merchant vessels were hidden, mostly behind pivotal false deck or side structures. For example, aboard another raider *Atlantis*, a phony crane and deckhouse on the after part of the ship screened two of her guns; the other four were concealed behind flaps in the hull that were raised when action was imminent. Torpedo tubes were similarly hidden from view; those of *Atlantis* were at her waterline.[13]

Under the command of Kapitän zur See Kurt Weyher, the 15,700-ton *Orion* had arrived in New Zealand waters on the afternoon of 13 June 1940. The raider was carrying 228 type moored-contact mines, with orders from the German Naval Command to lay them in the approaches to Auckland. At dusk, with good visibility under a cloudless sky, she began laying three rows of mines in the following areas:

- Across the eastern approach to the passage between Great Mercury and Cuvier Islands
- Across the approach to Cuvier Island, in a zigzag pattern which overlapped the southeast end of Great Barrier Island
- Across the northern approaches to Hauraki Gulf, with the large mine barrage extending from a point off the northern end of Great Barrier in a wide arc 6½-miles off Moko Hinau Islands, then in a straight line to the northwest, passing about six miles outside the Maro Tiri Islands to a point about five miles from the mainland[14]

*Orion* sighted three outland steamers and one inward-bound vessel, during the seven-hour long operation. She was either undetected or ignored. By 2300, the sky clouded over, helping maintain the secrecy of her actions. Upon dropping her last mine at 0236, the raider cleared the area, proceeding at full speed to the northeast.[15]

## UNEVENTUAL JOURNEY OF *WHANG PU*

*Whang Pu*, mentioned at the beginning of this chapter, arrived safely in Fremantle, on 1 March 1942. She had been fitted out as a submarine depot ship at Singapore for the RN, and continued to be assigned those type duties after arriving in Australia. *Wang Pu* spent the next year and a half at Fremantle, serving as an accommodation ship for Dutch submarine and minesweeper crews that had escaped the Netherlands East Indies. Meanwhile, the Australian and British naval authorities considered how best to use her for the remainder of the war.[16]

Photo 5-5

Former Chinese coastal steamer *Whang Pu*, on 20 March 1942, shortly after escaping from Singapore. She was later commissioned into the RAN as a mobile repair ship. Australian War Memorial photograph 301724

## REMAINING TWO CHINESE STEAMERS ACQUIRED FOR RAN SERVICE

Photo 5-6

Cargo steamer *Poyang*, on 5 January 1942, prior to her being requisitioned by the RAN for service as an armament stores-issuing ship.
Australian War Memorial photograph 301187

*Poyang* and *Yunnan* did not have to escape Singapore, they were already in Australian waters. *Poyang* had arrived in Broome from southern China, on 19 December 1941, and, four days later, *Yunnan*, arrived in Fremantle from Singapore, on 23 December. Both were requisitioned by the RAN, in February 1942. After being fitted out as armoured stores-issuing ships in Melbourne, they proceeded to Sydney at the end of April. The two ships operated primarily off the Australian east coast for the remainder of the year and most of 1943, while making occasional voyages to New Caledonia and New Guinea.[17]

Photo 5-7

Merchant vessel *Yunnan* at Fremantle, Western Australian, on 24 December 1941. She was requisitioned by the RAN, on 22 June 1942, for service as the ammunition stores-issuing ship HMAS *Yunnan*.
Australian War Memorial photograph 301779

More about the Chinese steamers, which were eventually all commissioned into the RAN, follows in later chapters of the book. The ammunition ships *Poyang* and *Yunnan*, oiler *Bishopdale*, and provisions ship *Merkur*, represented the RAN in Task Group 77.7, the Leyte Gulf Service Force of the Seventh Fleet. They were a part of the larger Leyte Gulf force of some 550 ships, consisting of battleships, cruisers, escort carriers, destroyers, destroyer escorts, attack transports, cargo ships, landing craft, survey vessels, minecraft, and supply ships. Within this force were thirteen RAN ships, including those previously cited:

- Task Force 74: Heavy cruisers HMAS *Australia* and *Shropshire*, and destroyers *Arunta* and *Warramunga*
- Supply ships *Bishopdale*, *Poyang*, *Yunnan*, and *Merkur*
- Landing ships HMAS *Manoora*, *Westralia*, and *Kanimbla*

- The frigate HMAS *Gascoyne* and motor launch *HDML 1074*, which were part of the minesweeping and hydrographic group[18]

## HMAS *BARALABA*'S SHORT RAN SERVICE

Photo 5-8

Stores carrier HMAS *Baralaba* under way, circa 1942-43.
RAN historical photograph 4981

HMAS *Baralaba*, also acquired by the RAN in 1942, was neither a former Chinese vessel, nor long serving as a stores ship, but was employed with the U.S. Army Small Ships Section, introduced in the previous chapter. *Baralaba* was built as the 998-ton steamer *Nurnburg* at Stettin, Germany, in 1921, by Stettiner Oderwerke, and subsequently underwent several changes of ownership. She was acquired by the Australasian United Steam Navigation Co. Ltd., in 1925, renamed *Baralaba*, and employed on the Queensland sugar trade. *Baralaba* was requisitioned by the RAN, on 31 May 1942, as a stores carrier, but returned to her owners less than nine months later, on 11 February 1943—perhaps because she became too hot below decks to operate in the tropics.[19]

Former RAN stoker John Serjeant served aboard *Baralaba* when she was running stores from Townsville to Port Moresby and other nearby towns. He recalled that *Baralaba*, being a natural-draft coal burner, "was so hot down below, her stokers worked 4 hours on and 12 hours off instead of the more usual, 4 hours on and 8 hours off." During his short stint aboard *Baralaba*, no formal training on coal firing was given RAN personnel. They fired alongside experienced merchant navy firemen who taught them the skills of coal-firing marine boilers. Shoveling coal into a boiler's firebox to produce steam for ship propulsion, required hard physical labor, and was hot, dirty, exhausting work.[20]

Photo 5-9

Four members of the "black gang" in the stokehold of the light cruiser HMAS *Sydney* in January 1919. Boy First Class Henry Wilson is shoveling coal into the furnace. Australian War Memorial photograph EN0171

Photo 5-10

Port Moresby, Papua, shipping facilities and a portion of the township, 11 July 1942. Australian War Memorial photograph 025876

## *BARALABA*'S DUTY WITH U.S. ARMY SMALL SHIPS SECTION, TRANSPORTATION SERVICE

In 1942, while assigned to the U.S. Army's Small Ships Section, *Baralaba* was camouflaged and fitted for a special voyage. In the latter part of 1942, she carried freight between North Queensland and New Guinea.[21]

Photo 5-11

Walter Edward Sayers, Small Ships Section, United States Army Services of Supply, Southwest Pacific Area (USASOS SWPA). Born in England, Sayers enlisted in the Royal Navy in 1904 at the age of 15. During the First World War, he served in the Dardanelles, Atlantic Ocean, and North Sea. In 1923 he immigrated to Australia. In 1940 during the Second World War, he attempted to enlist in the RAN, however, at the age of 51, was deemed to be too old. In July 1944, Sayers enlisted in the USASOS SWPA, serving until August 1945. He saw service in the Pacific region and was aboard the cargo ship USAT *Chippewa* during the invasion of the Philippines. Around 3,000 Australians enlisted in the Small Ships Section during the Second World War. They were not inducted members of the US Army, but were civilians (merchant seamen) who were attached to the Army.
Australian War Memorial photograph P08923.001

The Australian government allowed the U.S. Army to recruit men and boys who were either too young or too old, or medically unfit for service in the Australian military services. They ranged from age 15 to over 70 years in age. Among the older members were veterans of World War I and old hands in New Guinea's days of exploration, as well as men with only one arm and, in one or two cases, one leg. The Grace Building on the corner of York and King Street in Sydney was used by the Army's Small Ships Section, Transportation Service, for administrative offices and most hiring of crewmen was done there. Recruiting on a small scale was also carried out at Brisbane and Townsville.[22]

Photo 5-12

Gili Gili, Milne Bay, New Guinea, in July 1942. Drums of petrol are being moved by native labor from the Dutch transport SS *Bantam* to the beach.
Australian War Memorial photograph 148937

The author is not sure what ports *Baralaba* called at; cargo destined for water movement to Buna-Gona was transferred at Milne Bay to small boats and carried up along southeast New Guinea to Oro Bay. The 211-mile-long-coast from Milne Bay to Oro Bay had not been accurately charted. The most recent charts were dated 1895, and new coral formations, which grew rapidly in warm tropical waters, had formed since then. As the Allies pushed up the northern coast of New Guinea, small vessels able to navigate among the reefs sailed beyond Oro Bay and Buna-Gona to Biak and Sansaport, and thence Morotai.[23]

Photo 5-13

Liberty ship at Oro Bay, New Guinea, 21 July 1943. Large ocean ships did not venture up the coast northwest of Milne Bay in 1942.
Australian War Memorial photograph 054581

Photo 5-14

U.S. Small Ships Section vessels in their daytime hideout at Morobe, New Guinea, 13 August 1943. They operated at night to avoid attack by Japanese float planes.
Australian War Memorial photograph 055776

## *BARALABA*'S POST-RAN SERVICE

Following her return to her owners, in February 1943, *Baralaba* was sold to San Ernesto Steamship Co., Hong Kong. She changed hands again, in February 1952, when sold to Wallem & Co and renamed *Brenda* and, a few years later, yet again in 1956, when purchased by Cambodian interests and renamed *Bayon*.[24]

6

# Capture and Defense of Guadalcanal

*1832: Torpedo passed about 20 yards astern, running at high speed with wide wake, and ran up on Kukum beach. Ship's position: 2800 yards bearing 260 true from Lunga Point. Ship's head 055. Torpedo track approximately due south which would place source in open water of Lunga Roads – probably a submarine, judging from accuracy of aim although no sound contact was reported. Weather – misty with occasional rain; visibility about 7000 yards.*

—USS *Fomalhaut* (AK-22) war diary entry, regarding a Japanese torpedo attack made on the evening of 23 August 1942—the opening day of the Battle of the Eastern Solomons—while the ship was anchored off Kukum, Guadalcanal, discharging cargo to lighters.[1]

Recurring small-scale combat action followed the U.S. Marine landings on Guadalcanal and Tulagi, on 7-8 August 1942, and severe ship losses occurred in the night action off Savo Island on the 9th. During this "breathing spell" of about two weeks after the landings, U.S. naval forces carried out reinforcing operations for the Marines, while the Japanese employed surface units and aircraft in an attempt to prevent this strengthening action. During afternoon raids, Japanese bombers from Rabaul attacked the Marines' beachhead, and pounded the fledgling airstrip they were struggling to clear on Guadalcanal. At night, enemy cruisers and destroyers regularly raced down "The Slot"—a narrow waterway from Rabaul southeast through the islands—to harass with gunfire the Marine positions, withdrawing before daybreak. (See map on the next page.) In separate enemy actions daring attacks on the small U.S. convoys carrying supplies and munitions to the 16,000 increasingly beleaguered Marines, considerably retarded the strengthening of positions on Guadalcanal.[2]

Submarine and aircraft attacks were made against convoys en route to, present at Guadalcanal, and in return transit. One such involved the *Fomalhaut*, which fortunately escaped harm. On 23 August, Rear Adm. Yoshitomi Setsuzo (commander of Submarine Squadron 7 at Rabaul) ordered the Japanese submarine *RO-34* to intercept a supply convoy at

Lunga Roads, Guadalcanal. This was the cargo ships *Alhena* (AK-26) and *Fomalhaut* (AK-22), escorted by the destroyers *Blue* (DD-387), *Henley* (DD-391), and *Helm* (DD-388), with supplies for the Marines. The convoy had arrived on the 21st, with the escorts remaining overnight.[3]

The *RO-34*, commanded by Lt. Morinaga Masahiko, was a 960-ton, K5 submarine, commissioned on 31 October 1941. The sub left Rabaul, on 21 August, to reconnoiter Guadalcanal on her seventh war patrol. On the evening of 23 August, while searching the Lunga Roads area, Lieutenant Masahiko observed what he thought was a 10,000-ton transport being unloaded, as well as a destroyer and a corvette patrolling in the area. He then commenced an approach and, at 1827, fired two torpedoes at the *Fomalhaut.* One explosion was heard and Masahiko later reported his target as sunk. Thankfully, as related in the opening quotation, this proved erroneous.[4]

## AMERICAN CARRIERS INITIALLY SAFEGUARDED

Map 6-1

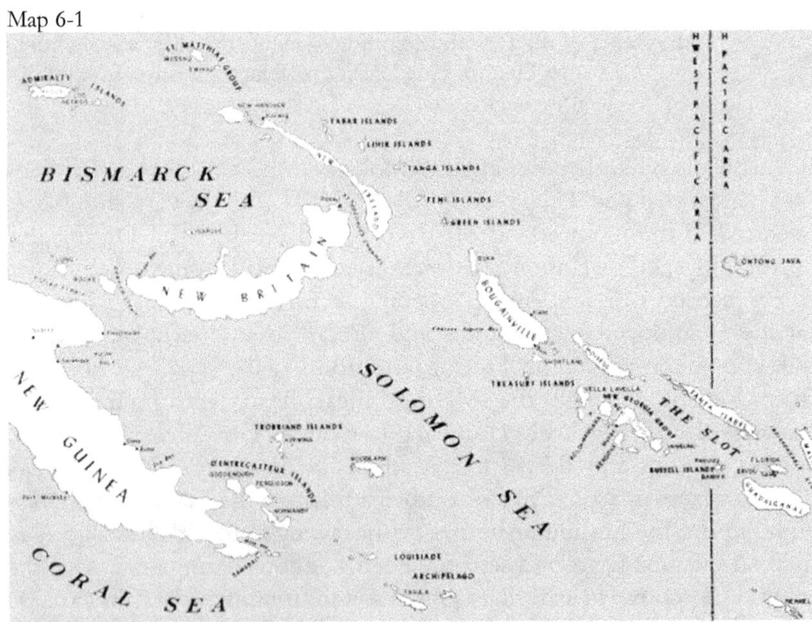


Rabaul Strategic Area (From Rabaul, on the northeast coast of New Britain Island, Japanese ships and aircraft could attack Allied forces on Guadalcanal following passage down New Georgia Sound, referred to as "the Slot," through the Solomons) http://ibiblio.org/hyperwar/USMC/II/maps/USMC-II-I.jpg

Between 9 and 23 August, as groups of Japanese cruisers and destroyers bombarded Guadalcanal with relative impunity almost nightly, three

USN carrier task forces operated well to the south of Guadalcanal. Keeping the carriers out of the range of hostile search planes, concealed their presence from the enemy. However, it also allowed the enemy in the passing days to inflict damage and seriously restrict the flow of Allied supplies, and bring its own reinforcements to the Rabaul area for a major attack on the Allies' footholds in the Solomons. U.S. Naval commanders were aware of the accumulation of Japanese strength in the Rabaul area, via air reconnaissance. By 23 August, whatever local fleet superiority the U.S. had enjoyed had vanished.[5]

For the Japanese, the Marines' toehold on Guadalcanal was intolerable. If the Allies were allowed to establish air bases in the eastern Solomons, their position at Rabaul would be directly threatened. Moreover, without control of the entire island chain, the Japanese were powerless to break the Allied supply line, stretching from Hawaii, south through Samoa and the Fiji Islands, and westward to Brisbane.[6]

## BATTLE OF THE EASTERN SOLOMONS

In the third week of August, in an effort to bolster their garrison on Guadalcanal, a reinforcement unit was dispatched from Rabaul by the Japanese. Commanded by Rear Adm. Raizo Tanaka, the unit—3 carriers, 3 battleships, 9 cruisers, 13 destroyers, 36 submarines, and several auxiliaries—was to run down the Slot and land the troops at night. Meanwhile, a force, under Vice Adm. Nobutake Kondo, centered around carriers *Shokaku* and *Zuikaku*, with two battleships and three heavy cruisers, arrived northeast of the Solomons to engage opposing American naval forces.[7]

The enemy's plan was for a detached force, commanded by Rear Adm. Chuichi Hara, to thrust ahead of the troop convoy bound for Guadalcanal. The dual-role of his light carrier *Ryujo*, accompanied by the heavy cruiser *Tone* and two destroyers, was to launch bombers to attack Henderson Field, and bait American carrier planes into attacking her, while air groups from the *Shokaku* and *Zuikaku* punished the carriers of Vice Adm. Frank Jack Fletcher with a resulting weakened air defence.[8]

Japanese submarines were a major part of this offense. Six proceeded ahead to scout, six others were positioned southwest of the Santa Cruz Islands, and four groups of three each were moving down to the south and west.[9]

Warned by Australian coastwatchers and American reconnaissance planes, Admiral Ghormley, commander, South Pacific Force, ordered Fletcher's carrier force (*Enterprise*, *Saratoga*, *Wasp*) northward to cover sea lanes into the Solomons. By daybreak on 23 August, this force was about 150 miles from Henderson Field. At the close of that day, Fletcher, misinformed by Pacific Fleet Intelligence that the Japanese force was north of Truk, and believing that there would be no battle for several days, sent the *Wasp* group to a fueling rendezvous. This decision left Fletcher with a two-carrier force.[10]

As in the Coral Sea and at Midway, U.S. and Japanese ships never sighted each other during the Battle of the Eastern Solomons—all attacks were carried out by carrier-based or shore-based aircraft. After a series of these, both sides withdrew their ships from the battle area. Neither achieved a clear victory. However, the Japanese lost significant numbers of aircraft and aircrew, and their ground reinforcements for Guadalcanal were delayed. Thus, the Allies had more time to prepare for ensuing enemy counter-thrusts and were able to prevent or interdict landings of heavy artillery, ammunition, and other supplies.[11]

A summary of Japanese and American ship losses/damage follows:

**Battle of the Eastern Solomons**

| Japanese Ships | Result | American Ship | Result |
|---|---|---|---|
| light carrier *Ryujo* | sunk by aircraft torpedoes | carrier *Enterprise* | damaged by aircraft bombs |
| light cruiser *Jintsu* | damaged by aircraft bombs | | |
| destroyer *Mutsuki* | sunk by USAF B-17 bombers | | |
| troop transport *Kinryu Maru* | set on fire by aircraft bombs, sunk by Japanese destroyer | | |
| seaplane carrier *Chitose* | damaged by bomb fragments[12] | | |

## TWO AIRCRAFT CARRIERS, A BATTLESHIP, AND A DESTROYER TORPEDOED; CARRIER *WASP* SUNK

Following the Battle of the Eastern Solomons that turned back the first major Japanese attempt to recover Guadalcanal, the Japanese and the Allies continued to move troops and supplies to the island. In support of the Allied effort, USN carrier forces operated southeast of the Solomons but in doing so suffered terrible loses. By 15 September, the carrier *Wasp* was sunk; the carrier *Saratoga*, battleship *North Carolina*, and destroyer *O'Brien* torpedoed and damaged; and the carrier *Hornet* barely missed being torpedoed—all within a period of fifteen days and in the same general area. Some details of these engagements follow.[13]

On 31 August, at a point northwest of Espiritu Santo, the Japanese submarine *I-26* was on the surface when her bridge team sighted with night binoculars a vessel resembling a large tanker. Comdr. Yokota Minoru identified the contact as a *Saratoga*-class carrier which, in fact she was—the *Saratoga* (CV-3). The submarine dove to periscope depth, but Yokota could not gain an attack position. Then the *Saratoga*'s task force changed course, reducing the distance to 1,100 yards with *I-26* at that time in a perfect position to fire a salvo at the carrier's starboard beam. However, her torpedomen were not prepared. Finally, at 0746, Yokota belatedly fired a six-torpedo salvo at the carrier that he estimated was making 13 knots.[14]

Alerted by a destroyer's "submarine" warning flag signal, *Saratoga* (Capt. DeWitt C. Ramsey) increased speed and began an evasive turn. One torpedo broached the surface because of a steering malfunction, and four others also missed, but one slammed into the ship's starboard side, just aft of the island. No personnel casualties resulted, and water flowing in through the breech in her hull only flooded the aftermost fireroom. However, the impact caused short circuits which damaged *Saratoga*'s turbo-electric propulsion system, leaving her "dead in the water." The cruiser *Minneapolis* (CA-36) took the carrier under tow and, by early afternoon, *Saratoga*'s engineers had restored propulsion. After patch-ups at Tongatabu, Fiji, from 6 to 12 September, the ship arrived at Pearl Harbor, on 21 September, for permanent repairs.[15]

A week after the submarine attack on *Saratoga*, Task Force 17—the carrier *Hornet* (CV-8), battleship *North Carolina* (BB-55), two heavy cruisers, one light cruiser, and six destroyers—was conducting a sweep toward the southern Solomons, when another deadly encounter with a Japanese submarine occurred. At 1149 on 6 September, their submarine *I-11* was patrolling submerged northwest of Espiritu Santo, when her soundman (sonar operator) detected the carrier group.[16]

Comdr. Shichiji Tsuneo commenced an approach and successfully penetrated the destroyer screen unobserved. After coming to periscope depth, he sighted a fleet carrier crossing dead ahead, 765 yards distant. At 1249 he fired a salvo of four torpedoes, one each from the four bow tubes, dove to 200 feet, and rigged for silent running. Sequentially, three minutes later, heavy explosions were heard, resulting from the detonation of two of the torpedoes churning toward *Hornet*, but also by a depth charge dropped by one of her planes. Apparently, two of the torpedoes intended for *Hornet* were exploded by the depth charge, before they could reach their intended target.[17]

Turning to what was happening on the surface, at 1250, three explosions in quick succession were sighted astern of the *Hornet* by shipboard personnel and airborne aircrews, while she was conducting flight operations. The carrier's after lookouts reported a wake passing across her stern from port to starboard, and an aviator stated that this wake ended coincidentally with the third explosion (the second of two torpedoes blowing up). Action by Ens. John Cresto, USNR, from Torpedo Squadron VT-6, was likely responsible for countermining the two torpedoes. In his account of the incident, he stated:

> The USS *Hornet* launched a patrol of TBF-1 airplanes at approximately 1245, September 6, 1942. I took off in a TBF-1 airplane with this patrol for the purpose of making an engine run-in and while making my first circle of the ship after taking off I sighted a torpedo on a parallel course with the ship, on her starboard quarter. I immediately started a turn when an explosion occurred at the position from which the torpedo appeared to have come. I followed the torpedo wake and dropped a MK 17 depth charge at the point where I through the submarine launching the torpedo was located....[18]

Photo 6-1

Grumman TBF-1 Avenger torpedo planes flying in formation, circa 1942-43.
National Archives photograph #80-G-K-16151

Ens. Harold J. W. Eppler, USNR, the pilot of the first TBF-1 off the deck, witnessed, while making a wide sweep around *Hornet*, the destruction of the final torpedo headed for her:

> On the starboard quarter [of the carrier] just outboard of my left wing I saw an explosion and immediately turned in that direction

> and armed my depth charge. Then I saw the wake of a torpedo. There was another explosion just aft of the torpedo, which caused the torpedo to broach, do a nose dive and then explode. The torpedo was headed directly at the stern of the USS *Hornet* when it exploded.[19]

Later in the war, Lieutenant Eppler was awarded the Navy Cross. His medal citation reads in part, "for extraordinary heroism in operations against the enemy while serving as Pilot of a carrier-based Navy Torpedo Plane in Torpedo Squadron SIX (VT-6), attached to the U.S.S. *HANCOCK* (CV-19), in the attack on major units of the Japanese Fleet in Kure Harbor, consisting of battleships, carriers and heavy cruisers, on 24 July 1945."[20]

## THREE TF 18 SHIPS TORPEDOED – *WASP* SUNK

> *All who witnessed the torpedoing agree on one point. A large explosion occurred in the vicinity of the hit under the forward gun gallery about four or five minutes after the hit and from then on flames appeared to be out of control. This explosion was accompanied by a column of dense, white smoke which shot up into the air which was painfully similar to the one seen when the* Arizona*'s forward magazine blew up on December 7th [1941]. It is believed that this explosion was the forward magazine of the* Wasp.
>
> —Commanding Officer, USS *San Francisco*, Action Report – torpedoing of USS *Wasp*, *North Carolina*, and *O'Brien*, September 15, 1942, forwarding of, 30 September 1942.

On the morning of 15 September 1942, the carriers *Hornet* and *Wasp* and their respective task forces were 250 miles SE of Guadalcanal, escorting a convoy of six transports carrying the 7th Marine Regiment from Espiritu Santo to reinforce Guadalcanal. The carrier groups were steaming in sight of each other about eight miles apart. The Japanese submarine *I-19* was running submerged nearby. At 1250, her soundman reported "heavy screw beats," and Lt. Comdr. Kinashi Takakazu ordered the sub brought up to periscope depth. He made a sweep with his "scope," scanning the horizon for ships, but none were in sight.[21]

An hour later, Kinashi raised the periscope again. This time, he saw a carrier, a heavy cruiser, and several destroyers (Task Force 18), bearing 045 degree true, 9 miles away. Kinashi estimated the task force was steering a northwest course and began a slow approach. The ships, zigzagging at 16 knots, changed course to west-northwest and then, at 1420, reversed course to the south-southeast. Then *Wasp* made a slow port turn into the wind to launch and recover her aircraft, putting the carrier on a perilous course toward the *I-19*.[22]

Kinashi calculated that *Wasp* was steering 130 degrees at 12 knots. At 1445, he fired a spread of six Type 95 oxygen-propelled torpedoes at the carrier from 985 yards. The carrier had just hauled down the two black day shapes from her mast displayed during flight operations, and was turning right to return to base course 280 and increase speed to 16 knots when she was hit by two torpedoes. The commanding officer of the heavy cruiser *San Francisco*, described in his report, the first evidence of the attack on the aircraft carrier:

> Those keeping station on the *Wasp* noticed a large column of dirty black water rise up on her starboard side forward of her island, and near the forward gun gallery on the starboard side. This column of water was followed by a large cloud of thin white smoke, which hung over the fore part of the vessel. The Navigator, who was sitting in the chair on the port wing of the Nav bridge, noticed a large flash of flame which appeared to go aft through the hanger and flight decks.[23]

As sea water surged into the *Wasp*, she took an almost immediate list of 15 degrees to starboard, and appeared down by the bow by about ten feet. Observers aboard the *San Francisco* could see personnel in the vicinity of the forward gun galleries, jettisoning ammunition and aircraft aboard. This was being done despite the flames raging in that area.[24]

Personnel aboard the *San Francisco* did not see any torpedo wakes near *Wasp*. The *San Francisco*'s commanding officer received a report of one wake crossing astern of his cruiser, from port to starboard, at 1455; and a report of a second wake at 1505. He rightly considered the second report somewhat doubtful.[25]

The battleship *North Carolina* (BB-55) received one of the spread of torpedoes fired by *I-19*, portside, twenty feet below her waterline, killing six crewmen. A 5.6 degree-list was righted in as many minutes, and she maintained her station in the formation at 26 knots.[26]

Another of the submarine-fired torpedoes found the destroyer *O'Brien* (DD-415). At about 1454, her lookouts spotted a torpedo closing the ship, forward of the port beam, 1,000 yards away. It missed close astern, but another "fish" struck the port bow, causing buckling at several places including the engine room. There were some injuries to personnel, none serious, and no killed or missing crew. *O'Brien* was also able to proceed under her own power.[27]

With the fires aboard *Wasp* out of control, her crew assembled aft to abandon ship, beginning about 1513. The destroyers *Duncan*, *Lardner*, *Laffey*, *Lansdowne*, and *Farenholt* collectively rescued 143 officers, and 1,767 men. Among the 1,910 survivors were the task force commander, Rear Adm. Leigh Noyes, and *Wasp*'s commanding officer, Capt. Forrest P. Sherman. There were also others who were in planes prior to the hit, and made it aboard *Hornet* or ashore. Upon orders of Noyes, *Lansdowne* scuttled the unsalvageable *Wasp* with three torpedoes. Her loss left *Hornet* the only operable U.S. fleet carrier in the theater.[28]

## *ALHENA* TORPEDOED BY SUBMARINE *I-4*

> *The torpedoing by submarines of four warships, with the loss of two of them was a serious blow that might possibly have been avoided. Carrier task forces are not to remain in submarine waters for long periods, should shift operating areas frequently and radically, must maintain higher speed, and must in other ways improve their tactics against submarine attack. Task forces have rendered invaluable support in maintaining the line of communications to Guadalcanal. They must and will continue this support but will venture into submarine waters only when necessary, and while there will operate in a suitable manner to reduce the submarine threat.*
>
> —Adm. Chester W. Nimitz, commander in chief, Pacific Fleet.[29]

The above guidance, promulgated by Admiral Nimitz, on 31 October 1942, in correspondence aptly titled, Solomon Islands Campaign – Torpedoing of *SARATOGA*, *WASP*, and *NORTH CAROLINA*, came after the torpedoing of the cargo ship *Alhena* (AK-26), on 29 September 1942, by the Japanese submarine *I-4*. However, *Alhena*, and other ships like her, were neither members of carrier task forces, nor able to proceed at high speeds, or avoid enemy submarine-patrolled waters, while delivering their desperately needed cargos to fighting forces ashore.

The *I-4* was laid down, on 17 April 1926, at Kawasaki's shipyard in Kobe as a Type J-1 submarine. She was completed, on 24 December 1929, and renumbered *I-4*. A unit of Submarine Division 7, she had left Truk, on 19 September, for an area south of San Cristobal Island, which lay southeast of Guadalcanal. Then currently under the command of Lt. Comdr. Kawasaki Rokuro, it was her fifth war patrol.[30]

On the night of 29 September, the submarine was twelve miles southwest of Cape Sidney, San Cristobal. At 2330, while patrolling on the surface, her lookouts reported a 7,000-ton cargo ship on the port bow heading east-southeast at 13 knots, escorted by a solitary destroyer.

*I-4* dove and began an approach at 5 knots. At 2344, Kawasaki fired two Type 96 torpedoes at the target, set at a depth of 10 feet. Seven minutes later, a hit to the ship's stern was observed, with the ensuing detonation creating a column of water and a plume of fire. He believed that the second torpedo also found its mark, but failed to explode.[31]

Comdr. Charles B. Hunt's *Alhena* (AK-26), was en route to Espiritu Santo from Guadalcanal, escorted by the destroyer USS *Monssen* (DD-436). The two ships had left Espiritu Santo on 23 September bound for the Guadalcanal beachhead with vital supplies. The Japanese still controlled the air by day, and sea by night, and the South Pacific high command believed that this small convoy might succeed where a larger one would surely attract Japanese resistance.[32]

Map 6-2

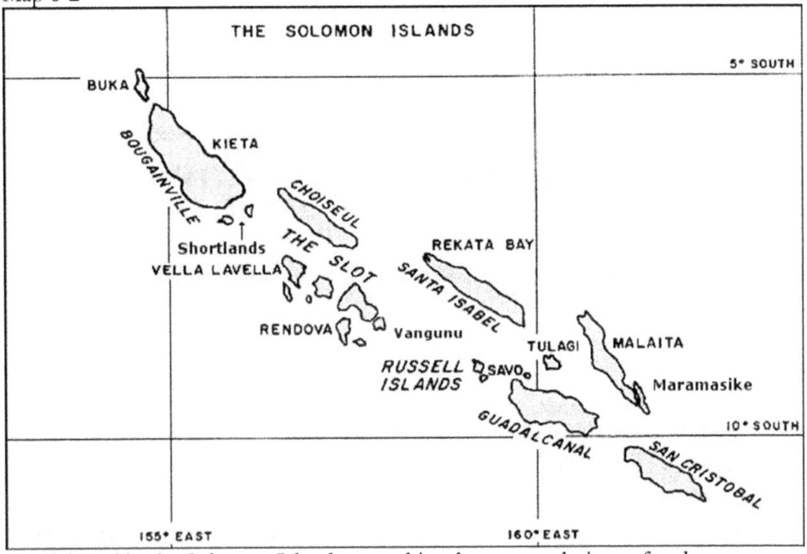

Guadalcanal in the Solomon Islands was a bitterly contested piece of real estate.
http://www.ibiblio.org/hyperwar/USN/ACTC/img/actc-35.jpg

*Alhena* and *Monssen* arrived off Lunga Point, on 26 September. While *Alhena* anchored as close to the beach as possible, the *Monssen* took up an anti-submarine patrol seaward. There were no shore dock facilities, so the discharge of cargo was slow; supplies had to be unloaded by hand on the beach from small boats. Moreover, while at Guadalcanal, several enemy air attacks interrupted operations. For safety, each day at dusk, the *Monssen* escorted *Alhena* eastward through Sealark Channel to an area out to sea, then accompanied her back to Lunga Point after daybreak. This pattern continued until the ships left for Espiritu Santo at 1145 on 29 September. Aboard the departing *Alhena* were forty-four wounded US servicemen and eight prisoners, for further transfer south.[33]

Twelve hours later, the cargo ship was 20 miles south of San Cristobal Island when a torpedo from the unseen *I-4* struck her, starboard side at frame 170 in the vicinity of No. 5 hold, just below the shaft alley flat. The torpedo opened both sides of her stern, causing extensive damage in the after part of the ship. (The resulting hole on the starboard side was approximately 25 feet long by 35 feet high; the port rent, 19 feet long and 35 feet high.) *Alhena* was zigzagging at 15 knots, on base course 125° true, steering 100°, with *Monssen* patrolling 1,500 yards ahead. The sky had been overcast since sunset, with visibility decreasing despite a fourth-quarter moon, because of approaching rain squalls which arrived coincidently with the attack.[34]

Photo 6-2

View of torpedo damage to the cargo ship USS *Alhena* (AK-26), showing gaping hole across her stern, which has been raised by shifting oil and armament forward.
Bureau of Ships Navy Department, War Damage Report No. 27, 24 May 1943

Fire broke out, but did not spread, and was extinguished in 3½ hours. *Alhena* with lost propulsion, settled 10 degrees by the stern, and drifted throughout the night and the next day. Casualties were: 4 killed, 1 missing, and 20 wounded. Believing the cargo ship doomed, Kawasaki dove to 165 feet and departed the area at 3 knots.[35]

Following the torpedo strike, *Monssen* circled the crippled ship prepared to attack, but no submarine was seen, or contact made with underwater sound gear. Fearing that the after bulkhead in lower four 'tween decks might carry away under pounding of the sea, a majority of the wounded were transferred to the destroyer by boat at daylight. (At least 41 of the wounded passengers were Marines.) When the extent of the damage to *Alhena* became known at first light, sharks could be seen swimming in and out of a gaping hole in her hull.[36]

On 1 October, a tow line was rigged astern of *Monssen* to the *Alhena*, and the destroyer then started towing on course 115°, "making turns" for 8 knots. This continued until arrival of the fleet tug *Navajo* (AT-64) and sub-chaser *PC-479* the following afternoon. The destroyer *Meredith* (DD-434) had joined the previous day and taken up patrol duties. *Monssen* turned over her tow to the *Navajo* and joined the screen. The group arrived at Espiritu Santo, on 7 October.[37]

On the 16th, after temporary repairs were made to *Alhena* there, *Navajo* took her in tow for Noumea, arriving on 20 October. On 8 November, following additional repairs, the tug *Sonoma* (AT-12) took the cargo ship under tow, and set course for Sydney, reaching the Australia port city on 20 November. *Alhena* remained in Sydney, until June 1943, undergoing repairs, during which she was redesignated the attack cargo ship AKA-9 and underwent a change of command. Captain Hunt was detached, and Comdr. Howard W. Bradbury, USN, the executive officer, assumed command.[38]

Charles Hunt, who would retire from the Navy as a rear admiral, was awarded the Navy Cross. His medal citation reads:

> The President of the United States of America takes pleasure in presenting the Navy Cross to Captain Charles Boardman Hunt, United States Navy, for extraordinary heroism and distinguished service in the line of his profession as Commanding Officer of the Cargo Ship U.S.S. *ALHENA* (AK-26), while engaged in transporting supplies, equipment, and reinforcements during and after the occupation of the Japanese-held island of Guadalcanal in August and September 1942. Though subjected to enemy high bombing, dive-bombing or torpedo attacks on each of the occasions on which the *ALHENA* entered this dangerous area, Captain Hunt coolly and courageously defied the threat of strong enemy surface forces in the vicinity to accomplish his vital and hazardous mission. On 30 September, when his ship was attacked and torpedoed while en route to base, Captain Hunt kept her afloat and prevented her grounding on a shoal near Rennel Island in spite of damage which flooded her aft, and put her propeller shaft out of action. His fine seamanship and gallant fighting spirit enabled him to bring his seriously damaged ship through for repairs and further useful service.

## BATTLE OF GUADALCANAL, 11-15 NOVEMBER

> *Perhaps the entire period between 7 August 1942, when the first Marines landed in the Solomons, and the final evacuation of Guadalcanal by the Japanese should be labeled "The Battle of Guadalcanal." Hardly a day went by during that 6 months which did not see action on land, in the air, or on the sea. Nevertheless,*

> *the climax and turning point of the campaign came with the shattering of the enemy's supreme effort to overwhelm the island between 11 and 15 November. During the desperate sea and air battles of those 5 days, Japanese losses as estimated by CINCPAC were 2 battleships sunk and 2 damaged, 4 cruisers sunk and 6 damaged, 8 destroyers sunk and 4 damaged, and 12 transports sunk or destroyed. Our losses consisted of 1 battleship damaged, 2 cruisers sunk and 3 damaged, 7 destroyers sunk and 4 damaged, and 3 cargo vessels damaged, 2 of these negligibly.*
>
> —*Combat Narratives Solomon Islands Campaign: VI Battle of Guadalcanal 11-15 November 1942* (Washington, DC: Office of Naval Intelligence, 1944). Sources title the period 11-15 November, the Battle of Guadalcanal, and that involving opposing naval forces (12-15 November), the Naval Battle of Guadalcanal.

As fighting continued on Guadalcanal, and in the skies overhead and adjacent waters, cargo ships continued to be dispatched to the island with ammunition and supplies for the Marines ashore. On 1 November, *Libra* was at anchor in Berth A-4, Segund Channel, Espiritu Santo, awaiting orders. These came in the early afternoon on the 8th, assigning her to Task Group 62.4.6—a part of Rear Adm. Richmond K. Turner's Task Force 67. At 0930 the following morning, the group stood out to sea, en route Guadalcanal. *Libra* took her place in the column formation, led by light cruiser *Atlanta* (CL-51) with the transport *Zeilin* (AP-9), and cargo ships *Libra* (AK-53) and *Betelgeuse* (AK-28) astern of her. Four escorting destroyers formed an anti-submarine screen.[39]

**Task Force 67: Rear Adm. Richmond K. Turner, USN**
**Task Group 62.4.6: Rear Adm. Norman Scott, USN**

| | |
|---|---|
| light cruiser *Atlanta* (CL-51) | Capt. Samuel P. Jenkins, USN |
| Destroyer Squadron Twelve | Capt. Robert G. Tobin, USN |
| destroyer *Aaron Ward* (DD-483) | Comdr. Orville F. Gregor, USN |
| destroyer *Fletcher* (DD-445) | Comdr. William M. Cole, USN |
| destroyer *Lardner* (DD-487) | Comdr. Willard M. Sweetser, USN |
| destroyer *McCalla* (DD-488) | Lt. Comdr. William G. Cooper, USN |
| cargo ship *Betelgeuse* (AK-28) | Comdr. Harry D. Power, USN |
| cargo ship *Libra* (AK-53) | Comdr. William B. Fletcher Jr., USN |
| transport *Zeilin* (AP-9) | Capt. Pat Buchanan, USN[40] |

While in transit, on 9 November, a sighting was made of the cargo ship *Alchiba* with barge in tow, escorted by the destroyer minesweeper *Hopkins* (DMS-13), on an opposite course. The two ships had departed Guadalcanal on the evening of the 6th, bound for Espiritu Santo. On 11 November, the task group passed Nuga Island abeam to port at 0127.

Continuing their approach to Guadalcanal, the ships entered Sealark Channel and steered various courses to the anchorage off Lunga Point. *Libra*, *Betelgeuse*, and *Zeilin* anchored east of the point at 0545, and began discharging troops and cargo into boats and lighters. *Zeilin* carried 1,599 troops, 33 naval passengers, and also cargo. She first debarked of troops, and then followed by discharging of No. 4, 2, 3, 7, and 8 holds.[41]

At 0905, Radio Guadalcanal issued an air warning that 10 bombers and 15 fighters were approaching from the northwest and should arrive at about 0930. Admiral Scott ordered the task group to get under way and form anti-aircraft disposition. *Atlanta*, *Zeilin*, *Betelgeuse*, and *Libra* were in column, with *Aaron Ward*, *Fletcher*, *Lardner*, and *McCalla* on the flanks of the column at a distance of 1,500 yards.[42]

The ensuing Japanese air raid opened the Battle of Guadalcanal (11-15 November 1942), the most complex naval engagement of the bloody Guadalcanal Campaign. Commencing on that day, 11 November 1942, it lasted for five days and involved three phases: a cruiser night action, a carrier action, and a battleship night action. Both forces suffered heavy losses in the battle, but the U.S. Navy was able to thwart a determined Japanese attempt to land reinforcements in order to seize Henderson Field on Guadalcanal. Further, the enemy failed to gain control of the seas around Guadalcanal or to significantly increase the strength of its force ashore. This action marked the last major offensive by the Imperial Japanese Navy in the Guadalcanal area.[43]

## PRELUDE TO ENEMY ACTION AT GUADALCANAL

Allied aircraft reconnaissance had identified at least 60 Japanese naval vessels massed in the Buin-Faisi-Tonolei anchorages, including 4 probable battleships, 6 cruisers, and 33 destroyers. By the afternoon of Monday, 9 November, there was no longer any doubt that the Japanese had set in motion a vast amphibious offensive aimed at Guadalcanal. Admiral Halsey's overmatched combatant forces thus had a two-fold responsibility to protect the Supply Group (TG 62.4.6), and Transport Group (TG 67.1) arriving at Guadalcanal a day later, and simultaneously counter the enemy offensive—lest the Allies be forced to retire from the Solomons.[44]

## FIRST AIR ATTACK ON 11 NOVEMBER

Further details of the air attacks, already mentioned, follow. At 0935 on 11 November, nine "Vals" (Aichi D3A Type 99 carrier dive bombers) emerged from the clouds over Henderson Field at about 10,000 feet. Shortly before the first one started its dive, the *Atlanta* and the other ships opened fire.[45]

The transport, *Zeilin*, opened fire at 0940 with all her 3-inch/50 guns two minutes before five planes attacked her, at which time her 20mm anti-aircraft guns joined the battle and began pouring out rounds. Two bombs fell close aboard her port side. One minute later, a third bomb glanced off her hull at frame 45, starboard side, and exploded underwater. Dive bombers then strafed the ship. Immediate results of the damage suffered by the transport were a flare back (combustion explosion in the firebox of a boiler) in No. 2 fire room; steam pressure lost on her steering gear; and flooding of her starboard shaft alley and No. 8 hold.[46]

By 0955, *Zeilin* had developed a starboard list; was also taking on water in No. 7 and 9 holds; and her port shaft alley was leaking badly, in addition to the existing flooding in her starboard one. Fortunately, there was also some good news. Only one crewmember had been wounded by bomb fragments, though gun crews claimed two planes shot down and another two damaged. Late that afternoon, after discharging all of her troops and nearly half her cargo, *Zeilin* departed Guadalcanal with the *Lardner* for escort, bound for Espiritu Santo, which she reached the morning of 14 November.[47]

## *BETELGEUSE* DOWNS TWO JAPANESE BOMBERS

At 0941, two bombs, dropped by an enemy plane that had dived from about 8,000 feet and leveled off at 1,500, landed close aboard *Betelgeuse*. Both detonated upon contact, spraying the ship with water and numerous metal and wood fragments. (Six personnel were wounded as a result of this attack, and two others that followed.) Despite heavy fire from the cargo ship's four 3-inch/23 guns, as well as her 20mm, .50, and .30-caliber machine guns, veering to port, the plane escaped with no apparent resultant damage.[48]

Two minutes later, another plane started a shallow dive from about 8,000 feet, leveled off at about 800-1,000 feet, and dropped a bomb which landed off *Betelgeuse*'s port quarter. The concussion from this near miss temporarily knocked her No. 3 and 4 guns out of commission, due to the upward jar jamming them. Unlike the other, this plane did not escape harm. Hit many times by still functioning 20mm AA guns, it caught fire, veered to port, and crashed about 300 yards from the ship.[49]

At 0945, a third plane made a bombing run on the *Betelgeuse*. Diving down from off the ship's port quarter toward her starboard bow, it released a bomb from about 1,200 feet. As it fell, the bomb slanted diagonally, following the plane's path, and landed about 20 feet clear of *Betelgeuse*'s starboard bow. Like the previous attacker, this dive bomber was downed by the cargo ship's 20mm gun crews:

> The plane was hit repeatedly by our 20mm and fire was seen to break out in the lower part of the fuselage directly between the wheels.... [It] then veered to starboard, then made a large circle to port and crashed in flames about 2,000 yards off the starboard bow. The plane was showing increasing flame and smoke from the time it was hit by our guns until it crashed into the sea.[50]

## *LIBRA* CLAIMS ONE PLANE SHOT DOWN

Concurrent with actions by the other ships, *Libra* opened fire at 0941, on two low-level dive bombers approaching her port quarter. Two dropped bombs missed her, one landing in the sea 100 feet off her port quarter and the other 20 feet clear of No. 2 hold. A strafing attack shot away her radio antennae and fore truck, and bomb detonations caused some distortion and slight leakage at frames 40-46, port side above the turn of the bilge. *Libra* ceased fire at 0945, after having shot down one of her tormentors which crashed off her port bow.[51]

## SECOND ENEMY AIR RAID OF 11 NOVEMBER

The Vals that escaped destruction by anti-aircraft fire during the attack, were downed by fighters during their withdrawal. After the first air raid, unloading resumed. At 1106, an air raid warning came of 27 enemy bombers with fighter escort approaching from the south over Guadalcanal. The bombers concentrated on Henderson Field. Shipboard anti-aircraft fire opened, but the planes were out of range, and consequently most bursts were short. At conclusion of the attack, the ships returned to anchorage, around noon, and resumed discharge of cargo.[52]

## JAPANESE AIR RAID OF 12 NOVEMBER

At 1327 on 12 November 1942, *Betelgeuse* received warning over the Guadalcanal control circuit that enemy planes were expected in the area. She and the *Libra* were anchored two miles east of Lunga Point, and the *Zeilin* off Kukum Beach, with combatant ships disposed about them in two protective semicircles. *Betelgeuse*'s anchor was under foot with the chain at short stay to facilitate getting under way rapidly, if necessary. Rear Adm. Richmond K. Turner (commander, Task Force 67 embarked in the transport *McCawley*) then made a general signal for ships to get under way immediately and form up.[53]

In response by 1340, all ships were formed up and proceeding north on course 340°, headed for open water. Turner maneuvered the formation via flag hoist signals directing turns. The transports and cargo ships were in two columns, screened by the cruisers *San Francisco*,

*Atlanta*, *Helena*, *Juneau*, and *Portland*; and destroyers *Buchanan*, *Shaw*, *O'Bannon*, *Cushing*, *Barton*, *Aaron Ward*, *McCalla*, *Sterett*, *Hovey*, *Fletcher*, *Southard*, and *Monssen*.

| **Left Column** | **Right Column** |
|---|---|
| *McCawley* (AP-10) | *President Jackson* (AP-37) |
| *Crescent City* (AP-40) | *President Adams* (AP-38) |
| *Betelgeuse* (AK-28) | *Libra* (AK-53)[54] |

As expected, the enemy aircraft appeared at 1405, after approaching low behind Florida Island, so as to remain out of sight as long as possible. There were 21 torpedo bombers and 8 Zero fighters. Only the bombers were immediately visible. As observed, flying at approximately 170 knots, they headed toward the transport and cargo ship anchorage in a long line abreast, skimming so close to the water that they occasionally dipped below the horizon. As they came within range, the screening ships to the north and east, opened fire, blasting several planes from the air and setting others ablaze. The remainder of this flight, split into two groups.[55]

At 1408, *Betelgeuse* sighted about twelve enemy planes to the east, about eight miles distant. The planes were divided into two groups: one bearing 060°T and headed westward toward the starboard bow of the ship formation; the other bearing 120°T, flying southward, apparently to make a simultaneous attack from different angles. Comdr. Harry D. Power, *Betelgeuse*'s commanding officer, described formation maneuvers taken to minimize the air threat, and his necessary focus on the second group of torpedo bombers:

> Our formation seemed to be turned toward the Northeastern group to force their torpedo attack before the Southeastern group could get in position and the Task Group Commander could turn to meet each attack in succession. This we were able to do, taking the Southern attack on our tail.
>
> Our screen had opened fire by this time. This ship's attention was held by the southern group as we were the rear ship in the left-hand column.[56]

## *BETELGEUSE* SHOOTS DOWN THREE PLANES

> *In addition to having 10 – 20mm guns firing on these two planes, all members of the crew and embarked construction battalion not otherwise engaged were firing*

> *.30 cal BAR's* [Browning automatic rifles], *rifles, machine guns and .45 cal pistols at the planes.*
>
> *The fire power was terrific and must have amazed the Japanese pilots who evidently thought that they were attacking a defenseless looking AK and that they would have easy pickings.*

—Comdr. Harry D. Power, commanding officer, USS *Betelgeuse* (AK-23), describing the barrage of fire responsible for downing the second and third of three enemy torpedo bombers destroyed by the cargo ship during a six-minute period on the first day of the Naval Battle of Guadalcanal (12-15 November 1942).[57]

Photo 6-3

USS *Betelgeuse* (AK-28), circa 1942, displaying her camouflage scheme.
National Archives photograph #80-G-31906

Aboard *Betelgeuse* at 1412, a single torpedo plane was sighted heading for her port quarter. Fire was held on the 20mm guns until the enemy was within range, then intense fire was opened on the aircraft which sheered away to the left at a distance of about 300 yards. Flames broke out in the wings of the plane as it passed the cargo ship, before crashing into the water about 1,000 yards on her port bow.[58]

Three minutes later, at 1415, two torpedo planes directly astern of *Betelgeuse* began a run on her. Their approach was steady, directly for the ship, flying about 60 feet off the water. The cargo ship's stern gun, a 5-inch/51, fired one round of shrapnel which burst near the two planes, likely contributing to an erratic torpedo attack. At a range of about

1,500 yards from *Betelgeuse*, they dropped their torpedoes which did not make successful runs.[59]

Comdr. Harry D. Power, *Betelgeuse*'s commanding officer, described witnessing the starboard plane's failed weapon employment, and the subsequent attacks by both planes on the vessel:

> The torpedo appeared to hang by the tail and entered the water in a plunging motion at an angle of about 60° and when it struck the water it bounced back out clear of the water like a pole which had been plunged end first into the water. The torpedo then plunged back into the water nose first and was not seen again.
>
> The planes then divided and came up each side of the ship at a height of about 50 feet and the wing tips about 75 feet from the side of the ship.
>
> The fire of the 20mm guns was held until the planes were within effective firing range. The planes were on a steady run up the side of the ship and made a perfect point of aim until they came up abreast of the ship when the rate of change of bearing increased rapidly.[60]

As expressed in the quoted material at the beginning of this section, all of *Betelgeuse*'s naval guns were firing, as were unoccupied men topside employing small arms. Two of the ship's officers, who were ashore on the beachhead and witnessed the entire action, watched *Betelgeuse* in particular. They stated that it appeared as if the ship was not going to defend herself at first, as her guns were silent and the two planes were closing rapidly. "Then when the planes came within close range it looked as if the entire ship was surrounded by a sheet of fire from all of the smaller calibre guns."[61]

Hits were made almost immediately on the two aircraft, and by the time the plane on the starboard side was opposite No. 4 hatch, it was aflame, and burning briskly at a distance of about 75 feet at the level of the flying bridge. This plane, with No. 318 painted on its tail, burst into full flame and crashed in the sea off the ship's starboard bow.[62]

*Betelgeuse*'s rudder was immediately put hard left, then shifted to hard right, to keep her bow and stern from running into the burning plane, the flames of which now reached higher than her masts. The heat generated was intense as the ship passed close aboard.[63]

Simultaneously, the second plane (tail number 817) was passing up *Betelgeuse*'s port side at about the same distance off as the first plane. As machine gun fire poured into it, the aircraft started to burn under the starboard wing, and then appeared to lose control either from damage, or injury to the pilot. Tracer rounds were seen going into the fuselage all around the pilot. The plane then swerved across the ship's bow, from port to starboard and pancaked into the sea. It sank to the top of the fuselage and when last sighted was still floating.[64]

At 1418, *Betelgeuse*'s guns fell silent. During the attack, the planes had raked exposed positions aboard the cargo ship with their machine guns, wounding twelve personnel and causing considerable minor material damage to topside areas.[65]

Photo 6-4

Smoke rising from two planes shot down during a Japanese air attack on U.S. Navy ships off Guadalcanal, 12 November 1942. Photographed from the USS *President Adams* (AP-38), the second of three units in the right column of the ship formation. USS *Betelgeuse* (AK-28), pictured above, was the third (last) unit in the left column. National Archives photograph #80-G-32376

## DESCRIPTION OF ATTACKING ENEMY AIRCRAFT

Commander in Chief, Pacific Fleet, in an action report about the air attack of 12 November, identified the enemy planes as "Nells" and described their formations, with results of the combat action:

> Twenty-one Mitsubishi, type 96, heavy bombers were sighted at 14,600 yards at 1410. They approached over and around Florida Island in three groups of 7, 9, and 5 planes. The first group of seven

> approached from the starboard bow of the formation and the other two groups from the starboard quarter. Their speed was about 170 knots and altitude was not more than fifty feet and they used the dark background of Florida and Tulagi Islands to reduce the silhouette effect.... The action lasted seven minutes and only one plane escaped.[66]

Photo 6-5

Lae, New Guinea, 2 October 1943, Japanese Mitsubishi G3M2 Navy type 96 bomber, gutted by fire, and recovered by the Allies for intelligence purposes.
Australian War Memorial photograph 133469

*Libra*'s war diary entries describing the combat action were brief:

1411: Sighted large formation of enemy torpedo planes close to the water approaching from northeast.
1412: Commenced firing.
1425: Ceased firing. About 8 enemy planes shot down by ships – burning on water.[67]

## RETURN TO ESPIRITU SANTO

*Libra* returned with the *Betelgeuse* and transports to anchorage, following the air attack, and resumed discharging cargo. At 1700, she received orders from commander, Task Force 67, to get under way and standout about 1800. In view of the developing situation, a buildup of enemy surface forces in the area stronger than available U.S. forces, Admiral Turner decided to withdraw all transports and cargo vessels from the area. Once formed, Task Group 67.1—the transports and cargo ships, screened by the *Buchanan*, *Shaw*, *McCalla*, *Southard*, and *Hovey*—proceeded eastward out of Savo Sound, and southward around the western edge of San Cristobal Island, bound for Espiritu Santo. The group arrived there on 15 November.[68]

## THIRD TRIP BRINGS MISFORTUNE TO *ALCHIBA*

> *She was one of the Navy regular "work horses" in the Atlantic putting in at almost all of the eastern seaports. She even had made a trip to Antofagasta, Chile to pick up some desperately needed ingot and electrolytic copper.*
>
> *But December 7th [1941] changed all that, and the* Alchiba *was transferred immediately to the South Pacific to partly alleviate the great shortage of ships in that area. Few of us appreciate the importance of the few auxiliaries in this early phase of the war, for had not those ships, probably less than a hundred at the time, brought up supplies to the defensive outlying bases, the Japs would probably have attacked down as far as Australia.*
>
> —From the War History of USS *Alchiba*, 17 December 1945.[69]

As previously mentioned, Task Group 62.4.6, while in transit to Guadalcanal on 9 November, had sighted the *Alchiba* (AK-23), a former merchant vessel, returning to Espiritu Santo from their destination. When the Navy acquired the MS *Mormacdove* (a type C-2, diesel propelled ship) on 2 June 1941, and commissioned her USS *Alchiba* thirteen days later on 15 June, little did it realize she would turn out to be the heroine of the auxiliary fleet. She was named *Alchiba* in conformity with a policy of naming cargo ships after stars, and one of the Navy's stars she soon became. The exploits of her crew in earning a Presidential Unit Citation, are detailed in Chapter One.[70]

Arriving at Wellington, New Zealand, *Alchiba* loaded with cargo for Noumea, New Caledonia. In August 1942, she loaded at Pago Pago, Samoa, to take part in the first offensive attack of the Pacific war—destination Guadalcanal. Following the landings on Guadalcanal and Tulagi, *Alchiba* was to depart to reload and return with more supplies. This she was able to do, but on her third trip to "the Canal" (a "fleet shorthand" used by Marines in reference to Guadalcanal), she was torpedoed, on 28 November, by a Japanese midget submarine.[71]

Temporary repairs were made, as described in the opening chapter, on 7 December 1942, before *Alchiba* was torpedoed a second time by a different midget submarine. Following additional temporary repairs, she limped across the Pacific to the Navy Yard at Mare Island, California. On 7 August 1943, then redesignated as attack cargo ship AKA-6, *Alchiba* rejoined the Pacific Fleet.[72]

## END OF THE GUADALCANAL CAMPAIGN

*Before Guadalcanal the enemy advanced at his pleasure—after Guadalcanal he retreated at ours.*

—Adm. William F. Halsey Jr., USN, commander, South Pacific Force and South Pacific Area, who had succeeded Vice Adm. Robert L. Ghormley, USN, on 18 October 1942.

Photo 6-6

World War II poster, featuring Adm. William F. Halsey, USN.
Naval History and Heritage Command photograph #NH 76342-KN

Over the course of the hard-fought Guadalcanal campaign, U.S. forces suffered 7,100 dead and nearly 8,000 wounded. Japanese losses were at least 19,200 dead and an unknown number of wounded. In conjunction with the Battle of Midway, the Allied victory on Guadalcanal was likely the turning point of the Pacific war.[73]

By the end of December 1942, the Japanese Imperial High Command had evaluated and accepted recommendations for the evacuation of Japanese forces on Guadalcanal. In preparation for Operation KE, the Japanese 17th Army, the bulk of the enemy's ground forces on Guadalcanal, withdrew to the island's western shore while staging counterattacks and fighting rearguard skirmishes. The last

Japanese troops were evacuated by a destroyer task force during the night of 7 February 1943.[74]

## AWARDING OF BATTLE STARS

The eligibility period for battle stars associated with the Capture and Defense of Guadalcanal was from 10 August 1942 to 8 February 43. Listing of the cargo ships awarded stars for this portion of the Guadalcanal Campaign as well as for the Naval Battle for Guadalcanal are provided in the tables.

**Capture and Defense of Guadalcanal**

| | |
|---|---|
| USS *Fomalhaut* (AK-22) | 23 Aug-27 Dec 42 |
| USS *Bellatrix* (AK-20) | 8 Sep-25 Oct 42 |
| USS *Alhena* (AK-26) | 29 Sep 42 |
| USS *Alchiba* (AK-23) | 15 Oct-28 Nov 42 |
| USS *Betelgeuse* (AK-28) | 11 Nov 42 |
| USS *Libra* (AK-53) | 11-12 Nov 42 |
| USS *Carina* (AK-74) | 1 Feb 43 |

**Guadalcanal – Naval Battle of Guadalcanal**

| | |
|---|---|
| USS *Betelgeuse* (AK-28) | 12-15 Nov 42 |
| USS *Libra* (AK-53) | 12-15 Nov 42 |

*Carina* (AK-74) earned a battle star on 1 February 1943, the last one awarded a cargo ship for Capture and Defense of Guadalcanal. She was anchored off Guadalcanal late that morning, and got under way at 1133, following warning of an impending air raid. The captain had the conn (was personally giving rudder and engine orders), as the navigator observed enemy bombers to port. The ship opened fire at 1135 and, in the anti-climactic action that followed, expended two rounds each from her 3-inch and 5-inch guns. There were no casualties.[75]

# 7

# HMAS *Patricia Cam* sunk by Japanese Aircraft

Photo 7-1

The trawler *Patricia Cam* prior to her commissioning by the RAN in 1941.
Australian War Memorial photograph 301155

In January 1943, beleaguered Japanese troops on Guadalcanal were preparing to flee farther up the Solomons island chain; Allied and Japanese forces were facing each other in New Guinea, the Americans and Australians having ejected the Japanese from Papua; and enemy floatplanes were attacking Allied vessels off Northern Australia, and those working the Papua-New Guinea coast. One victim, sunk by an aircraft bomb on 29 January, was a wooden-hulled, modestly built and armed, 120-foot stores-issuing ship, proudly flying the RAN White Ensign.

She was the motor vessel *Patricia Cam*, built in 1940 at Beatties shipyard, Daleys Point (Brisbane Water, NSW), where she began life as a tuna fishing boat. She was requisitioned by the Royal Australian Navy from her owners, Cam & Sons Pty. Ltd, Sydney, on 9 February 1942. The newly acquired vessel was commissioned at Sydney the following month as the auxiliary minesweeper HMAS *Patricia Cam*. Minelaying by

German surface raiders in 1940-1941 had demonstrated the shortage of suitable vessels to keep Australian sea lanes clear of this threat, but she was not employed in this role even though it was her naval designation.[1]

*Patricia Cam* sailed for Darwin, on 8 March, under the command of Lt. John A. Grant RANR(S), with a crew of 2 officers and 12-17 naval ratings. Thereafter, she worked out of that port in Northern Australia as a general-purpose vessel, armed with one 20mm Oerlikon, two .303 Vickers machine guns, and one Browning machine gun.[2]

Ten months later, on the morning of 13 January 1943, *Patricia Cam* left Darwin, carrying stores and personnel for several outlying stations. Embarked aboard her were the Reverend Leonard N. Kentish, chairman of the Methodist Northern Australian Mission District, and five native passengers. The latter individuals included Paddy Babawun Wanambi of Milingimbi, a native pilot from the Yolngu tribe, one of a number of coastal aborigines regularly carried to assist with navigation among the area's uncharted reefs and shoals. The other four Yolngu were Djimanbuy and Djinipula Yunupingu, Narritjin Maymuru, and Milirrma Marika.[3]

Following a stop at Goulburn Island, *Patricia Cam* left there, on 19 January, for Millingimbi Island. She arrived the following afternoon and departed, on the 22nd, for Elcho Island.[4]

Photo 7-2

Darwin and its harbour, Northern Territory, 19 October 1942.
Australian War Memorial photograph 027335

In early afternoon that day, a Japanese float plane of the 734th Kokatai attacked the *Patricia Cam* from out of the sun. (More about this enemy air group, and Millingimbi Island in the following pages.) The Japanese Aichi E13A passed down the length of the ship from stem to stern, at an elevation of less than 100 feet above the water, and dropped a large bomb. It landed amidships and exploded, opening the hull planking. *Patricia Cam* sank within a minute at position 11°19'S, 136°23'E.[5]

Unlike purpose-built naval ships, the former fishing vessel did not have the compartmentation that naval architects required in their plans, to help control flooding or spread of fire in the event of combat, or non-combat damage to such ships. Sub Lt. John Leggoe, RANVR, who joined *Patricia Cam*, in January 1943, as her first lieutenant, described the challenges associated with the ship, while praising her crew:

> *Patricia Cam* was really nothing more than a vast wooden hold with an engine room right aft. The low after superstructure consisted of two cabins, mine and one shared by the engine room artificers (ERAs) and the coxswain, a tiny wardroom mess just big enough to take a table and a couple of benches, and the galley. On the deck above were the wheelhouse and chartroom and the Captain's cabin. Above the wheelhouse was a compass platform and just forward of the wheelhouse was the mainmast.
>
> Right forward on the forecastle was a tall foremast carrying the topping lift for an enormous boom which was used for working cargo at the various coastwatching and mission stations around the coast. The winch on the forecastle was driven by a rowdy diesel engine which was also used for weighing the anchor. A point-five machine gun, which swayed from side to side on its mounting, seemed to be something of an afterthought.
>
> The ship's company consisted of eight seamen, three stokers, a cook, a steward, a telegraphist, a coxswain and three ERAs. They lived in a fairly capacious but cockroach-ridden forecastle. When the tropical rain squalls of the Wet swept over the ship the forecastle leaked like a sieve. The crew, a magnificent bunch clad only in brief shorts and tanned like bronze statues, lived most of their time on top of the huge hatch.[6]

Several crewmembers who were sitting on the forward hatch at the time of the attack, were propelled by the blast down into the hold of the ship. They were almost immediately washed out by the inrush of water up through the hull, before the *Patricia Cam* sank. Ordinary Seaman Neil G. Penglase and Djinipula Yunupingu died at once. The survivors took to a single raft and floating debris, the ship's two rescue boats having been destroyed in the attack.[7]

The Japanese floatplane returned and dropped its second bomb on survivors in the water, killing Able Seaman Edward D. Nobes and

Djimanbuy Yunupingu. The plane then circled about for a half hour, directing machine gun fire at the survivors without managing to inflict any new injuries. It flew away to the north, returned five minutes later, and landed near the men in the water.[8]

Photo 7-3

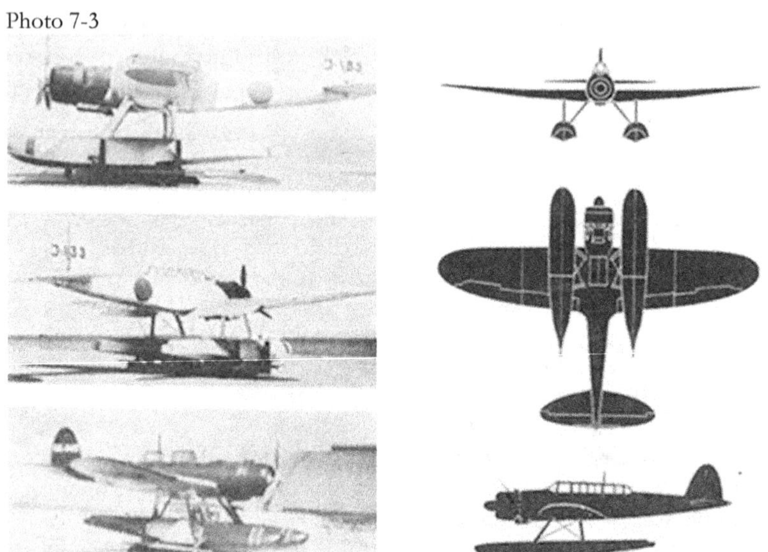

Japanese "Jake" long-range reconnaissance seaplane with crew of three; one flexible, rearward-firing 7.7mm machine gun; and which could carry a bombload of 550 lbs. Japanese Operational Aircraft CinCPOA 105-45 "Know Your Enemy!"

Leggoe watched as the rear gunner in the aircraft inserted a new magazine into his machine gun and fired a few rounds at the survivors in the water. The floatplane then taxied around to the other side of the group of survivors, where Reverend Kentish and a rating were afloat on a hatch cover, about 50 yards away from the main group. One of the Japanese aircrewman jumped down onto a float, pointed his revolver at Kentish and ordered him to swim over to the aircraft.[9]

Kentish complied, was hauled up onto the float, given a drink from a flask, then ordered into the aircraft. The plane then took off and disappeared to the north. This might well have been the only case of an Australian being taken prisoner by the Japanese in mainland Australian waters during World War II.[10]

Ordinary Seaman Andrew Johnston and Chief Engine Room Artificer William Moffitt floated away that night and were not seen again. Eighteen survivors (those in the raft, and others using improvised rafts, or clinging to a hatch cover) landed on the small rocky islet

Gurraka at 0330 the following morning. Gurraka lay off Guluwuru Island, west of Cumberland Strait, a waterway between the two principal Wessel Islands. The men were barely ashore before Paddy started a fire and found fresh water. Stoker Percival J. Cameron and Milirrma Marika died of their wounds that day. They were buried side by side, and their graves covered with rocks.[11]

Searching island natives found the survivors, on 25 January 1943. They took the commanding officer, Lt. Alexander C. Meldrum RANR(S), by canoe across Cumberland Strait to Marchinbar Island, to seek further assistance. Meldrum made a 34-hour barefoot trek, more than 25 miles up Marchinbar Island, to the coastwatcher station on Jensen Bay. Wet season storms prevented radio contact with Darwin.[12]

Meanwhile, Flight Sergeant Len Gairns, of 7 Squadron RAAF based at Horn Island, was on an operational flight, on 27 January, when he spotted a man waving furiously from an island beach below. The man wrote a message in the sand to identify he was from HMAS *Patricia Cam*, "Patcam bombed no food have water." The survivors were dropped some food and first aid kits the next day, and were rescued by the auxiliary patrol boat HMAS *Kuru* from Darwin, on 29 January.[13]

Photo 7-4

Group portrait of survivors of HMAS *Patricia Cam* made on 17 February 1943. Front row (left to right): Stoker Arthur Bennett, Able Seaman (AB) Greg Durrington, Telegraphist Bert R. L. Stevens, Steward Alfred R. Tanner. Standing: AB Don Brun, Stoker George W. Williams, Ordinary Seaman (OS) D. Murray, Sub Lt. John Leggoe, Cook John R. Hawkins, Lt. Alexander C. Meldrum (commanding officer), Petty Officer Hulbert Challender, AB Aubery R. White, and Engine Room Artificer John D. McKimmie.
Australian War Memorial photograph 014329

It was later determined that Reverend Kentish was taken to Dobo in the Aru Islands (now a part of Indonesia), and beheaded there, on 4 May 1943. After the war, Sub Lt. Sagejima Maugan, 24 Naval Base Force Det. Dobo, who had carried out the execution, was arrested as a war criminal and sentenced to death. He was hung in Hong Kong's Stanley Gaol, on 23 August 1948. Petty Officer Hoyama Kenzo and Civilian Administrator Kohama Shozuke received a sentence of life imprisonment for their part in the execution of Leonard Kentish.[14]

| **HMAS *Patricia Cam* Crewmember and Passenger Casualties** |
|---|
| Stoker 2nd Class Percival James Cameron |
| Ordinary Seaman Andrew Alexander Johnston |
| Chief Engine Room Artificer William Robson Moffitt |
| Able Seaman Edward David Nobes |
| Ordinary Seaman Neil Gray Penglase |
| Milirrma Marika (passenger) |
| Djinipula Yunupingu (passenger) |
| Djimanbuy Yunupingu (passenger) |
| Reverend Leonard N. Kentish (killed by Japanese while in captivity) |

# 8

# Consolidation of Southern Solomons

*KILL OR BE KILLED*

—Declaration in the name of Vice Adm. William F. Halsey, USN, posted on a sign at Tulagi in 1943, reminding Navy servicemen of the realities of war.

Photo 8-1

Cargo ship USS *Carina* (AK-74), location and date unknown.
Naval History and Heritage Command #NH 84609

"Consolidation of Southern Solomons" is a phrase used to describe the period from 8 February to 20 June 1943, following the completion of the enemy's evacuation of all its forces from Guadalcanal, on 7 February 1943. Thereafter, the Japanese strengthened their positions in the upper Solomons, while still carrying out air raids and submarine attacks on shipping in the southern islands.[1]

The first cargo ship to earn a battle star in support of the Allied strengthening operations was USS *Carina* (AK-74), launched on 6 November 1942, at Permanente Metals Corp., Yard No. 1, Richmond, California as the merchant ship SS *David Davis*. She was transferred to the Navy on 20 November, and commissioned on 1 December 1942, with Lt. Comdr. James Ian MacPherson, USNR, in command.[2]

*Carina* stood out of San Francisco, on 14 December, laden with cargo for Espiritu Santo and Guadalcanal. She discharged her carried cargo, between 23 January and 4 February 1943, during the last phases of the bitter campaign for Guadalcanal. Then she began plying between Espiritu Santo and Purvis Bay, Florida Island; Tagoma Point, Guadalcanal; and Tongatabu, Tonga. On the night of 3 March starting at 2227, while anchored at Tulagi Harbor and discharging cargo, five or six bombs fell off her starboard beam at a distance of 20 to 50 feet.[3]

At 2323, a second flight of bombers passed overhead, and another group of six bombs dropped into the sea about 50 to 100 feet away. There had been no "Condition Red" forewarning given of the approach of enemy aircraft. At 0115, Condition Green signaled "all clear." As a result of these attacks, six crewmembers suffered shrapnel wounds, and there were about twenty other cases of men who suffered cuts and bruises from being thrown down by concussions from bomb explosions. These blasts were responsible for about sixty holes, varying from one to four inches in size, on the starboard side of the ship. On the upper deck, there were a considerable number of small shrapnel holes, and other damage which included a large number of lines (boom guys and hatch runners) cut, a lifeboat badly damaged, a masthead light smashed, and ventilators holed.[4]

*Carina* was subsequently repaired at Espiritu Santo.[5]

## *ARIDED* SUBJECTED TO AIRCRAFT ATTACKS

About one week after Japanese bombers achieved near misses on *Carina* at Tulagi, the cargo ship *Arided* (AK-73) was subjected to a similar attack at Guadalcanal. On 11 March, while discharging cargo, she was the target of enemy land-based bombers. No casualties were sustained then, nor in another attack, weeks later, on 23 May 1943.[6]

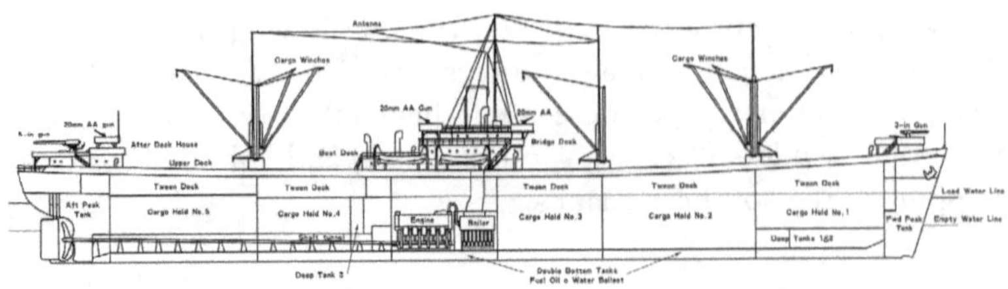

The *Carina* and *Arided* were both 441-foot Liberty ships of the EC2-S-C1 cargo ship design, and products of the same shipyard in Richmond, California. *Arided* was built as the *Noah H. Swayne* for the War Shipping Administration, and was delivered, on 12 November 1942. Valued at $1,750,000 she had taken only two months to build. Immediately after delivery, she was transferred to the US Navy and taken to Pier 36 in San Francisco, where she was converted for Navy use by the Matson Maintenance Company. On 26 November, she was placed in commission under Lt. Comdr. John B. Blaine, USNR, with a complement of 12 officers and 154 crewmembers.[7]

Following commissioning, *Arided* reported for duty with the U.S. Pacific Fleet assigned to Service Squadron Eight. With a full cargo of ammunition, gasoline, and other war materiel, she sailed from San Francisco, on 4 December, for Espiritu Santo, via Pago Pago and American Samoa. Arriving at "Santo," the cargo ship began operating as part of the South Pacific Force under Vice Admiral Halsey. (Halsey's command was redesignated the U.S. 3rd Fleet, on 15 March 1943). In January 1943, *Arided* was assigned duty transporting cargos of ammunition and gasoline from New Zealand ports and Noumea, New Caledonia, to Guadalcanal and other southern Solomon Islands.[8]

Photo 8-2

Pago Pago Harbor, Samoa, 1899.
Naval History and Heritage Command photograph #NH 1457

As background to the supply missions of the ships, descriptions of warehousing and fuel storage facilities is provided. By late autumn of 1942, ammunition depots had been established at Noumea, and at Espiritu Santo and Efate in the New Hebrides. Fuel-oil supply storage had been constructed on Ducos Peninsula at Noumea, with a capacity of 370,000 barrels of black and 30,000 barrels of diesel fuel, together with a pier at which ships could be unloaded and supplied. Ships sailed on water but were propelled by oil, and the demand never ceased. At Vila on Efate, there were seven 1,000-barrel steel tanks for aviation gasoline, two 10,000-gallon diesel tanks, and four buried 5,000-gallon aviation-gasoline tanks. At Port Havannah, eight other buried tanks held 5,000 gallons each.[9]

The consumption of fuel and lubricants was tremendous, and storage, like demand, mounted steadily. At Tulagi alone, during the early part of 1943, the motor torpedo boats based there used up to 3,000-7,000 gallons a day and aircraft about a thousand. To support this requirement, there were, in the Tulagi area, ten 1,000-barrel tanks, 12,000 barrels of aviation gasoline, a 60,000-barrel diesel-oil storage, and a 280,000-barrel fuel-oil farm. Guadalcanal had storage for 1,300,000 gallons of aviation gasoline.[10]

## *CASSIOPEIA* EARNS BATTLE STAR

> *The USS* Cassiopeia *AK 75 spent her time during the war months, carrying cargo among the islands in the South Pacific. She was one of the many "Liberties" taken over completely by the Navy to serve during the wartime emergency.... She was a clumsy looking ship with her high straight sides; slow moving, 10 knots; and dull gray in color. It was one of the first "K" ships in the Pacific and out of her class of ten ships, one of the few lucky enough to return unscathed. Of the other ten, two were sunk, one damaged by bombs, and three damaged having to return to the states for repairs.*
>
> —David A Friederich, former crewmember aboard USS *Cassiopeia*.[11]

Photo 8-3

Cargo ship USS *Cassiopeia* (AK-75), date and location unknown.
Courtesy of NavSource and David Friederich

The cargo ship USS *Cassiopeia* (AK-75) earned a battle star, on 11 March 1943, during a Japanese enemy air raid on Guadalcanal. She was anchored off Lunga Point that evening, discharging cargo while the destroyers *Sterett* (DD-407), *Morris* (DD-417), and *Smith* (DD-378) patrolled Lunga Roads. Details of how *Cassiopeia* earned the star are scarce, however it is known that *Sterett* (which had escorted *Cassiopeia* from Espiritu Santo, arriving at Guadalcanal that morning) went to General Quarters at 1954, as enemy planes passed overhead, and later released all hands from battle stations at 2135.[12]

## TEN "LIBERTY SHIPS" TO SOUTH PACIFIC FORCE

USS *Cassiopeia*, like *Carina* and *Arided*, was a recently converted Liberty Ship that had arrived in the South Pacific. She was part of a detachment resulting from earlier action by Adm. Ernest J. King, USN (commander in chief, U.S. Fleet). On 3 October 1942, he had directed the Auxiliary Vessels Board to consider a dispatch from commander, South Pacific Force, urgently requesting that ten Liberty (EC2) type ships be acquired, given a minimum conversion, and be manned by Navy crews for use in the logistic support of forces in his area. A week later, the Board recommended that ten EC2 type hulls then being built in the San Francisco area be acquired. Having a construction rate averaging one every ten days, these ships were chosen to fulfill this requirement. They were to be assigned to commander, South Pacific Force.[13]

The ships identified in the table, were built by Permanente Metals Corporation No. 1 and 2 Yards, Richmond, California. Details about the torpedoing and loss of *Aludra* and *Deimos*, on 23 June 1943, are provided later in the chapter.

**USS *Crater*-class (MC EC2-S-C1 "Liberty Ship")**

| Ship/<br>ex-Merchant | Com/<br>Decom | Disposition |
|---|---|---|
| USS *Crater* (AK-70)<br>*John James Audubon* | 31 Oct 42<br>24 Jun 46 | |
| USS *Adhara* (AK-71)<br>*G. H. Corliss* | 16 Nov 42<br>7 Dec 45 | |
| USS *Aludra* (AK-72)<br>*Robert T. Lincoln* | 26 Dec 42<br>lost | Sunk by Japanese submarine *RO-103* off San Cristobol Island with AK-78. |
| USS *Arided* (AK-73)<br>*Noah H. Swayne* | 23 Nov 42<br>12 Jan 46 | |
| USS *Carina* (AK-74)<br>*David Davis* | 1 Dec 42<br>17 Oct 45 | Badly damaged by Japanese suicide boat off Okinawa 31 May 45 |
| USS *Cassiopeia* (AK-75)<br>*Melville W. Fuller* | 8 Dec 42<br>21 Nov 45 | |

| | | |
|---|---|---|
| USS *Celeno* (AK-76)<br>*Redfield Proctor* | 2 Jan 43<br>1 Mar 46 | |
| USS *Cetus* (AK-77)<br>*George B. Cortelyou* | 17 Jan 43<br>20 Nov 45 | |
| USS *Deimos* (AK-78)<br>*Chief Ouray* | 23 Jan 43<br>lost | Sunk by Japanese submarine RO-103 off San Cristobol Island with AK-72. |
| USS *Draco* (AK-79)<br>*John M. Palmer*[14] | 16 Feb 43<br>28 Nov 45 | |

## ENEMY AIR ATTACKS ON SHIPPING CONTINUE

The next enemy air attack on a convoy plying between Espiritu Santo and Guadalcanal came on 20 March 1943, carried out by a single plane that did little harm. Comprising Task Group 32.4.5 were the cargo ship *Adhara* (AK-71), and merchant vessels *Day Star*, *Lyman Beecher*, and *William J. Turner*, escorted by the destroyer *Aaron Ward* (DD-483), high-speed transport *Ward* (APD-16), and minelayer *Gamble* (DM-15).[15]

At 1117, lookouts aboard *Adhara* sighted a large twin-engine plane, at high altitude, going into clouds and sun. *Fletcher* opened fire a minute later, and *Adhara* ordered "Emergency turn 45 degrees to starboard" for the convoyed ships. At 1126, four bombs fell, two each off the port quarters of *Lyman Beecher* and *Adhara*. It appeared that two of the bombs were delayed action, and the others, fragmentation.[16]

At 1127, *Adhara* ordered "Emergency turn 45 degrees to port." The plane appeared to cross the formation, and disappeared in a northerly direction. *Fletcher* then ceased fire. The escorted ships had not opened fire, because the plane was too high and out of reach of their guns. No casualties resulted from the attack.[17]

## *ADHARA*, *LIBRA* ATTACKED OFF GUADALCANAL

> *Three [bombs] which 'greased' the ship's side, two to starboard and one to port, abreast the bulkhead separating No. 2 and No. 3 holds were delayed action. These bombs caused the ship to lift bodily out of the water, to vibrate most violently, engines stopped, steering-gear 'went out' and a great deal of movement and minor damage to portable appliances, fixtures and stores was evident, cargo booms snapped.*
>
> —Lt. Comdr. William W. Ball, USNR, commanding officer, USS *Adhara* (AK-71) describing in a war diary entry, damage to the cargo ship from six of the eight aircraft-dropped bombs which straddled her, on 7 April 1943. During two successive attacks, she was also strafed with machine gun fire.[18]

Three minutes before noon, on 7 April 1943, the first warning of an impending air attack on Guadalcanal was received by the Senior Officer Present Afloat (SOPA) aboard the attack transport USS *Hunter Liggett* (APA-14). There were four US Navy attack transports, two US Navy cargo ships, one US Navy oiler, and three civilian merchant vessels unloading or loading along the shore from Kukum to Koli Point, a distance spanning about nine miles.[19]

On signal from *Hunter Liggett*, all ships got under way and retired eastward through Lengo Channel, the transports forming Task Unit 32.4.1 and the auxiliaries, 32.4.4. The cargo ship USS *Adhara* initially followed toward the channel in the wake of *Hunter Liggett*, and merchant vessels *John Penn*, *George S. Clymer*, and *Fuller*. She overtook the merchant vessel *Dona Nati* off Koli, but the US transports rapidly drew away. At 1330, *Aludra* received a message from the cargo ship *Libra* (AKA-12) eight miles astern of her, "Suggest you join my formation that will provide two destroyers for screen." This she did.[20]

At 1411, radio warning of enemy approach off the eastern tip of Guadalcanal was received. By 1430, the fleeing USN and merchant ships had formed up with *Libra*, *Louis Joliet*, and *William Williams* in starboard column; and *Adhara* and *Dona Nati* in port column. *Taylor* (DD-468), *Sterett* (DD-407), and *Ward* (APD-16) were screening the formation. At 1515, the oiler USS *Tappahannock*, with destroyer escort, overtook and quickly passed the formation, well inshore and to starboard—apparently complying with an old Navy adage, advising that "speed is life, and more is better."[21]

Four minutes later, at 1515, flagship *Libra* ordered commencement of two sequential 45 degree turns to starboard, and all ships opened fire at approaching enemy aircraft. At 1520-1521, *Adhara* was straddled by eight bombs from three dive bombers from a formation of five, approaching from starboard. After being bombed and strafed by these aircraft, *Adhara* was attacked by a different dive bomber approaching from astern. It straddled her with two bombs and also strafed her with machine gun fire. A fragmentation bomb exploded close on her port bow. Shrapnel from it penetrated her hull below and above No. 1 tween-decks, bulwarks, and splinter protection, and ready ammunition boxes at No. 1 Gun, severely injuring two personnel (one of whom later died).[22]

The emergency turns to starboard, and combined fire of formation ships (two AKs and the civilian merchant ships), greatly aided mutual defense and minimized damage. The *William Williams* shot down one plane while suffering a near miss from a bomb, The *Louis Joliet* claimed a hit on a plane, and the *Dona Nati* one aircraft destroyed and another one damaged. She reported that ten bombs were dropped, some close to her. Defense aboard the merchant ships was provided by Navy Armed Guardsmen. Commander Transport Group, South Pacific Force, highlighted as noteworthy in his action report:

- "The tactical control exercised by the Task Unit Commander over these cargo vessels and the defensive maneuvers executed during the attack"
- "The large volume of AA [anti-aircraft] fire, particularly by the SS *Louis Joliet*"[23]

## NAVY ARMED GUARD SERVICE

The U.S. Navy Armed Guard was a branch of the United States Navy in World War II, responsible for defending U.S. and Allied merchant ships from attack by enemy aircraft, submarines and surface ships. The complement for a ship armed with a 5"/38 dual purpose stern gun, a 3"/50 AA gun, and eight 20mm machine guns was one officer and 24 gunners, plus normally about three communications men comprising a total of 28 Armed Guards. However, many ships operated in the early days with less than the armament desired and with smaller Armed Guard crews.[24]

Three basic Armed Guard schools prepared officers and enlisted men for their duties. These were located at Little Creek (later moved to Shelton), Virginia; Chicago, Illinois; and San Diego, California. The school at Chicago was closed because winter conditions at the nearby Great Lakes Naval Base were not suited to training, and a replacement school was opened, in the fall of 1942, at Gulfport, Mississippi. Upon completion of their training, graduates were ordered to an Armed Guard Center at Brooklyn, New York; New Orleans, Louisiana; or Treasure Island, San Francisco, California, for assignment to ships.[25]

Armed Guardsmen lived aboard ship with highly paid merchant seamen, and had to get along with men whose highest form of discipline probably came from the regulations of their labor union. The officers in charge of Armed Guard crews were expected to have the usual traits of leadership necessary for all naval officers, but experience soon showed that a certain ability to get along with masters of merchant ships was an important characteristic. The ideal Armed Guard officer was a tactful individual able to look after the interests of his men, and concurrently facilitate smooth relations between the Navy complement and the master, officers, and crew of the merchant vessel.[26]

## *TITANIA* PRESENT FOR RAID ON GUADALCANAL

On 13 May 1943, USS *Titania* (AKA-13) was anchored off Lunga Point discharging cargo. Following receipt of a Condition Red warning at 1230, she set General Quarters, and let go all barges with their crews. At 1236 with anchor aweigh, *Titania* was proceeding on varying courses and speeds, standing out to sea. At 1241, a sighting was made of an enemy plane approaching from the west, 25 degrees above the horizon. Meanwhile by 1258, a large number of planes were closing in on Henderson Field. Five minutes later, personnel aboard *Titania* observed American planes engaging enemy planes over Cape Esperance, the northernmost point on Guadalcanal. Condition Green was received at 1350, upon which the cargo ship turned around to return to Lunga. Anchoring at 1518, she commenced discharging as before.[27]

## CONVOY ATTACKED, NO RESULTANT DAMAGE

One week later, on 20 May, the cargo ship USS *Fomalhaut* (AKA-5) was in convoy with the civilian transport ships SS *Dashing Wave* and SS *Santa Ana*, proceeding from Noumea, New Caledonia, to Guadalcanal, and screened by USS *Pringle* (DD-477) and *McCalla* (DD-488).[28]

At 1138 aboard flag ship *Fomalhaut*, Comdr. Henry C. Flanagan, received two reports and observed two bomb splashes approximately 100 yards off the starboard bow of the ship. He immediately ordered the convoy to execute emergency turns, and called his crew to their battle stations. The escorts opened fire on enemy aircraft then overhead.[29]

Photo 8-4

Merchant vessel SS *Dashing Wave*, a C-2 type cargo ship which had been converted to a troopship a few months after her construction in 1943.
Naval History and Heritage Command photograph #NH 102230

A sighting was made at 1146 of a twin-engine Japanese bomber, proceeding in the same direction as *Fomalhaut*, on a parallel course, altitude about 16,000 feet. The cargo ship's 3"/50 guns opened fire. One minute later, two bombs fell close aboard the starboard bow of the *Santa Ana*. The plane then passed out of sight into a cloud formation. *Fomalhaut* ceased fire, having expended 11 rounds at the high-flying target. No hits were observed on it, and no damage done to any vessel of the task unit by dropped bombs.[30]

## *CELENO* DOWNS JAPANESE PLANES, SUFFERS 14 CREWMEN KILLED, 2 MISSING, 22 WOUNDED

> *On 16 June [1943], the largest enemy force since 7 April attacked our shipping in the GUADALCANAL area. Enemy forces consisted of at least 60 VB [bombers], screened by a like number of fighters. 104 U.S. fighters were scrambled in defense, and 74 made contact with the enemy. There were numerous U.S. ships*

*in the transport area off LUNGA Point and in TULAGI. The attack lasted from 1315 to 1513. ComSoPac [commander, South Pacific Force] credits his VFs [fighters] and ships' A/A [anti-aircraft fire] with the destruction of 107 enemy planes. Six of our VF were lost, two pilots were recovered.*

—Commander in Chief, U.S. Pacific Fleet, Operations in the Pacific Ocean Areas – June 1943, 6 September 1943.

Photo 8-5

Cargo ship USS *Celeno* (AK-76), 27 November 1943.
Naval History and Heritage Command photograph #NH 84964

On the morning of 9 June 1943, a group of six ships stood out of Noumea Harbor, New Caledonia, passed through the anti-submarine net, then leaving Maitre Island to starboard, set a course for Guadalcanal. Designated Task Unit 32.4.4, the cargo ships USS *Deimos* (AK-78), USS *Aludra* (AK-72), and USS *Celeno* (AK-76), and the merchant Liberty ship SS *Nathaniel Currier* formed a convoy, with the high-speed transport USS *Waters* (APD-8) and minesweeper USS *Skylark* (AM-63) serving as escorts. A third escort ship, the destroyer USS *O'Bannon* (DD-450), arriving from Espiritu Santo, joined the task unit later on the morning of 12 June.[31]

The task unit arrived at Guadalcanal around noon, on 14 June, and as the escorted ships proceeded to their anchorages off Lunga Point, the escorts took up patrols offshore, carrying out anti-submarine screening duties. Subsequent unloading operations were regularly interrupted by Japanese air raids. At 1305 on 16 June, *Celeno* and the other ships received a preliminary alert warning of Condition "Yellow" from the shore station, and began making all necessary preparation for getting under way. *Celeno* was particularly vulnerable, having aboard a cargo of bombs, ammunition, gasoline, etc., loaded at Noumea.[32]

At 1352, a Condition "Red" warning was received. *Celeno* sounded General Quarters, calling her crew to battle stations, and got under way at 1355 on course 070°T, with *Deimos* following about 1,000 yards astern, and *Skylark* screening both vessels. Fifteen minutes later, at 1410, *Celeno* was attacked by eight Japanese dive bombers, and all gun batteries that could be brought to bear on the enemy began firing. The attacking planes were part of a group of eighteen fighters and dive bombers sighted by the ship's lookouts at a range of five miles, coming in fast at an intermediate altitude.[33]

Aboard *Celeno*, fifteen guns (3"/50, 20mm, and .30-cal. machine guns) were soon pouring out rounds. At 1411, one bomb, a near miss, landed off the cargo ship's starboard bow, and three more near misses fell off her port bow and beam. *Celeno*'s gunners scored their first hit on an attacking plane a minute later. A bomb then made a hit on the ship's stern, forward of the five-inch gun platform, knocking the gun out of commission, and disabling the steering with full right rudder on. Another near miss, abreast of No. 5 hold on the starboard quarter, set ablaze the deck cargo of gasoline and diesel oil.[34]

As attempts made to utilize the emergency steering gear aft, proved unsuccessful, all available crew were detailed to fight the spreading fire. When the enemy attack ended at 1425, the horrors of its effect became known. *Celeno*'s commanding officer, Lt. Comdr. Niles E. Lanphere, reported that the bombs (believed to be 550-lb armor piercing) were dropped from approximately 600 feet. Ship's history cites 14 men killed, 2 missing, and 22 wounded. The Naval History and Heritage Command's *Directory of American Naval Fighting Ships* history for the *Celeno* records that fifteen of her valiant crew were killed and nineteen wounded in the attack.[35]

As *Celeno* was being bombed and thoroughly strafed, her gun crews shot down three planes confirmed, and possibly another three. One dive bomber was destroyed with 3-inch/50 fire, two by 20mm, and the three other possibles were hit hard by 20mm fire; if not forced down, they were seriously damaged. In total, the cargo ship expended seven rounds of 3"/50, 860 rounds 20mm, and 970 rounds .30-caliber.[36]

*Celeno* was left with jagged holes in her decks and sides, her hull seams opened by bomb blasts, and her rudder useless. At 1500, the fleet tugs USS *Rail* (AT-139) and USS *Vireo* (AT-144)—former World War I vintage minesweepers—and the tank landing craft *LCT-322* came alongside the cargo ship to move her toward Lunga Point. *Celeno* was settling by the stern (her no. 5 hold, after steering station compartment, and magazines were flooded), she was capable of only 35 propeller shaft RPM, and her rudder was jammed hard right. Consequently, the task of getting her to the beach was a slow and difficult one.[37]

The cargo ship was finally run aground, just barely, at the mouth of the Lunga River; with only her bow touching bottom, she risked sliding off into deeper water. It was believed that the bulkhead between No. 4 and No. 5 holds had collapsed, flooding No. 4 hold, and leaving the ship settling dangerously by the stern, and threatening to drag her down slope. Salvage operations were begun and, on 24 June, with preliminary repairs made, the fleet tug USS *Menominee* (AT-73) took her in tow to an anchorage in Purvis Bay.[38]

*Celeno* lay in the bay, until 27 July, when she was taken under tow to Espiritu Santo for docking, and removal of her screw and rudder. Following this work, she was towed to San Francisco, and docked at the Mare Island Navy Yard for overhaul, damage repair, and conversion to a troop transport. In her new role, *Celeno* returned to the war zone arriving at Espiritu Santo, on 3 January 1944, with cargo and troops loaded at Port Hueneme, California.[39]

Photo 8-6

Minesweeper USS *Rail* (AM-26), prior to her redesignation as a fleet tug, 1941. Naval History and Heritage Command photograph #NH 81002

## *ALUDRA* AND *DEIMOS* SUNK BY SUBMARINE

> *The moon was full and was directly overhead, the sky was completely overcast, the sea was choppy, the wind was force about 3 from southeast. The indirect illumination caused by overcast sky with a full moon caused our ships to stand out very clearly at a range of 4,000 yards.*

> *It is definitely felt that for three large vessels steaming at such slow speed the screen was inadequate; this coupled with the fact that the screening units had never worked together before.*
>
> —Comdr. Donald J. MacDonald, USN, commanding officer, USS *O'Bannon* (DD-450), and senior officer of the screening ships assigned to Task Unit 32.4.4.[40]

Task Unit 32.4.4 departed Guadalcanal on 22 June 1943, en route to Espiritu Santo, without *Celeno* after her combat damage and subsequent beaching, and with USS *Ward* (APD-16) replaced by USS *Waters* (APD-8) as one of the escorts. By early morning the following day, the ships were passing south of San Cristobal Island.[41]

The cargo ship USS *Aludra* (AK-72) was the guide, followed by the merchant ship SS *Nathaniel Currier* at a distance of about 700 yards, with USS *Deimos* (AK-78) abeam of *Aludra*, 1,000 yards to starboard. The destroyer USS *O'Bannon* (DD-450) was stationed 4,000 yards, 45 degrees on the starboard bow of the convoy. The minesweeper USS *Skylark* (AM-63) was 4,000 yards on the port bow of it, and the high-speed transport USS *Ward* (APD-16) 3,000 yards astern.[42]

Minesweepers frequently served as "maids of all duties" when not performing their primary mission. The 221-foot *Skylark*, commissioned on 25 November 1942, was armed with one 3"/50 gun mount, two 40mm mounts, and eight 20mm guns, fittingly enabling her to be pressed into escort duties. She also boasted two depth charge tracks, and five depth charge projectors for use against submarines.[43]

Photo 8-7

Minesweeper USS *Skylark* (AM-63), location and date unknown.
Naval History and Heritage Command photograph #NH 89202

At about 0445, with the convoy zig-zagging in accordance with Plan #38, steering course 124°T, speed 10 knots, the two leading ships in convoy—*Aludra* and *Deimos*—were each hit with a torpedo on their port sides. These strikes occurred almost simultaneously, four minutes after the vessels changed course 10° to the left of the base course, on which they had been for ten minutes. (Greater ship speed and more dramatic course changes reduce the likelihood that a submarine can gain an attack position and accurately calculate the target's projected position for a successful torpedo attack. However, cargo ships, being slow, could not deviate much from their base course, lest their overall speed of advance be even further reduced.)[44]

Photo 8-8

Cargo ship USS *Aludra* (AK-72), location and date unknown.
Naval History and Heritage Command photograph #NH104025

The torpedoes were fired by a Japanese submarine from a position off the convoy's port quarter. The *RO-103*, under the command of Lt. Ichimura Rikinosuke, had departed Rabaul on 12 June, en route the Gatukai-San Cristobal area. Eleven days later, when fifty miles south of the eastern tip of San Cristobal Island, Ichimura had sighted what he believed to be a convoy of three transports protected by three destroyers, and he made a torpedo attack. Surprisingly, the three escort vessels gained no radar or underwater sound contacts, before or after the attack.[45]

*Aludra* was hit in No. 1 hold, and both No. 1 and 2 holds flooded in about twenty minutes. Number 3 hold was flooding when the ship sank about 4 hours and 49 minutes after being torpedoed. Two of Aludra's crew were killed and 12 were wounded.[46]

*Deimos* was hit aft, resulting in the flooding of No. 4 and 5 holds, as well as the steering engine room, and buckling and leakage of the engine room bulkhead. *Deimos* settled slowly until sunk by destroyer gunfire, 7 hours and 48 minutes after being struck by the torpedo fired at her.[47]

Aboard *Deimos*, Lt. Comdr. Walter Louis Sorenson gave the order, "Stand by to abandon ship" at 0450. This action followed an assessment that salvage was not possible, and a thorough search of the vessel for survivors by himself, the executive officer, the first lieutenant, the navigator, and six men. When "abandon ship" was passed (five minutes after the warning was given to standby), *Deimos* was settling by the stern. At 0555, the commanding officer ordered all remaining crew off the ship, and then left himself. At noon, *O'Bannon*'s commanding officer, observing the stern of *Deimos* settling rapidly and her after gun mount disappearing underwater, likewise decided that salvage was impossible and ordered her sunk with naval gunfire.[48]

## BATTLE STARS

The following cargo ships were awarded a battle star for the campaign titled, Consolidation of Southern Solomons. Per Navy practice, a ship could only earn one battle star for any single campaign, no matter how many qualifying actions. As shown in the table, the recipients were mostly new *Crater*-class cargo ships, which had recently joined the South Pacific Forces, but some older ships also garnered a star.

**Consolidation of Southern Solomons**

| Ship | Ship Class | Qualifying Date or Period for Star |
|---|---|---|
| USS *Carina* (AK-74) | *Crater*-class (Liberty) | 3-5 Mar 43 |
| USS *Cassiopeia* (AK-75) | *Crater*-class (Liberty) | 11 Mar 43 |
| USS *Arided* (AK-73) | *Crater*-class (Liberty) | 11 Mar-23 May 43 |
| USS *Adhara* (AK-71) | *Crater*-class (Liberty) | 20 Mar-7 Apr 43 |
| USS *Libra* (AKA-12) | *Arcturus*-class (C2-F) | 7 Apr 43 |
| USS *Titania* (AKA-13) | *Arcturus*-class (C2-F) | 13 May 43 |
| USS *Fomalhaut* (AKA-5) | *Fomalhaut*-class (C1-A) | 20 May 43 |
| USS *Aludra* (AK-72) lost | *Crater*-class (Liberty) | 16 Jun 43 |
| USS *Celeno* (AK-76) | *Crater*-class (Liberty) | 16 Jun 43 |
| USS *Deimos* (AK-78) lost | *Crater*-class (Liberty) | 16 Jun 43 |

# 9

# HMAS *Maroubra* Falls Victim to Japanese Zeros

Photo 9-1

Millingimbi Island with wire mesh matting laid in the sand to enable transports to get down to the water's edge, to obtain barge-delivered, ship cargo, 18 November 1943. Australian War Memorial photograph 060715

While Japanese aircraft continued to strike Allied bases and shipping in the southern Solomons, they were similarly engaged in New Guinea and Northern Australia.

On 10 May 1943, Japanese aircraft struck the Royal Australian Air Force Airfield on Millingimbi Island, which had been pressed into wartime service as 59 Operational Base Unit (59 OBU). Millingimbi (also known as Yurruwi), is the largest of the Crocodile Islands Group off the coast of Arnhem Land. Located in the Northern Territory, approximately 270 miles east of Darwin and 120 miles west of Nhulunbuy, it is situated between the Blyth and Glyde River mouths, separated from the mainland by a narrow channel.[1]

As a primitive RAAF airfield, on a low-lying island of clay, sand and mud, and surrounded by convoluted mangrove waterways, it served a vital purpose. It was part of a chain of austere bases used to provide Australian Hudson patrol bombers greater reach to the east and west of Darwin in their coastal patrols along the "Top End" shipping routes. In addition to these patrols, marauding Bristol Beaufighter aircraft of No. 31 Squadron, operating from RAAF Coomalie Creek Airfield, Northern Territory, were challenging Japanese dominance over the Arafura Sea in the Netherlands East Indies.[2]

The expanded ocean coverage hindered operations of the Imperial Navy's 934 Air Group, an Ambon-based floatplane unit charged with reconnoitering and attacking Australian coastal shipping that plied the supply route between Queensland and Darwin. As previously described, one of its aircraft (an Aichi E13A three-seater) had surprised and sunk the wooden-hulled supply vessel HMAS *Patricia Cam* on 22 January 1943.[3]

In response to greater RAAF activity in the Arafura Sea, the enemy 934 Air Group had moved its seaplane base to Taberfane, on the west coast of Tarangan Island (also in the Aru Islands). The new, forward base, provided the floatplanes with additional reach eastward. They could then operate right up to the edge of the Gulf of Carpentaria—a large, shallow sea enclosed on three sides by Northern Australia and bounded on the north by the eastern Arafura Sea.[4]

## REPRESENTATIVE OPERATIONS OF *MAROUBRA*

Photo 9-2

HMAS *Maroubra* off Millingimbi, on 10 May 1943, before the air attack. Australian War Memorial photograph 300993

HMAS *Maroubra* was a 60-ton vessel built in 1930 at Brisbane by Norman Wright. The small vessel boasted a flush deck, raised foc's'le, straight stem, and pucked stern. She had been requisitioned by the RAN, on 20 March 1942. Propelled by sail and/or auxiliary engine, she could make a modest 7.5 knots while serving as a coastal stores ship.[5]

In representative operations, from February through May 1942, *Maroubra* usually called at Port Moresby, Thursday Island, Cairns, and Townsville. Then, in a variation of that run, Cairns, Thursday Island, Darwin, Millingimbi, and Darwin once again.[6]

## ENEMY AIR ATTACK ON MILLINGIMBI RAAF FIELD

Photo 9-3

The coastal stores ship HMAS *Maroubra* on fire and sinking off Millingimbi Island, Northern Territory, after being attacked by Japanese Zeros, on 10 May 1943.
Australian War Memorial photograph 300992

On 10 May 1943, HMAS *Maroubra* was lying at anchor at Millingimbi, when the RAAF field came under enemy air attack. While six of a group of Japanese Zeros were engaged by Spitfires at altitude, three others descended for a strafing attack. Approaching the island from the east, the fighter aircraft found the coastal stores ship as their first target. Hit repeatedly by 20mm explosive rounds, *Maroubra*'s POL (petroleum, oil, and lubricant) cargo ignited; she burned with the loss of 30-tons of Royal Australian Air Force stores.[7]

Although likely a satisfying attack for the enemy pilots, the ruin of *Maroubra* greatly reduced their remaining ammunition and undoubtedly saved several Beaufighters from destruction. The Zeros arrived over the island base during busy runway operations by 31 Squadron, with planes on the ground and vulnerable.[8]

Photo 9-4

Ground crews with Beaufighter aircraft of No. 31 Squadron RAAF in World War II. Australian War Memorial photograph 100448

# 10

# North Pacific Action in the Aleutians

Photo 10-1

Cargo ship USS *Spica* (AK-16) off Boston, 26 April 1940.
Naval History and Heritage Command photograph #NH 86626

As the Solomons and New Guinea campaigns continued to be waged, there was another Allied offensive far to the north. In the summer of 1942, the Japanese had occupied Attu and Kiska in the Aleutian Islands. By the spring of 1943, U.S. and Canadian forces were prepared to take them back. On 11 May, the United States invaded Attu. Gunfire support was provided by the battleships *Nevada* (BB-36), *Pennsylvania* (BB-38), and *Idaho* (BB-42), with air support by the escort carrier *Nassau* (CVE-16) and U.S. Army Air Force aircraft. An additional three heavy cruisers, three light cruisers, nineteen destroyers, and four attack transports, as well as oilers, minesweepers, and small craft supported the troop landing. Later a similar force was deployed to reclaim Kiska but was not needed when landing forces found that the enemy had fled.[1]

Space does not permit coverage of all U.S. vessels involved in the Aleutian campaign, instead, this chapter singles out a cargo ship, whose presence during the landings and related air attacks earned a battle star for the engagement.

Oddly, the cargo ship USS *Spica* (AK-16), one of the ships that received a battle star for Attu, was not designated in plans as part of the support forces for occupation of the island. She was pressed into this role because of an unplanned requirement of the invasion, and her availability in the area. An advanced group of "Seabees" (from the 23rd Naval Construction Battalion) was needed at Attu to build a base and airfield, and *Spica* was accessible to ferry them there.[2]

With the advent of war, on 7 December 1941, it immediately became apparent that the services of civilian contractors and their employees that were formerly used on US naval construction projects throughout the world could not be employed adequately for construction work outside the continental United States in potential combat zones. Accordingly, the first Navy Seabee detachment, consisting of some 300 officers and men was authorized the following day for such purposes.[3]

Map 10-1

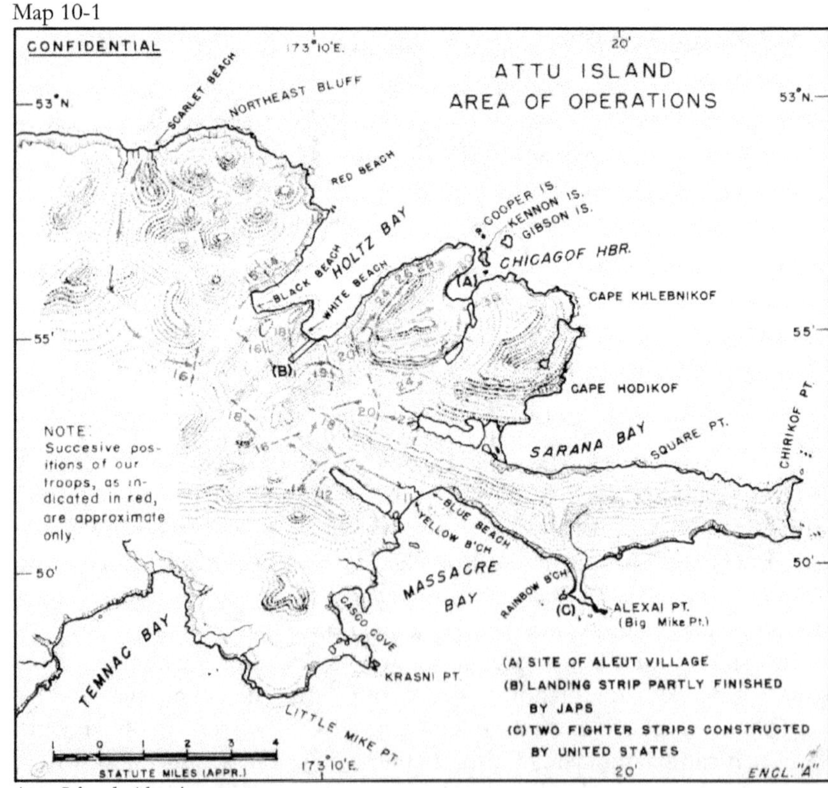


Attu Island, Aleutians
Commander in Chief, U.S. Pacific Fleet, Operations in the Pacific Ocean Areas – May 1943, 13 August 1943

Attu Island lay at the extreme western end of the Aleutian chain, at a distance of 740 miles from Dutch Harbor and only 208 miles shy of the Russian Commander Islands to the west. The bleak island was 14 miles wide and 38 miles long, with terrain comprising a rugged coastline, tundra-padded valleys, and wind-swept mountains devoid of trees, similar to that of others in the Aleutians. The Attu area was subject to violent and unpredictable weather, with wind a prevailing factor, and fog, snow, and sleet, any or all, possible at any time.[4]

When Japanese forces occupied Attu and Kiska, in June of 1942, approximately 2,000 troops comprised their Attu garrison. The primary installation at Kiska was several times larger. Bypassing Kiska, two assault forces of the U.S. 7th Infantry Division landed on Attu, on 11 May, one going ashore at Massacre Bay and the other landing west of Holtz Bay. Although little resistance was encountered on the beaches, the Japanese made strong defensive stands inland. On 17 May, the American forces were able to join and, over the next two weeks, as a result of ferocious and bloody fighting, continued to inch forward and slowly boxed the Japanese into the eastern sector of the island.[5]

## *SPICA*'S SERVICE

Stretching 410 feet in length, *Spica* was the former SS *Shannock*, a cargo ship built in 1919 by American International Shipbuilding Corp., at Hog Island, Pennsylvania. The Navy acquired the freighter from the United States Shipping Board, on 16 November 1921, and renamed her *Spica* (AK-16). Following several years of service, she was decommissioned and remained laid up in "mothballs" in the Reserve Fleet, until 1 March 1940, when she was recommissioned at Norfolk, Virginia, with Comdr. Edwin D. Gibb, USN, in command.[6]

*Spica* was initially employed running cargo on return voyages from the Brooklyn Navy Yard south to San Juan, Puerto Rico, and Guantanamo Bay, Cuba, in the Caribbean. By mid-1941, she was assigned to the 13th Naval District, headquartered at Seattle, Washington. She would ply between the Bremerton Navy Yard and Alaskan waters, carrying supplies to American outposts in Alaska and the Aleutians, including Sitka, Kodiak, Dutch Harbor, and Adak, until late 1943.[7]

On 19 May 1943, *Spica* was laying at anchor in Kuluk Bay, at Adak Island in the Aleutians. (She had been operating in the Alaskan-Puget Sound area, under the commandant, 13th Naval District, until assigned to ComAlSec, on 22 April 1943.) Also present at the bay were various units of Task Force 16 (North Pacific Force) and Task Force 51 (Assault Force), as well as USS *Vega* (AK-17), which was also a Hog Island product and a sister ship to *Spica.* In accordance with commander, Alaskan Sector (ComAlSec) orders, *Spica* stood out of the bay that afternoon and joined Task Unit 16.8.2—a component of the larger Attu Reinforcement Group (Task Group 16.8).[8]

After she joined, the task unit consisted of the *Spica* (AK-16), Liberty merchant ship SS *David D. Field*, and transports merchant ship SS *David W. Branch* and USAT *President Fillmore* in convoy; screened by the destroyer USS *Dewey* (DD-349), destroyer minesweeper USS *Long* (DMS-12), and sub-chasers USS *PC-485* and *PC-572.* Upon arrival of the task group at Massacre Bay, on the morning of 21 May, the convoy ships and subchasers anchored in their assigned berths, while the destroyers departed for other duties.[9]

Photo 10-2

Recruiting poster for Navy Seabee reservists.
Naval History and Heritage Command #NH 78817

As the convoy ships began unloading, the Seabees embarked aboard *Spica* left her for the shore, to begin their assigned construction tasks. As an aside pertaining to the work of the Seabees, it is noted that, by the latter part of June, all of the 23rd Naval Construction Battalion had arrived and were occupied with construction work. Building a base and airstrip on the inhospitable Aleutian island would prove a "tough slog." Fortunately, it was soon bolstered by the 22nd Battalion which arrived, on 15 July, and the 68th Battalion, on 29 July 1943.[10]

## WARNING OF IMPENDING JAPANESE AIR ATTACK

At 1530 on the afternoon 22 May, *Spica*'s second day at Massacre Bay, she received a warning of enemy aircraft in the vicinity. Comdr. Joseph W. Long, her commanding officer, ordered General Quarters set, and her anchor brought up. By 1550, with anchor aweigh, the cargo ship was headed outbound in the channel, per visual orders (flag hoist) from the transport USS *Heywood* (AP-12) whose commanding officer was the Senior Officer Present Afloat (SOPA). *Spica*'s captain "had the conn" (meaning he was personally giving all engine speed and rudder orders), and her navigator was on the bridge.[11]

At 1650, after all ships leaving port had cleared the channel, they formed two columns and, continuing seaward, passed Chirikof Point, on the easternmost peninsula of Attu Island, abeam to port. At 1810, *Spica* received orders from *Heywood* for all ships to proceed independently back into Massacre Bay.[12]

## ATTACKS ON USS *PHELPS* AND USS *CHARLESTON*

> *1548 Patrolling entrance to Holtz Bay. Weather: visibility less than 5 miles. 18 Japanese twin-engined heavy bombers (Navy, Mitsubishi OB-01) made torpedo and strafing attack on* PHELPS *and* CHARLESTON *respectively. No personnel casualties. 8 machine gun bullets penetrated ship during attack doing very minor material damage.*
>
> —USS *Phelps* (DD-360) War Diary entry for 22 May 1943.

> *1600 Sighted 12 to 15 enemy type 01 Mitsubishi heavy bomber torpedo type planes. About 6 planes commenced firing torpedoes and delivering strafing attacks at this vessel. Planes were fired on continuously while in range by AA battery 1 plane shot down. All torpedoes fired missed ship and exploded in water at end of their run. By 1610 Enemy planes departed toward westward.*
>
> —USS *Charleston* (PG-51) War Diary entry for 22 May 1943.

Good fortune smiled on Naval forces during the Japanese air attack that prompted *Spica* and the other ships to get under way, in order that they would have sea room to maneuver and not be easy stationary targets. The bombers arriving off Attu either didn't see the two columns of ships nearby, or ignored them to focus their attacks on the patrol gunboat USS *Charleston* (PG-51) and destroyer USS *Phelps* (DD-360), which were patrolling off the Chicagof – Holtz Bay areas. It was normally overcast/foggy in the Aleutians and, as cited in *Phelp*'s war diary entry, restricted visibility conditions existed that day.[13]

Addressing actions by the two ships attacked, in his endorsement to *Charleston*'s report of Anti-Aircraft Action, Rear Adm. Thomas C. Kinkaid (commander North Pacific Force) noted:

> That neither ship received a torpedo hit was most fortunate. It was undoubtedly due to the fact that the attacks were not aggressively forced home by the torpedo planes, and the fact that the PHELPS and CHARLESTON were handled most effectively during the attack. The concentrated fire of both ships definitely accounted for one plane, definitely damaged another so that its return to base was unlikely, and inflicted minor damage on a third.[14]

The inclement weather was particularly important to the safety of ships, because there had been a failure of warning net communications. Unidentified planes were detected by *Dewey* on radar at 1515, range 74 miles, 45 minutes prior to the attack. She reported the contact to SOPA in *Heywood* by TBS (talk-between-ships, a very high frequency shipborne radio equipment of medium power). This warning was not received by *Phelps* and *Charleston*, as mountains blocked the transmission from reaching the two ships. Kinkaid highlighted that such reports must also be passed on CW (continuous wave) radio circuits.[15]

## CONCLUSION OF COMBAT OPERATIONS ON ATTU

On 29 May 1943, with his defense force reduced to about 1,000 men, Col. Yasuyo Yamasaki mounted a counterattack. In the early morning hours, he led hundreds of screaming soldiers in a desperate banzai charge. The human wave penetrated deep into the American lines, but was soon routed in fierce hand-to-hand combat. When it was over, Japanese casualties, including those from this recent action, totaled over 2,300, with only 28 enemy soldiers taken alive. U.S. casualties amounted to 549 killed and 1,148 wounded (and another 1,814 casualties owing to cold injuries and disease.)[16]

## *SPICA* AWARDED BATTLE STAR

USS *Spica* (AK-16) earned a battle star for "Aleutians operation: Attu occupation," for the period 20 May to 4 June 1943. Later that year, the cargo ship departed San Francisco, on 5 December, for Kwajalein Atoll via Funafuti, and returned with a stop at Pearl Harbor, en route to Seattle, on 22 March 1944. For the next six months, she resumed her Alaska-Aleutian circuit before, beginning a series of voyages from the west coast to Hawaii in mid-September. This continued until mid-March 1945, when she returned to Seattle and once again took up the northern Pacific supply runs through 14 September. *Spica* was decommissioned at Seattle, on 18 January 1946.[17]

## CAPTURE OF KISKA NEXT OBJECTIVE

Map 10-2

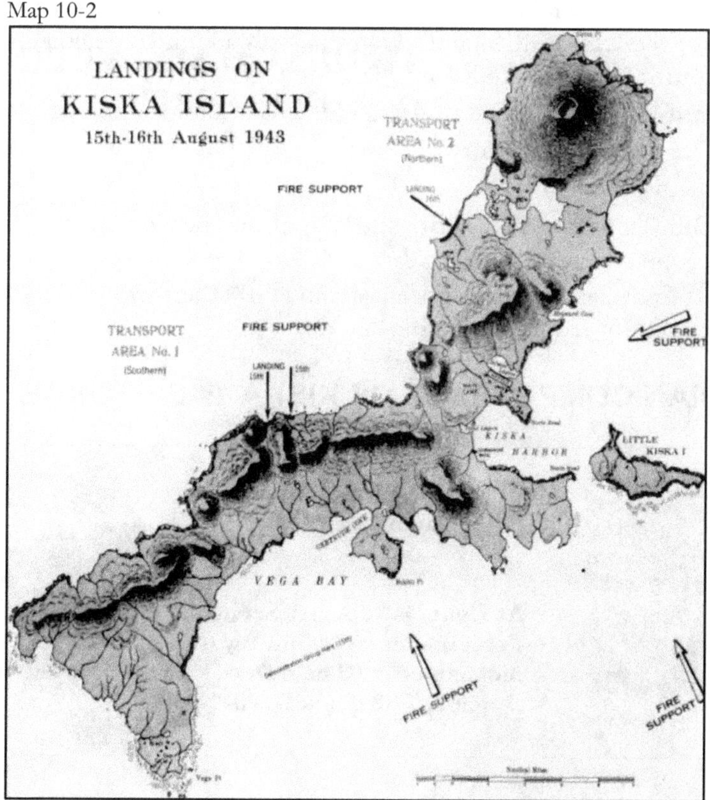


Kiska Island
*Combat Narratives: The Aleutians Campaign June 1942-August 1943* (Washington, DC: Office of Naval Intelligence), 1945.

With Attu secure, attention was focused on Kiska Island, located about 45 miles northwest of Amchitka and 165 miles southeast of Attu. Kiska was about 22 miles long, with a maximum width of 6 miles, and a mountainous northern end, sloped southward into a flat tableland. There were several lagoons, lakes, and ponds; numerous small streams; and tundra and muskeg bogs in the lowlands. The shoreline was steep and rocky, with a few sandy beaches.[18]

## AMERICAN / CANADIAN ASSAULT FORCES

Considerably larger forces were allotted to the assault on Kiska than had been used at Attu, since the garrison of the former island was known to have been several times as large as the latter. Rear Adm. Francis W. Rockwell, USN (commander, Amphibious Force, North Pacific Area) was in charge of the attack force (Task Force 16), and Maj. Gen. Charles H. Corlett, USA, the landing force (Task Force 9). The troops employed consisted of the U.S. 17th Infantry Division with additional regiments attached for the invasion of Kiska (Operation COTTAGE):

- 53rd Infantry (formed of Alaska Defense Command units)
- 87th Mountain Infantry
- 184th Infantry
- 13th Canadian Infantry Brigade Group, and headquarters troops
- 1st Canadian Special Service Battalion (700 Canadian troops trained for special operations)[19]

## CANADIAN COMPONENTS OF KISKA TASK FORCE

**At left: US Army Kiska Task Force patch**

**At right: 1st Special Service Force patch. This unit was nicknamed, "The Devil's Brigade, The Black Devils"**

Canada is most known for her World War II contributions related to the Allied effort in Europe. Like the United States, she was committed to a "Germany first" policy and, having a small population and limited forces, the bulk of her effort was devoted to the Battle of the Atlantic and ground combat in Europe. However, she also contributed forces to the Pacific Theater.

This involvement began following a request by the British Government in September 1941, that she assist in the defense of Hong Kong. Canada sent 1,975 soldiers from The Royal Rifles of Canada and The Winnipeg Grenadiers to garrison Hong Kong, therefore freeing up troops for other British possessions in the Far East. After the Japanese invasion, on 8 December 1941, the island held out until Christmas against superior enemy forces. Those men not killed in the fighting were captured, many dying of mistreatment in captivity. Company Sergeant Major John Osborne of the Grenadiers was awarded a posthumous Victoria Cross for actions in the fighting, including that of covering a Japanese grenade with his body, thereby saving those of several other soldiers nearby.[20]

Photo 10-3

Infantrymen of the 13th Canadian Infantry Brigade embarking aboard U.S. transports at British Columbia, Canada, July 1943, to take part in the invasion of Kiska. Note Canadian troops used U.S. helmets and uniforms for this campaign.
Library and Archives Canada photograph 3262661

The 13th Canadian Infantry Brigade Group (under Brigadier Harry Foster) was formed using the headquarters staff of the 13th Canadian Infantry Brigade, a unit raised for home defense, and three infantry battalions in Pacific Command that were numerically strongest. These were:

- The Canadian Fusiliers
- The Winnipeg Grenadiers
- The Rocky Mountain Rangers[21]

On 13 August 1943, 5,300 Canadians (including the Canadian component of the 1st Special Service Force, known administratively as the 1st Canadian Special Service Battalion), sailed from Adak for Kiska, part of a force of 34,426 soldiers. The so-called "Devil's Brigade," properly designated as the 1st Special Service Force, was a joint World War II American-Canadian commando unit initially trained at Fort Harrison near Helena, Montana, in the United States.[22]

The 1st Special Service Force had been officially activated, on 20 July 1942, under the command of US Army Lt. Col. Robert T. Frederick. Its volunteer members consisted primarily of enlisted soldiers recruited by advertising at Army posts, with preference given to men previously employed as lumberjacks, forest rangers, hunters, game wardens, and the like. The moniker "The Black Devils" was adopted by the unit after discovery of a German officer's personal diary referring to *die schwarzen Teufeln* (the Black Devils), who were much feared for their fighting prowess.[23]

## KISKA ASSAULT LANDINGS / AFTERMATH

In early morning on 15 August, minesweepers cleared the approach channels, as the transports took up their position's northwest of the waist of the island in Transport Area 1 (see map on page 103). The attack cargo ship USS *Thuban* (AKA-19), under Comdr. James C. Campbell, was a part of this group. At 0621, the first assault wave—covered by Special Service troops, which were already ashore after stealthily landing in small boats at night—came ashore in the left section on the west coast of the island, against, surprisingly, no opposition. By late afternoon, 6,500 troops were ashore. The lack of contact with enemy forces was not unexpected; intelligence had suggested that the Japanese might withdraw to prepared positions on high ground, as they had done at Attu.[24]

The following day, remaining assault troops, including the 13th Canadian Brigade, landed in the right section, on the northwest coast from Transport Area 2. By noon, 3,100 troops were ashore in this sector. When Ranger Hill was occupied, indications of recent Japanese evacuation were found. By 1600, all Canadian troops had landed, making a total of about 7,000 men in the sector.[25]

Photo 10-4

Infantrymen of the 13th Infantry Brigade Group disembarking from a landing craft during the invasion of Kiska, Aleutian Islands, 16 August 1943.
Library and Archives Canada photograph 3211083

Further evidence of hasty abandonment was found, on 17 August. While the fighting on Attu was still in progress, the Japanese high command had decided to withdraw the Kiska garrison and use it to strengthen the Kuriles, an island group further to the west and close to Japan. An early attempt was made to remove the troops by submarine, but this was done at the cost of four submarines lost to the U.S. Navy or navigational hazards. Orders were then issued for this task to be accomplished by surface forces. On 28 July, while one light cruiser stood off south of Kiska, two others and a flotilla of destroyers made their way through the fog to the island. The remaining garrison members were jammed aboard the cruisers and six of the destroyers, and the force reached Paramushiro in the Kuriles, now the site of Japan's northernmost naval base, in safety.[26]

Photo 10-5

Canadian and American soldiers examining abandoned Japanese submarines on Kiska, Aleutian Islands, 1 September 1943.
Library and Archives Canada photograph 3599747

With the departure of the last Japanese from Kiska, the Aleutians Campaign effectively ended. The withdrawal of the enemy without a fight was a false presentation of what might be expected when the odds were hopelessly against him. Attu, where the Japanese had fought to the death, was to be the pattern of the future, not Kiska, where they had faded into the fog without a struggle.[27]

## CANADA'S CONTINUED CONTRIBUTIONS

In addition to providing assault troops and two corvettes for anti-submarine duties to the Aleutian campaign, the Royal Canadian Airforce (RCAF) played a significant role in the defense of Alaska and the north west coast by the provision of patrol and fighter aircraft to the extent of actually basing aircraft in Alaska to cover for hard-pressed US air forces.

On 25 September 1942, Squadron Leader Kenneth Arthur Boomer shot down a Japanese Nakajima A6M2-N "Rufe" (a Zero converted into a seaplane via the addition of floats) over Kiska. Boomer was the commanding officer of RCAF No. 111 (Fighter) Squadron (Curtiss P-40K Kittyhawks) based at Unmak in the Aleutians. His victory over the Japanese fighter was the only WWII kill credited to the Home War Establishment (HWE). For this action, Boomer was awarded both the Distinguished Flying Cross and the US Army Air Medal.[28]

The strike force of which he was a part, consisted of B-24D Liberator heavy bombers of the 21st Bombardment Squadron, escorted by Eleventh Air Force Bell P-39 Airacobras and Curtiss P-40 Warhawks of the 343rd Fighter Group, plus four P-40s from No. 111 Squadron, Royal Canadian Air Force. The mission set shore facilities ablaze, and downed two Rufes, the other by Maj. John S. Chennault, the son of Maj. Gen. Claire Chennault, leader of the American Volunteer Group (AVG) in China—the famous Flying Tigers.[29]

Photo 10-6

RCAF No. 111 Squadron group photograph, 3 March 1942.
Department of National Defence photograph PBG 1657

Canada remained active in the Pacific Theater until the end of the war, on 2 September 1945, through naval and air patrols along the west coast, provision of ships to the British Pacific Fleet (BPF) and through the augmentation of Commonwealth forces. The various roles of Canadian individual service members included: signalers in Australia, aviators and sailors with the British Pacific Fleet (BPF), RCAF airmen in Burma, and naval aviators with the Royal Navy's Fleet Air Arm. Additionally, Chinese-Canadian soldiers were recruited for service in Malaya as spies and trainers of local guerillas. Prior to the war's end, preparations were well underway to provide a large naval component to the BPF to augment the work already provided by the auxiliary cruiser HMCS *Prince Robert* and the light cruiser HMCS *Uganda*.[30]

The term "signalers" referred to No. 1 Special Wireless Group of the Royal Canadian Corps of Signals (RCCS), which was employed intercepting enemy radio traffic. Force 136, a branch of the British Special Operations Executive (SOE), recruited Southeast Asians and about 150 Chinese-Canadians to support and train local resistance movements to sabotage Japanese supply lines and equipment. Earlier in the war, many of the latter individuals had volunteered their services to Canada but were either turned away or recruited and sidelined. Force 136 provided the Chinese-Canadian men opportunity to demonstrate their courage and skills and especially their loyalty to Canada.[31]

Canada's contribution to the prosecution of the Pacific war and provision of security for Alaska, also included three magnificent civil engineering projects that were accomplished in unprecedented time through close cooperation with their U.S. Allies. The first was the expansion of the system of inland airports that Canada had established that ultimately became known as the "North Western Staging Route." The second was the construction of the Alaska Highway pioneered by the U.S. Army for first convoys by fall of 1942. The third was the Canol pipe line constructed by the U.S. Army from 1942-1944 to carry crude oil from established fields at Norman Wells, Northwest Territories, in Canada to Whitehorse, Yukon, on the Alaska Highway.[32]

# 11

# Central Solomons Campaign

Photo 11-1

American troops take cover on a Rendova Island Beach, while landing in a heavy rainstorm during the early morning of 30 June 1943.
National Archives photograph #80-G-52573

Except for continuing air battles, a comparative lull existed in the Solomon Islands for several months, in early 1943, following the bloody battles around Guadalcanal, in late 1942. Execution of planned U.S. offensive operations up the Solomon Islands was delayed and severely hampered by a paucity of resources well into 1943. The European theater of operations still had priority in accordance with the Allied strategy of "Germany First." In compliance with this policy, the Allied invasion of Sicily began in early July 1943. Measured by the strength of the initial assault, it broke all records for amphibious operations; even the Normandy landings the following year, if follow-on echelons of shipping are excluded from the consideration. As a result, resources of any kind, particularly landing craft, were in exceedingly short supply everywhere else.[1]

Although a massive U.S. shipbuilding program was underway, except for several new *Cleveland*-class light cruisers and *Fletcher*-class destroyers, naval forces already present in the Solomon Islands were limited to ships built prior to the war. Almost all the heavy cruisers in the Pacific had been sunk or damaged. Several new fast battleships were available, but Admiral Halsey wisely decided not to risk them in the constrained and poorly charted waters of the Central Solomons.[2]

Map 11-1

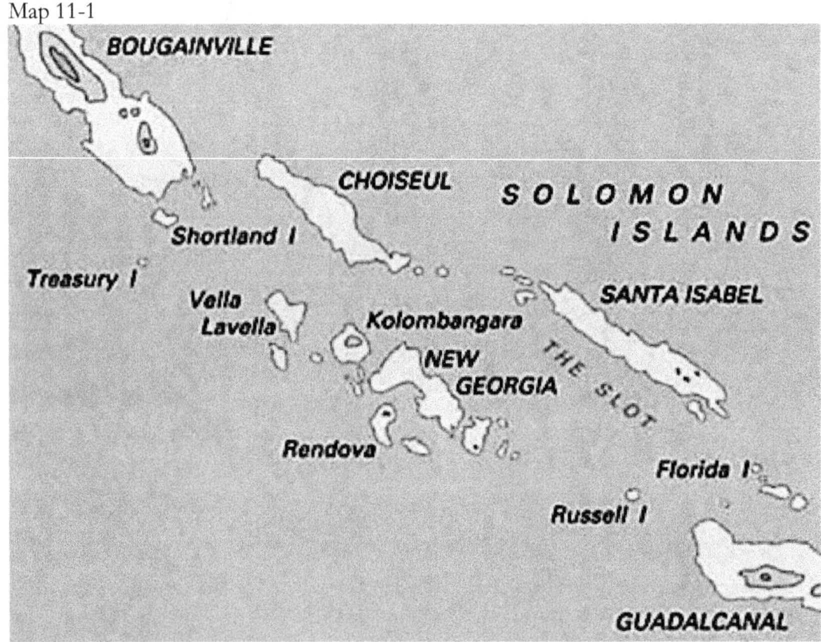


Central Solomon Islands

## ENEMY PRESENCE IN THE NEW GEORGIA GROUP

As early as August 1942, there had been reports of Japanese activity in the New Georgia Islands, 200 miles northwest of Guadalcanal. At first, the Japanese had made use of numerous hideouts and dispersal anchorages in this area, and had used them for establishing staging points for the purpose of ferrying troops and supplies to Guadalcanal. Later, in November 1942, after their failure to capture the airfield on Guadalcanal, they had begun the construction of an aerodrome near Munda Point on the southwest corner of New Georgia Island.[3]

So cleverly did the enemy conceal the construction of the aerodrome, Allied air reconnaissance failed to detect any activity until the project was well advanced. To camouflage work on the ground from passing aircraft, the Japanese crisscrossed heavy wire cables between the tops of palm trees in such a way as to form a net. The trunks of interior trees were then cut out from under their branches, which remained in place, held aloft by the cables. After the runway was completed, the camouflage was removed, ending concealment efforts. On 29 December, shortly before the completion of the Munda Airfield, work was begun on a second airbase, near the mouth of the Vila River, on the southern tip of adjacent Kolombangara Island.[4]

## RUSSELL ISLANDS OCCUPIED AS A PRELUDE TO INVASION OF THE NEW GEORGIA ISLANDS

Photo 11-2

U.S. Marine Raiders paddle their rubber assault boats ashore on Pavuvu Island, Russell Islands, 23 February 1943. They are leaving the USS *Humphreys* (APD-12).
National Archives photograph #USMC 54767

Immediately after the Japanese evacuation of their remaining troops from Guadalcanal, in February 1942, Admiral Halsey set in motion plans for occupation of the Russell Islands, so that they might be used as a staging point for movement into New Georgia. At about dawn on

21 February, some 9,000 infantrymen and Marines from New Caledonia landed unopposed at three different points in the Russells. Accompanying the landing forces was a major portion of the Navy's 33rd Construction Battalion, the "Seabees," who immediately began construction of a radar station, PT boat base, and an airstrip. By month's end, in addition to Construction Battalion, there were units of the 43rd Army Infantry Division, the 3rd Marine Raider Battalion, the 10th Marine Defense Battalion, and Naval Base personnel in the Russells.[5]

On 3 June, Halsey issued the basic operation plan for the invasion of New Georgia. Day-D was set for 30 June, on which assault forces were to make simultaneous landings at several points on Rendova Island, and on New Georgia at Viru Harbor, Segi Point, and Wickham Anchorage. Although supplies were moved forward to the Russell Islands from stockpiles in the Guadalcanal area, in preparation for the New Georgia Campaign, the primary bases for the staging of forces, remained at Noumea in New Caledonia and Espiritu Santo in the New Hebrides.[6]

Map 11-2

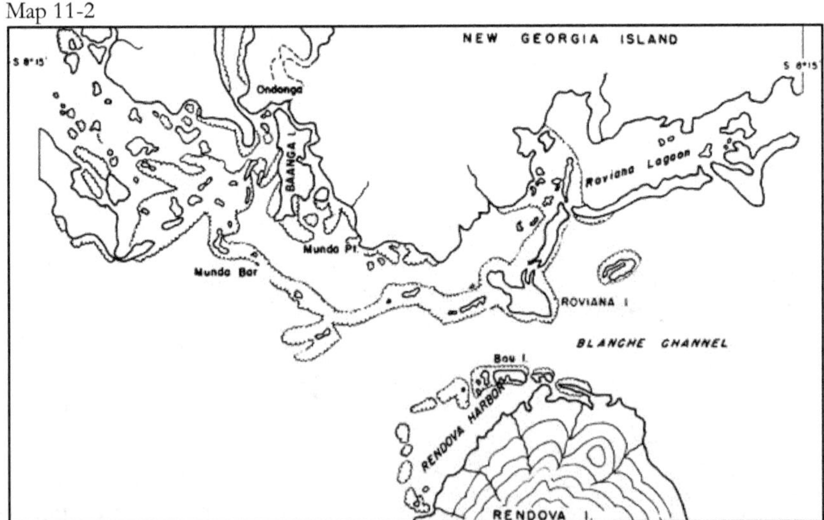


Coastal area of larger New Georgia Island, and the northern coast of Rendova Island
*Building the Navy's Bases in World War II: History of the Bureau of Yards and Docks and the Civil Engineer Corps 1940-1946 Volume II*

The plan called for three major task forces. Task Force Tare (under the command of Rear Adm. Richmond K. Turner, until relieved, on 15 July, by Rear Adm. Theodore S. Wilkinson, following the main landing operations) was subdivided into two groups. The Western Group carried out the operations in the Rendova-Munda area, while the Eastern Group was responsible for the landing and subsequent operations at Viru, Segi, and Wickham Anchorage. The second task force, under the direct command of Vice Adm. William Halsey, was responsible for covering (protecting) the operation and furnishing fire support. The primary air support was provided by Task Force Fox (South Pacific Air Force), under Vice Adm. Aubrey W. Fitch.[7]

A landing at Segi Point was carried out in advance of the main New Georgia operations by Task Force Tare units, to counter possible Japanese interference with the landings, as a result of reports that they were moving into that area. On 21 June, Companies O and P of the 4th Marine Raider Battalion made an unopposed landing there from the high-speed transports *Dent* (APD-9) and *Waters* (APD-8).[8]

## LANDING AT RENDOVA

The second landing in the New Georgia Campaign took place as planned, on 30 June, on the north side of Rendova Island, which is separated from New Georgia by only a few miles. Troops from the advance unit—APDs *Dent* and *Waters*—landed at dawn to secure the beachhead for the main occupation forces which followed one hour later. Picking their way through the reefs, the succeeding transport group entered the harbor and, at 0642, began debarking men and materiel.

**Transport Group: Capt. Paul T. Theiss (ComTransDiv Two)**
**Rear Adm. Richmond K. Turner embarked aboard USS *McCawley***

| | |
|---|---|
| transport USS *McCawley* (APA-4) | Comdr. Robert H. Rodgers |
| transport SS *President Jackson* | Capt. Charles W. Weitzel |
| transport SS *President Adams* | Capt. Frank H. Dean |
| transport SS *President Hayes* | Capt. Francis W. Benson |
| cargo ship USS *Algorab* (AKA-8) | Capt. Joseph R. Lannom |
| cargo ship USS *Libra* (AKA-12) | Capt. William B. Fletcher Jr.[9] |

Two groups of destroyers patrolled the entrances of the harbor, as others screened the transports, ready to silence any shore guns that went into action. The boats leaving the *McCawley* were cautioned, "You are the first to land. Expect opposition." This warning proved true, as boats making for the shore faced machine gun fire from the beach. A little after 0700, batteries at Munda Point on New Georgia opened fire. The first salvo scored a hit on the engine room of the destroyer *Gwin* (DD-433), and a near miss on the *Buchanan* (DD-484). Then, two enemy batteries on Baanga Island, and one or two at Lokuloku to the east, joined in.[10]

*Buchanan* and *Farenholt* (DD-491) silenced the enemy guns with counterbattery fire, while maneuvering skillfully to deny the Japanese easy targets. Exchanges of fire between still operational shore batteries and the destroyers continued intermittently throughout the day.[11]

On two occasions that morning, unloading was interrupted when the transports and cargo ships had to get under way and proceed eastward into Blanche Channel, to avoid the threat of air attack. In each instance, a 32-plane combat air patrol, maintained by fighters from bases in the Russells and Guadalcanal, drove off enemy planes threatening the ships. The unloading was completed by 1500. Flagship *McCawley* had discharged her cargo at the rate of 157 tons per hour, while placing 1,100 troops on the shore. Orders were then given to the transport group to form a cruising formation, and proceed southeast through the channel for return to Guadalcanal.[12]

## TORPEDO PLANE ATTACK ON TRANSPORT GROUP

> ALGORAB *was on the opposite of the task force from the point of attack, and only five (5) of the enemy planes succeeded in coming within range of this ship's guns. All were shot down by* ALGORAB, *but since two (2) had been hit prior to being taken under fire by this vessel credit for only three (3) was officially claimed.*
>
> —Commanding Officer, USS *Algorab*, History of the USS *Algorab*, 19 November 1945.

> *This vessel made hits on four enemy torpedo planes, which were seen to crash.*
>
> —USS *Libra* War Diary, June 1943.

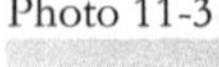
Photo 11-3

Attack cargo ship USS *Algorab* (AKA-8), location and date unknown. Naval History and Heritage Command photograph #NH 78541

Map 11-3

Route of flight of enemy torpedo planes prior to attack on the transport group. *Combat Narratives Solomon Islands Campaign X, Operations in the New Georgia Area 21 June-5 August 1943* (Washington, DC: Office of Naval Intelligence 1944).

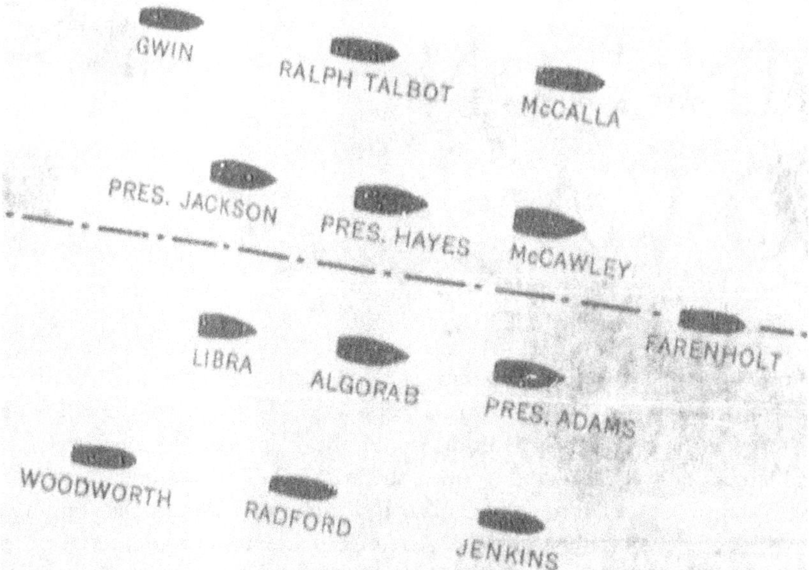


Expanded view of the disposition of ships at the time of attack, led by the destroyer *Farenholt.* Transports and cargo ships were formed in interior columns, and screening destroyers in the left and right columns of the formation.

Within an hour of the transport group departing Rendova, 24-28 Nells (Mitsubishi type 96 torpedo bombers), escorted by an unknown number of Zero fighters, were sighted coming in over the northwest corner of New Georgia Island near Munda Point. Setting up for an attack, the bombers circled the ship formation, using land background for masking, then made a very low approach at nearly 250 knots. All ships opened fire, and immediately scored hits. The surviving bombers ignoring losses of other formation members, pressed home attacks with great determination, releasing their torpedoes at ranges approximating 500 yards.[13]

Three planes, in succession, made drops off destroyer *Farenholt*'s port beam. The first torpedo passed ahead of her, and the second astern. The pilot of the remaining plane dropped his torpedo, and sheered-off. As *Farenholt*'s commanding officer ordered right full rudder to avoid it, a thud was felt on the bridge. All hands waited nervously for an expected explosion, but none occurred. Either the "fish" was a dud, or it had run too short a distance to arm the warhead.[14]

Photo 11-4

Transport USS *McCawley* (AP-10), 3 February 1941.
Naval History and Heritage Command photograph #NH 97720

The USS *McCawley*, flagship of Rear Adm. Richmond K. Turner, was similarly bracketed by three torpedoes. One passed ahead, one astern, and the other apparently under the ship. Flank speed and full right rudder were used to "comb the torpedoes," meaning bring the ship's course parallel to their tracks, and thus provide a smaller target.[15]

Immediately after the ship formation made a ninety-degree turn to starboard, a torpedo was seen churning toward the portside of the *McCawley*. Despite her rudder being put over hard right, and starboard engine backed full, in an effort to twist her beam away from the weapon, and present only the stern, the torpedo struck the transport amidships in the vicinity of her engine room. The resulting explosion blew a hole in her side about 18-20 feet in diameter, extending from about frame 87 to 93. The frames between them on the port side were carried away, and shell plating blown outward with very ragged edges.[16]

With the engines of the damaged transport stopped, and her rudder jammed hard over, Turner ordered USS *Libra* to take *McCawley* in tow, and the destroyers *Ralph Talbot* (DD-390) and *McCalla* (DD-488) to stand by to assist. Rear Admiral Turner and his staff then boarded the *Farenholt.* After only eight minutes at 1558, the air attack was over. Only two enemy planes survived the engagements of protecting fighters, and AA fire of the ships. However, these two aircraft were also shot down during retirement of the Allied planes.[17]

At 1640, upon orders from Turner, *Ralph Talbot* came alongside *McCawley* and removed all personnel from her except for a salvage crew. Rear Admiral Wilkinson, Turner's relief who was also aboard *McCawley*, remained aboard her in charge of salvage efforts.[18]

At 1718, twelve to fifteen Aichi dive bombers broke through the overcast at an elevation of about 1,000 feet and attacked the salvage group. *Libra* had just taken *McCawley* in tow, and was "in irons," while attempting to swing her to the right. Aboard the transport, salvage crew members manned nine 20mm and .50-caliber machine guns, and took the planes under heavy fire. Capt. William Hawkins, USMCR, on No. 2 .50-cal. MG, shot down one dive bomber. No damage was sustained by the salvage group and in total three enemy planes were destroyed. *Libra* expended 16 rounds of 3-inch and 1,020 of 20mm ammunition, with possible hits on two planes.[19]

At 1722, *Libra* was pointed fair and proceeding at 5 knots with the *McCawley* in tow. Despite success of the salvage party in bringing the transport's jammed rudder amidships and shoring her transverse bulkheads, it was apparent that water was rising in No. 4 and 5 holds. At 1850, with the draft aft thirty-eight feet, Wilkinson summoned the *McCalla* alongside, then gave the order for all hands to abandon ship.[20]

*McCalla* cleared *McCawley*'s side at 1930. Less than an hour later, the abandoned transport was hit by three torpedoes, then believed fired by a Japanese submarine, and sank stern-first in thirty seconds at position 08°25'05"S, 157°28'05'E, but it was another instance of friendly fire. It was later learned that *McCawley* was a victim of PT boats of Motor Torpedo Boat Squadron Nine, which erroneously mistook her for an enemy ship. Casualties aboard *McCawley* from the earlier actions were two officers and thirteen men missing in action, and seven wounded.[21]

## AFTERMATH

Map 11-4

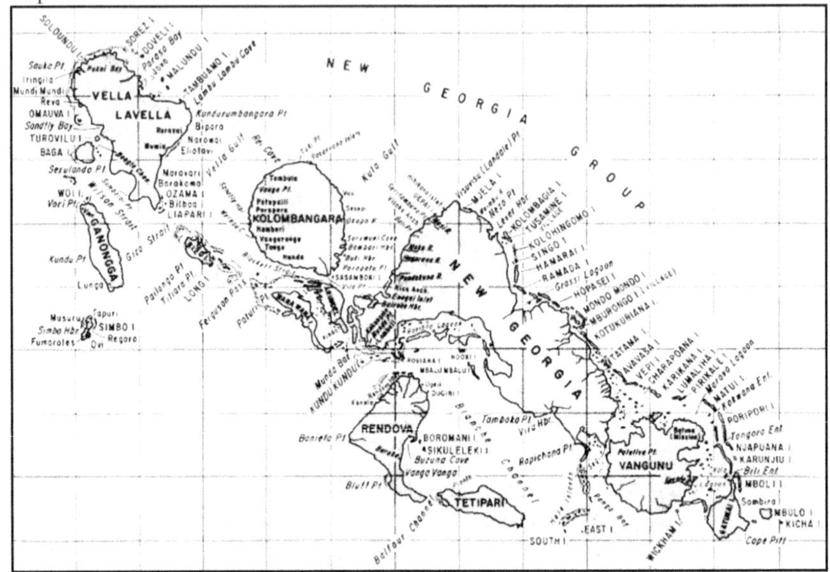


New Georgia Islands
Dyer, *The Amphibians Came to Conquer: The Story of Admiral Richmond Kelly Turner*, 1972

Starting on 30 June and continuing for several days, assisted by their artillery emplaced on Rendova, Allied landings were made on New Georgia near Munda. An attack was launched westward in the direction of the Munda Airfield, but progress was slow due to the thick jungle, and stout resistance from strong Japanese positions. The airfield was finally seized, on 5 August, but after its capture, the enemy still held the island of Kolombangara (immediately adjacent to, and northwest of New Georgia) with its airfield at Vila on its south coast. It was determined, however, to bypass positions on this island, and capture Vella LaVella, another island in the New Georgia Group, which lay northwest of Kolombangara. This was done 15-30 August.[22]

## BATTLE STARS

The attack cargo ships *Algorab* (AKA-8) and *Libra* (AKA-12) received a battle star for their participation, on 30 June 1943, in the "New Georgia Group: New Georgia-Rendova-Vangunu occupation." Many other ships earned battle stars as well; only USN cargo ships receiving such are listed here and elsewhere in the book, owing to space limitations.

# 12

# Occupation of Cape Torokina

*There were centipedes three fingers wide whose bite caused excruciating pain for a day, butterflies as big as little birds, thick and nearly impenetrable jungles, bottomless mangrove swamps, man-eating-crocodile-infested rivers, millions of insects, four types of rats larger than house cats, and heavy torrents of rain bringing enervating humidity. And sacred skull shrines, reminders of days of cannibalism and head-hunting.*

—James Brady and Ron Powers, describing in the book *Flags of our Fathers* the conditions encountered by U.S. Marines on Bougainville in the Solomons.

In 1943, the American military strategy in the Southwest Pacific was simple: advance northwest through the Solomon Islands and capture the powerful Japanese military base at Rabaul, New Britain. This action would make it possible to open a direct route to the Philippines, and retake those islands as General MacArthur had promised. By late autumn 1942, the 1st Marines had crushed the Japanese on Guadalcanal, and relieved of their duties by the 23rd Infantry Division, had left for Australia. (The 23rd was known as the "Americal" Division. Activated, 27 May 1942, on the island of New Caledonia, the division's commander had requested that the new unit be known as the Americal Division, a contraction of "American, New Caledonian Division.") While the Allied effort had given them use of Henderson Airfield on Guadalcanal, it did not relieve the threat of enemy air attack. The enemy stronghold at Rabaul, well-buttressed by other aerodromes (sites from which aircraft operations take place) located at Bougainville Island, the Shortland Islands, and New Guinea could mount formidable attacks on the Allies.[1]

The Allies soon mounted attacks to neutralize such airfields. Following the capture of Munda Airfield as well as Bairoko Harbor along the northwest coast of New Georgia, their next objective was to seize an area on Bougainville from which their bombers could fly airstrikes against Rabaul. Bougainville, the largest island of the Solomons, lay near the northwestern end of the island chain, some 190 miles east of Rabaul. It is a mountainous island dominated by the Emperor and Crown Prince Ranges, which include two active volcanoes, the largest of which, Mt. Balbi, rises to 10,171 feet. The dense jungle on the lower mountain slopes and coastal plains, and the swamp areas immediately inland from the beaches are precipitated by an annual rainfall averaging 100 inches. These conditions earned the island the moniker, as one Marine termed it, "a wet hell;" malaria and other tropical diseases were prevalent.[2]

Inhabited by approximately 54,000 natives, Bougainville was one of the few places in the South Pacific still possessing credible reports of headhunting taking place. The Japanese had invaded this island and the adjacent smaller one, Buka, at the tip of the chain in early 1942, when fewer than twenty Australian troops and a few native coast watchers were based there. Until evacuated, the soldiers withdrew into the jungle to observe the enemy, while the coast watchers remained to report enemy air and sea movements. Their warnings of convoys and air raids coming down "the Slot" had helped the Marines to achieve victory on Guadalcanal.[3]

Map 12-1

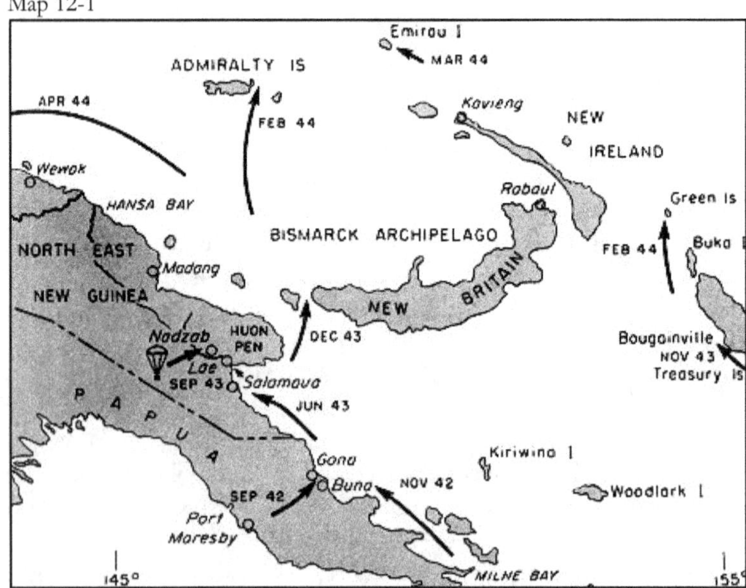


Bougainville, at the top of the Solomon Islands chain. To the northwest lay the Bismarck Archipelago, through which MacArthur and Halsey's forces would have to pass to advance into the Philippines.
(http://www.history.army.mil/books/AMH/Map23-43.jpg)

## AMPHIBIOUS LANDING AT CAPE TOROKINA

On 1 November 1943, Rear Adm. Theodore Wilkinson, commander, Third Fleet Amphibious Forces, directed the landings on Bougainville Island from his flagship, the attack transport USS *George Clymer* (APA–27). Maj. Gen. Alexander Vandegrift's I Marine Amphibious Corps—the 3rd Marine Division (reinforced), the 37th U.S. Infantry Division, and the Advance Naval Base Unit No. 7—constituted the landing force. The troops went ashore halfway up the island's western shore at Cape Torokina, just north of Empress Augusta Bay.[4]

Task Force 39, commanded by Rear Adm. A. Stanton Merrill, was assigned to cover the landings. His force, consisting of Cruiser Division Twelve—light cruisers *Montpelier* (CL–57), *Cleveland* (CL–55), *Columbia* (CL–56), and *Denver* (CL–58)—supported by Destroyer Divisions 45 and 46, was assigned to screen the transports and support vessels from Japanese air and surface attack. After the landings, Merrill's force was engaged in battle, following the interception of a Japanese force under Rear Adm. Sentaro Omori en route to attack the transports. The landing, which proved difficult owing to poor beach conditions and enemy resistance, was accomplished on 1 November when over 14,000 men and 6,200 tons of materiel were put ashore in eight hours.[5]

The naval force of the first echelon was comprised of eight transports, four cargo ships, seven destroyers, four destroyer minesweepers, and two fleet tugs. Designated the Main Body, Northern Force, it was organized around three divisions of the Transport Group, under the command of Commodore Lawrence F. Reifsnider.[6]

Transport Divisions Able and Baker were each comprised of four attack transport ships, screened by a destroyer division. Making up the remaining division, Charlie, were the four attack cargo ships:

**Transport Division Charlie: Capt. Harry E. Thornhill, USN**

| Ship | Commanding Officer |
|---|---|
| *Alchiba* (AKA-6) | Comdr. Howard Rutherford Shaw, USN |
| *Alhena* (AKA-9) | Comdr. Howard William Bradbury, USN |
| *Libra* (AKA-12) | Capt. Floyd Franklin Ferris, USN |
| *Titania* (AKA-13) | Comdr. Herbert Everett Berger, USN[7] |

## REPAIRED CARGO SHIPS RESUME THEIR DUTIES

The attack cargo ship *Alchiba*, torpedoed twice by Japanese submarines at Guadalcanal, had entered the Mare Island Navy Yard, at Vallejo, California, on 2 June 1943, for permanent repairs. After they had been completed in early August, she left the yard for sea trials off San Francisco, then moved down the coast to Port Hueneme, to load cargo. On 19 August, she took her leave of the U.S. West Coast, bound for the

South Pacific. Supply runs to New Caledonia and Guadalcanal, preceded her use in the landings on Bougainville.[8]

*Alhena*, another victim of a Japanese submarine attack, had returned to fleet service a little earlier. Following restorative repairs and conversion to an attack cargo ship at Sydney, she had left there, on 10 June 1943, shaping a course for Noumea. During the next few months, she plied between Noumea and Guadalcanal, making calls at Auckland, New Zealand, to take on cargo.[9]

## MOVEMENT TO THE OBJECTIVE

As a precaution to prematurely alerting the enemy during their transit to Bougainville, the three transport divisions with their embarked troops, equipment and supplies avoided merging into attack formation until late morning of the day prior to the landing. TransDiv Baker left Efate in the New Hebrides, in the late afternoon on 28 October, and proceeded directly to a point about twenty miles south of the eastern end of San Cristobal Island. TransDiv Charlie, which had departed from Guadalcanal the morning of 30 October, rendezvoused with Baker before dark later that day. Charlie was accompanied by the fleet tug *Sioux* (AT-75) and Comdr. Wayne R. Loud's, USN, Mine Squadron Two—*Hopkins* (DMS-13), *Hovey* (DMS-11), *Dorsey* (DMS-1), and *Southard* (DMS-10). During the night, these the two divisions and their escorts proceeded that night south of Guadalcanal, on a western course. They were joined by TransDiv Able (which had sailed from Espiritu Santo in the late afternoon on 28 October) west of Guadalcanal.[10]

Once all three divisions were consolidated, Commodore Reifsnider took tactical command of the transport group and the screen. The main body proceeded on a northwest course, with the transports and cargo ships formed in three division columns. At 0422 in early morning darkness, on 1 November, the group changed to a northeast course to make the approach to Cape Torokina. Speed was reduced from 15 to 12 knots, and the minesweepers were sent ahead to clear an approach channel. (No mines were found). Arrival in the transport area was set for daylight to permit visual detection of uncharted shoals believed to be offshore.[11]

## LANDINGS AT CAPE TOROKINA

*Beach Yellow Four assigned to the* Alchiba *proved almost entirely unworkable. Effective measures to stop landing on this beach, however, were not taken until practically all the LCVP's and tank lighters carried by the* Alchiba *were hopelessly broached to or otherwise out of commission.*

—*Combat Narratives Solomon Islands Campaign: XII The Bougainville Landing and the Battle of Empress Augusta Bay 27 October- 1 November 1943* (Washington, DC: Office of Naval Intelligence, 1945).

Map 12-2

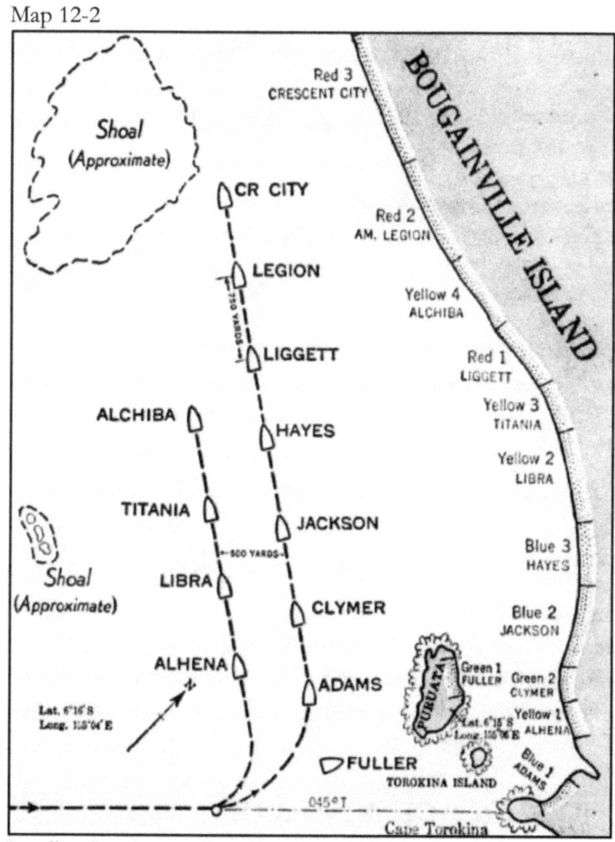


Landing Beaches at Cape Torokina, 1 November 1943
*Combat Narratives Solomon Islands Campaign: XII The Bougainville Landing and the Battle of Empress Augusta Bay 27 October- 1 November 1943*, ONI 1945

By 0645 all transports and cargo ships (often identified in Navy reports as transports) were in the transport area, and the signal, "land the landing force" was executed. Aboard transports, amphibious craft were lowered away, scrambling nets positioned, and Marines swarmed over the side to board the LCVPs and LCMs.[12]

Photo 12-1

A Marine Corps LVT-1 churns toward the shore at Cape Torokina.
National Archives photograph #80-G-54390

It was clear to those aboard ship, even from the transport area, that surf on the narrow, steep shore was bad, but boat crews found conditions worse than anticipated. This was particularly true of the four northernmost landing beaches. Admiral Halsey later termed the beach and terrain conditions, "worse than anything ever encountered before in the South Pacific." Coxswains reported that the steepness of the beach prevented safe grounding of boats along the length of their keels. Moreover, the beaches were so narrow in most places, that two bulldozers could not pass abreast between jungle and sea. In spite of these difficulties all of the troops were successfully landed. Once ashore, assault troops were confronted with swamp extending for more than a mile inland, except for two narrow corridors of land rising only a few inches above swamp level.[13]

## ENEMY AIR AND SURFACE ATTACKS

> *The brilliant performance of our fighter cover and Fighter Director Group who successfully turned back or completely broke up the concerted attacks of four separate groups of enemy planes. Without such effective air cover, severe losses or even failure of the operation may have resulted.*
>
> *The timely interception by Task Force 39 and its annihilation of the Japanese surface force which was closing in on the transports during their return to Empress Augusta Bay the night of 1-2 November.*

—Highlights identified by Commodore Lawrence F. Reifsnider, in his report of the landings in the Empress Augusta Bay area.[14]

At 0725, shortly after the first assault wave landed on the beach, the destroyer escort screen reported unidentified aircraft, twenty-one miles distant. Three minutes later, the convoy received orders to get under way, and to form "column open order." This type formation is designed to maximize firepower of its members in a small area. With the lead ship (in this case, *Alchiba*) acting as the guide, the second ship was typically 4 degrees off the port quarter of the guide at the standard distance of 2,000 yards, the third ship 2 degrees off the starboard quarter of the guide at twice the standard distance, and the remainder alternating in the same manner from port to starboard.[15]

After forming up, the disposition began making radical maneuvers, in accordance with ordered emergency turns. At 0745, enemy aircraft closed in to attack beach positions and the rear ships of the formation. A single enemy plane came in low, flying seaward from the beach, and dropped a bomb near a screening destroyer. A second plane dove on the convoy from its port side.[16]

Both were shot down; one fell in flames toward the beach, the other was observed to burn on the horizon, apparently a victim of *Titania*'s guns. The latter one, a "Kate" (one of a group of about six Nakajima B5N Navy Type 97 carrier attack bombers), had passed diagonally over the attack cargo ship, as her two 3-inch and ten 20mm guns sent up a barrage of fire. Hit by 20mm rounds, the plane's starboard wing began smoking, and a person or object was seen to bail or fall out of the aircraft before it crashed in the water several miles away.[17]

After the attack ended at 0820, the convoy returned to the transport area, finished lowering away all boats, and resumed unloading troops and cargo. At 1250, the escort reported three groups of unidentified aircraft to the northwest, at a distance of 65 miles. Once again, the convoy got under way from the transport area, to gain sea room and, hopefully, increase ship survivability. At 1335, the escort screen reported planes in sight on the port beam of the ship formation. Three bombs were dropped well clear of it, two on the port bow at a distance of 4,000 yards, and one on the starboard beam at 1,500 yards. One plane was seen to crash on the horizon to the northwest.[18]

The convoy returned to the transport area at 1445, and resumed unloading operations. Although some of the ships were not empty, the convoy departed Empress Augusta Bay at 1750 that evening, 1

November. At 2300, three transports, *Alchiba*, the fleet tug *Sioux*, and a screen of five destroyers detached to return to the bay to complete the unloading. The remainder of the convoy proceeded to Guadalcanal.[19]

In early morning darkness, on 2 November, the group en route back to the bay witnessed evidence of the Battle of Empress Augusta Bay in progress, as noted by Commodore Reifsnider in his report:

- 0245: Sighted gunfire bearing 335T
- 0345: Observed naval gunfire bearing 320 on starboard quarter. Gunfire ceased at 0440. Believed to be Task Force 39 in contact with enemy
- 0455: Intermittent gunfire of about 5 minutes duration bearing 315
- 0515: Intermittent gunfire observed bearing 320. Our course 308. Gunfire ceased at 0522[20]

At 0807, as the convoy was approaching the bay, anti-aircraft fire was observed on the horizon, on the port quarter of the formation. Five minutes later, the escort reported that Task Force 39 was under air attack on the port quarter, and seven columns of black smoke were visible on the horizon.[21]

The transports and cargo ship anchored in Empress August Bay at 0912, and began unloading once again. Following the discharge of all cargo, the convoy reformed at 1500 and left for Guadalcanal; arriving there a little after midnight, on 4 November.[22]

## BATTLE STARS

The attack cargo ships received battle stars for the periods noted in the table; the end dates of 9 and 13 November, denote additional qualifying actions on those dates, following the landings on 1 November.

**Treasury-Bougainville: Occupation of Cape Torokina**

| Ship | Battle Star | Commanding Officer |
|---|---|---|
| *Alchiba* (AKA-6) | 1-13 Nov 43 | Comdr. Howard Rutherford Shaw, USN |
| *Alhena* (AKA-9) | 1-13 Nov 43 | Comdr. Howard William Bradbury, USN |
| *Libra* (AKA-12) | 1-9 Nov 43 | Capt. Floyd Franklin Ferris, USN |
| *Titania* (AKA-13) | 1-9 Nov 43 | Comdr. Herbert Everett Berger, USN |

# 13

# Central Pacific Campaign

*The capture of Tarawa knocked down the front door to the Japanese defenses in the Central Pacific.*

—Observation by Adm. Chester Nimitz regarding the acquisition of the Gilbert Islands, which came at a high cost in terms of lives lost.

Photo 13-1

Adm. Raymond A. Spruance, USN, commander, Central Pacific Force, U.S. Pacific Fleet, on 23 April 1944, days before it was redesignated the 5th Fleet. Navy Photograph, now in the collections of the U.S. National Archives.

By October 1943, the Pacific war against Japan had progressed to a point where landings in a new theater, the Central Pacific, could be undertaken. The Japanese had been driven from the Aleutians in the spring and summer of 1943. In the South Pacific, the drive up the Solomons had progressed to Bougainville by late in that month. MacArthur's forces had turned the tide in New Guinea, and a start had been made in breaking through the Bismarck Sea and along the road leading to the Philippines from the south. This was accomplished by gaining control of Vitiaz Strait and, soon after, the adjacent Dampier Strait, so as to be able to dominate all sea passages between New Britain and New Guinea. At this juncture, it was time to push another spearhead along the main route to Japan from the east, by means of "island hopping" across the vast Central Pacific.[1]

In preparation for this progression, on 5 August 1943, Adm. Chester Nimitz established the Central Pacific Force (Task Force 10) with Vice Adm. Raymond Spruance in charge. Its title would be short-lived. After overseeing its first success, the battle of Tarawa in November 1943, Spruance would guide his force as it advanced through the Gilbert Islands, and assaulted Kwajalein and Majuro atolls in the Marshall Islands. For these accomplishments, he was promoted to admiral, in February 1944. Subsequently, on 26 April 1944, the Central Pacific Force became the U.S. Fifth Fleet, which Spruance would command through war's end.[2]

Map 13-1

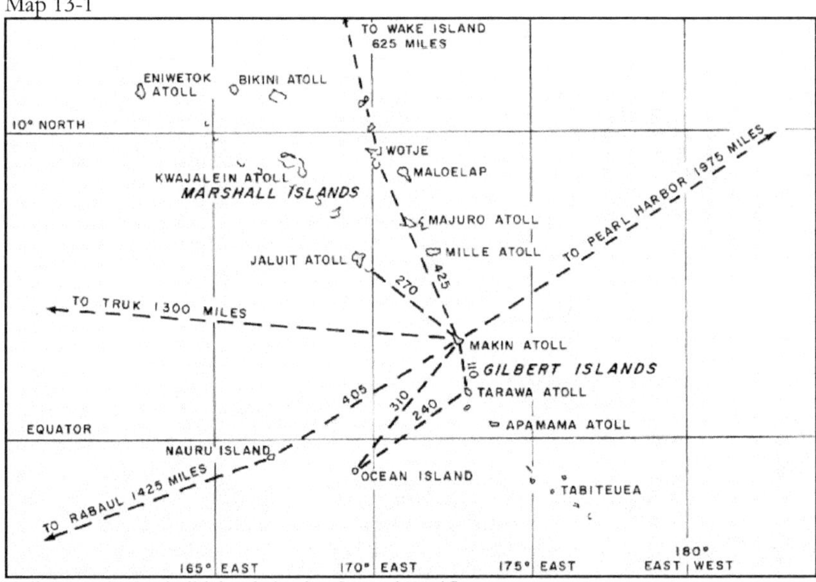


Gilbert and Marshall islands in the Central Pacific
http://www.ibiblio.org/hyperwar/USN/ACTC/maps/actc-16-p612.jpg

The Gilbert Islands, which included Tarawa, Apamama, and Makin, were a group of coral atolls oriented in a roughly north-south line near and north of the equator. The Japanese had invaded these British-held islands on the same day as the attack on Pearl Harbor and occupied them by 10 December 1941. The location of the islands immediately south and east of other important Japanese bases in the Carolines and Marshalls added to their strategic importance to the enemy. For the Allies, the Gilberts offered sites for airfields necessary for progress along the sea-road through the Central Pacific toward Japan. The planned assault and capture of Tarawa and Makin, in November 1943, were a part of the overall American strategy of conducting an offensive through Micronesia—the Gilbert, Marshall, and Caroline islands—at the same time as MacArthur's New Guinea-Mindanao approach to Japan.[3]

Map 13-2

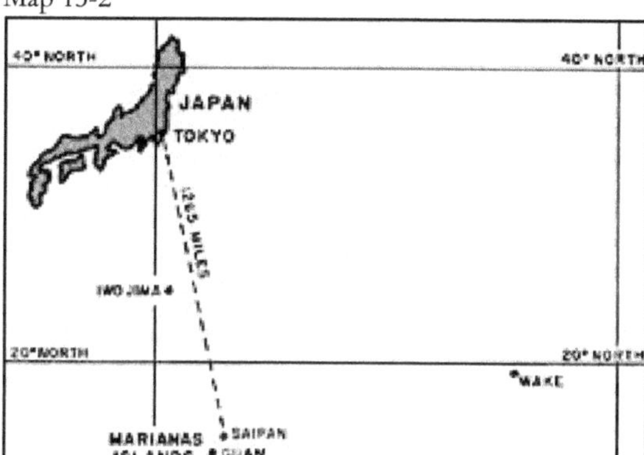

Principal route of the Central Pacific forces toward Japan
http://www.ibiblio.org/hyperwar/USN/ACTC/maps/actc-p735.jpg

In order to gain airbases capable of supporting operations across the Central Pacific to the Philippines and on to Japan, the U.S. needed to take the heavily defended Mariana Islands. Use of land-based aircraft, as part of this strategy, to weaken enemy defenses and to provide some measure of protection for the water-borne invasion forces, necessitated capturing the Marshall Islands, northeast of Guadalcanal, to provide for operation of such aircraft. But before any of this could be unrolled, first an enemy garrison and air base on Betio (one of the islands of Tarawa Atoll in the Gilberts) had to be captured, as it guarded against invasion forces arriving from Hawaii. Thus, the starting point for the planned invasion of the Marianas lay far to the east, at Tarawa.[4]

## CENTRAL PACIFIC FORCE

The occupation of the Gilbert Islands began with amphibious landings on Tarawa and Makin atolls, on 20 November 1943. The Central Pacific

Force under Spruance—consisting of some 116 combatant vessels and 75 auxiliaries—was, in general, organized into three major groups:

- Assault Force (TF 54): Rear Adm. Richmond K. Turner, USN
- Carrier Force (TF 50): Rear Adm. Charles A. Pownall, USN
- Defense Forces and Shore Based Air (TF 57): Rear Adm. John H. Hoover, USN[5]

The Assault Force was composed of the Northern Attack Force (Makin Atoll) and Southern Attack Force (Tarawa Atoll). Each attack force was made up of a transport group, a fire support group, an air support or carrier group, a minesweeper group, a landing force, and LST (tank landing ship) and garrison groups. The transports and cargo ships of these attack forces, and of the garrison group, are shown in the table.

| **Southern Attack Force (TF 53)**<br>**Rear Adm. Harry W. Hill** | **Northern Attack Force (TF 52)**<br>**Rear Adm. Richmond K. Turner** |
|---|---|
| Transport Division Four:<br>Capt. John B. McGovern, USN | Assault Transport Division |
| USS *Zeilin* (APA-3)<br>USS *Harry Lee* (APA-10)<br>USS *William P. Biddle* (APA-8)<br>USS *Arthur Middleton* (APA-25)<br>USS *Thuban* (AKA-19)<br>USS *Heywood* (APA-6) | USS *Leonard Wood* (APA-12)<br>USS *Calvert* (APA-32)<br>USS *Pierce* (APA-50)<br>USS *Alcyone* (AKA-7)<br>Reserve Transport Division<br>USS *Neville* (APA-9)<br>USS *Belle Grove* (LSD-2) |
| Transport Division Eighteen:<br>Capt. Herbert B. Knowles, USN | |
| USS *Monrovia* (APA-31)<br>USS *Doyen* (APA-1)<br>USS *Ashland* (LSD-1)<br>USS *Sheridan* (APA-51)<br>USS *Virgo* (AKA-20)<br>USS *LaSalle* (AP-102) | **Tarawa Garrison Group (TG 54.9)**<br>USS *President Polk* (AP-103)<br>SS *Dashing Wave*<br>USS *Jupiter* (AK-43)<br>SS *Cape Fear* |
| Transport Division Six:<br>Capt. Thomas B. Brittain, USN | |
| USS *Harris* (APA-2)<br>USS *J. Franklin Bell* (APA-16)<br>USS *Ormsry* (APA-49)<br>USS *Feland* (APA-11)<br>USS *Bellatrix* (AKA-3)[6] | |

The Gilberts operation was the U.S. Navy's first amphibious assault against defended enemy atolls, and was expected to be strongly contested. Small atolls were relatively easy to fortify by the enemy and, when garrisoned by a determined foe, their capture, as proved by experience, was to be difficult and cost heavily in casualties. These operations differed greatly from landing on a large island, where, with dispersed enemy defenders, lightly defended beaches were often found, and where the enemy might be surprised by an attack on his flank or rear. Here, the element of surprise was expected to be non-existent, except as to, possibly, the exact time of the attack.[7]

## MOVEMENT TO THE OBJECTIVE

Prior to start of Gilberts attacks, the forces designated for the operation were widely dispersed. At the beginning of the assembly and training period, the Southern Attack Force was in the New Hebrides area, and the Northern Attack Force, even more distant, was in the Hawaiian area. Part of the designated troops, the 2nd Marine Division was in New Zealand, and the Army 27th Infantry Division was on Oahu. The Marine Defense Battalions were in Samoa; Wallis Island; Nanumea Island (Ellice Islands); and Nukufetau (Motulalo Island, Ellice Islands). The Army defense battalions were yet to be organized, while service units (Seabees, etc.) were still in the United States, and the shipping and fleet units were scattered far and wide. All were brought together to join either the Southern or Northern Attack Force.[8]

The Northern Attack Force left Pearl Harbor, on 9 November, and the Southern Attack Force from Efate, New Hebrides, on 12 November. The Carrier Groups sortied at about the same time and, after rendezvousing with the fast battleships, proceeded as planned to carry out their strikes and bombardments. At 2243 on 19 November, the various task groups after sailing together were released to proceed to their initial attack positions. Passage between Tarawa and Maina atolls (separated by only 17 miles) was made near the southern part, where ships would be obscured by land masses in order to avoid detection by enemy radar on Tarawa.[9]

## FORMATION OF TRANSPORT DIVISIONS

Turning specifically to transport vessels, it is noted that several weeks earlier, on 5 October 1943, commander, Fifth Amphibious Force, Rear Admiral Turner, had named captains John B. McGovern, Thomas B. Brittain, and Herbert B. Knowles, commander of Transport Divisions Four, Six, and Eighteen, respectively. Turner had directed that they form nucleus staffs from ships' officers on the spot, drawing on the most experienced, and dispersing them throughout the whole flotilla, and to request additional officers to complete their allowances. All three division commanders drew heavily on USS *Heywood* and *J. Franklin Bell*, the two attack transports with the most experienced personnel, for this purpose. As shown in the preceding table, three to five attack transports

(APA) and one attack cargo ship (AKA) comprised each division, with the latter carrying amphibious craft as well as cargo.[10]

Ad hoc training of Southern Attack Force troops and ships' crews for amphibious landings on a simulated coral atoll had been carried out at Hawke's Bay, a region on the east coast of New Zealand's North Island. Complications arose because, as ships reported throughout the month, none of the transport divisions were able to operate as individual units, and troops were unable to work with the same shipboard personnel they would be teamed with during the assault. Improvised craft, manned by New Zealand Navy personnel, served as control vessels, used to direct assault craft to the beach; and mock air support was simulated by the New Zealand Air Force. The New Zealanders carried out these duties cheerfully and effectively, but they were a temporary, rather than an integrated element.[11]

## ASSAULTS ON TARAWA AND MAKIN

*The fighting ashore was bitter, and became a matter of destroying the enemy in small groups entrenched in pill-boxes, machine gun nests, and other strong fortifications. By nightfall of D-Day, 5 battalion landing teams had been committed, but the situation ashore was not clear to the Task Force Commander. It was later determined that a fair beachhead was held on the western beach, but that the two beaches to the eastward had practically no beachheads.*

—Description of ground combat on Tarawa, from Commander in Chief, U.S. Pacific Fleet and Pacific Ocean Areas, Operations in Pacific Ocean Areas – November 1943, 18 February 1944.

Map 13-3

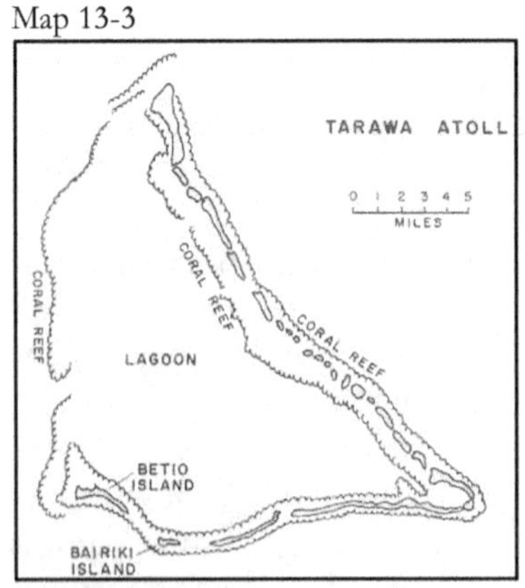


Tarawa Atoll
https://www.ibiblio.org/hyperwar/USN/ACTC/maps/actc-p683.jpg

After passing between Tarawa and Maina atolls, the transports of the Southern Attack Force continued on a westward track, partially shielded during the final approach to their unloading area, by being "up moon," moonrise being at 0058. The landing beaches at Tarawa were inside the sheltered lagoon on the northern side of Betio Island, located on the southwest corner of the triangle-shaped atoll. The Japanese had placed land mines and erected formidable barriers on the island's southern beaches. Many other mines were later found in storage ashore, indicating that the enemy had likely been in the process of completely fortifying the island perimeter with barriers and beach mines.[12]

Diagram 13-4

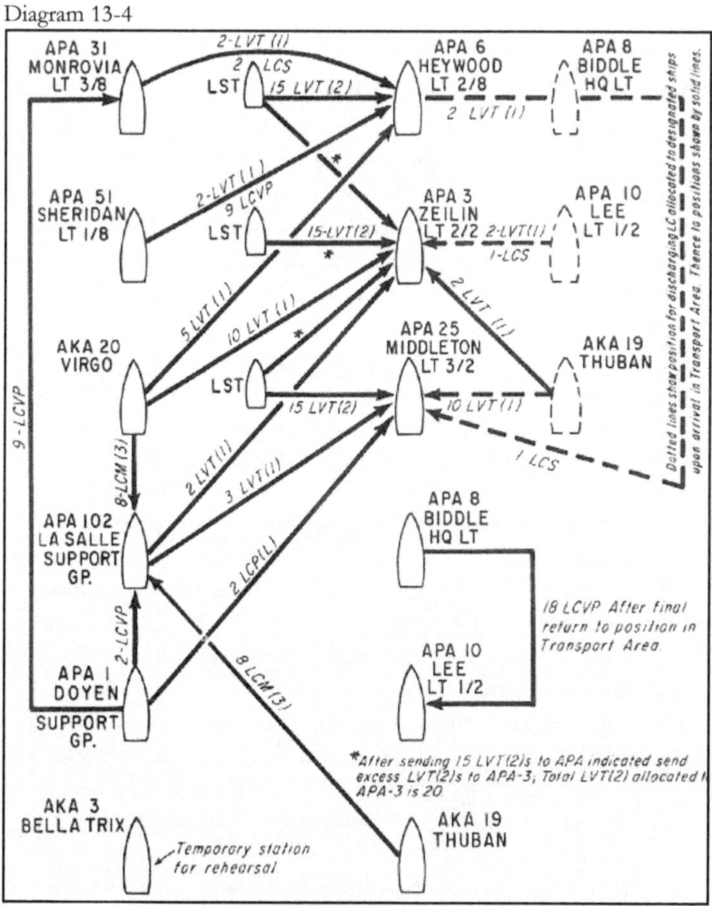


Relative positions of ships in the transport area, and landing craft allocation, based on 50 LVT-1s aboard APAs and AKAs, and 50 LVT-2s aboard LSTs https://www.ibiblio.org/hyperwar/USN/ACTC/charts/actc-p695.jpg

At 0355 on 20 November, the transports and cargo ships of the Southern Attack Force arrived in their designated area, some eight miles from the northwest corner of Betio Island, and began lowering their LVT-1 tracked amphibious vehicles (referred to as Alligators) for the first assault wave. As the amphibious craft were launched, they made their way to the transports *Zeilin*, *Heywood*, and *Middleton*, which were carrying the assault battalion landing teams of the 2nd Marine Regiment, and troops began to board them in the pale light of the third-quarter moon. About 0415, three tank landing ships carrying LVT-2 Buffalo tracked amphibious vehicles hove into position near the appropriate transports, and began loading the LVT-2s with the designated Marines aboard the APAs.[13]

## ENEMY SHORE GUNS OPEN FIRE

The first assault wave, composed of fifty LVT-1s, left the transport area for the rendezvous area at 0550. Immediately after, the second and third waves, comprising fifty LVT-2s, were formed and dispatched to follow the first wave. After these waves formed up, they proceeded from their rendezvous area, into the lagoon to the Line of Departure. At this point the craft were directed by a control ship, the minesweeper USS *Pursuit* (AM-108), to their landing beaches.[14]

Shortly before sunrise, at about the time the first wave departed the transport area, heavy enemy coast defense guns on the southwestern end of Betio opened fire on the transports. The Japanese had brought the 8-inch guns to Tarawa from Singapore, following their capture of the British Colony. In reply, the battleship *Maryland* (BB-46) fired ten salvos from her 16-inch main battery. The fifth salvo scored a hit on the battery, resulting in an explosion, and a cessation of firing by the Japanese. Other heavy U.S. support ships opened counter-battery fire as other enemy guns opened up on the Assault Force and continued shooting until about 0542, three minutes before the initial D-Day air strike was scheduled to start.[15]

During the initial part of this exchange, the cargo ships *Thuban* and *Virgo*, and transports *LaSalle* and *William P. Biddle* experienced near hits, ranging from 25 feet to 100 yards. At 0551, a round landed between the *Thuban* and the transport *Doyen*. *Thuban* notified the officer in tactical command via TBS (talk between ships), and as a result all transports stood farther out to sea. As this was happening, rounds continued to land among ships before they could move out of the range of the shore batteries, a distance of approximately 18,000 yards.[16]

## OPPOSED LANDINGS / FIERCE COMBAT ASHORE

Photo 13-2

U.S. Marines on Tarawa grab a "break" alongside a LVT amphibian tractor (Amtrak). Naval History and Heritage Command photograph #UA 55.01.07

The first wave of assault craft did not reach the Betio beaches until 0917. This event, designated H-Hour, had been tentatively set for 0830. The delay resulted from a combination of factors: overloading, wind, adverse sea conditions, an ebb tide, and the poor mechanical condition of a number of the LVTs, which slowed the speed of the first wave to 2½ knots when 3 knots had been expected. The first three waves of LVTs had no difficulty in crossing the reef which extended from 500 to 1,000 yards offshore. However, all were under heavy enemy fire. The third wave suffered the most casualties, believed a result of enemy reinforcements arriving from the south side of the island just before it landed.[17]

The 3/2 (3rd Battalion of the 2nd Marines) was the first unit to reach its assigned beach. The fourth and succeeding waves, comprised of wooden LCVPs (known as "Higgins boats," after George Higgins, their designer) and LCMs (steel landing craft), could not pass over the reef. Troops and equipment of these craft made their way ashore in LVTs or rafts, or by landing at a pier which extended out across the reef to deeper water. Troops attempting to wade ashore crossing over the reef, other than along the pier, met intense fire and suffered heavy casualties.[18]

Photo 13-3

LCVPs at Slapton Sands, England, in 1944, practicing for the Normandy landings. National Archives photograph #80-G-252326

Coming ashore, the 2nd Marines found Betio Island a mass of concrete pill boxes, heavily reinforced with concrete and coral sand, and well dug into the ground. Pre-assault naval gunfire and aerial bombing had succeeded in destroying only a fraction of the redoubts (a breastwork outside a fortification) along the narrow strip (island), only a few hundred yards wide, and a mile and a half long. Worse, the Marines landed in the center of this bristling beach defense, a point at which the enemy could concentrate their automatic weapons fire from both sides. Japanese mortars also kept up a murderous fire on the beach, and snipers used an old hulk in the lagoon and abandoned LVTs, as cover for firing at the landing assault forces.[19]

Despite heavy losses, in coming ashore and in the following land combat, which included the U.S. infantrymen using flame throwers and hand grenades to force the enemy out of their fortifications, the landing force managed to secure Betio within three days. Col. David Monroe Shoup, the commander of the Second Marine Regiment, and a future commandant of the U.S. Marine Corps, was awarded the Medal of Honor. His associated citation read in part:

> Although severely shocked by an exploding enemy shell soon after landing at the pier and suffering from a serious, painful leg wound which had become infected, Colonel Shoup fearlessly exposed himself to the terrific and relentless artillery, machinegun, and rifle fire from hostile shore emplacements. Rallying his hesitant troops by his own inspiring heroism, he gallantly led them across the fringing reefs to charge the heavily fortified island and reinforce our hard-pressed, thinly held lines. Upon arrival on shore, he assumed command of all landed troops and, working without rest under constant, withering enemy fire during the next two days, conducted smashing attacks against unbelievably strong and fanatically defended Japanese positions despite innumerable obstacles and heavy casualties.[20]

Providing added testament to the bravery of Shoup and his Marines was the totality of diverse weapons emplacements found on Betio:

- 14 coast defense guns, from 80mm to 8-inch in caliber
- 12 anti-aircraft guns of 75mm to 127mm in caliber
- 31 anti-aircraft 13mm guns, 4 being twin-mount
- 70 beach defense and anti-boat guns of from 13mm to 75mm
- A large number of emplacements for smaller guns and mortars
- Anti-personnel and anti-vehicle mines placed in some areas on the beaches and reefs[21]

## OCCUPATION OF MAKIN

As at Tarawa, one island in the Makin Atoll was defended by the Japanese: the long, narrow Butaritari Island. Commencing at 0436 on 20 November, the Northern Attack Force split up to form separate units. The Fire Support Units made their way to their respective support areas, the transports into Transport Area No. 1, and the Air Support Group to the southeast of Makin. At 0700, the LST group joined the transports, which included the cargo ship USS *Alcyone* (AKA-7).[22]

The landing at Makin was made under heavy enemy cross fire, during which *Alcyone* suffered her first fatality of the war. Seaman 1/c Jasper Anthony Murray, USN, was killed in action against the enemy while serving as coxswain of LCVP #15 in the fourth assault wave from USS *Neville.* (Cargo ships assigned to transport divisions carried boats used in assault landings, and supplied ship's personnel for their crews.)[23]

Two regiments from the U.S. Army 27th Infantry Division landed at Butaritari and, three days later following light losses—64 killed and 150 wounded—signaled "Makin taken." The 2nd Marines on Tarawa, facing a much larger enemy force, suffered heavy casualties—871 killed, an additional 124 men who would later succumb to their wounds, and 2,306 wounded or missing in action.[24]

Map 13-5

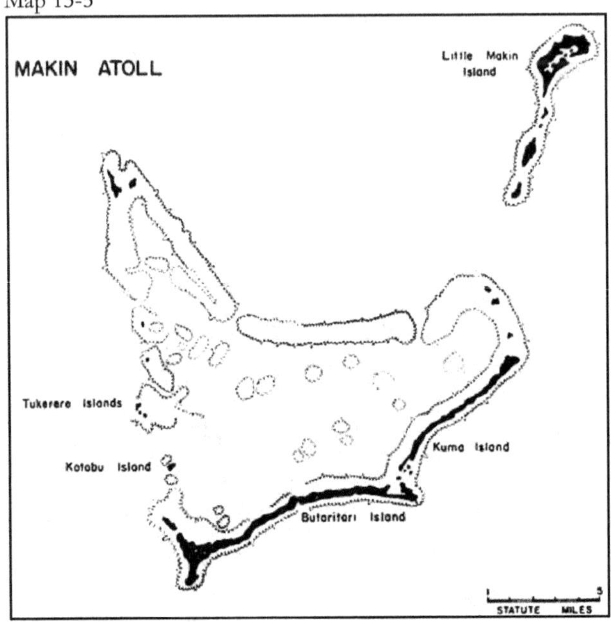


Makin Atoll
http://webdoc.sub.gwdg.de/ebook/p/2005/CMH_2/www.army.mil/cmh-pg/books/wwii/makin/m3.jpg

## BATTLE STARS FOR CARGO SHIPS

Eight cargo ships would earn battle stars for Tarawa, Makin, and other operations associated with assault and occupation of the Gilbert Islands.

**Gilbert Islands Operations**

| Ship | Award Period | Commanding Officer |
|---|---|---|
| *Auriga* (AK-98) | 13 Nov-8 Dec 43 | Lt. Comdr. John Charles Hart, USNR |
| *Hercules* (AK-41) | 17 Nov-8 Dec 43 | Comdr. William H. Turnquist, USNR |
| *Jupiter* (AK-43) | 21 Nov-8 Dec 43 | Lt. Comdr. Duncan Scott Baker, USN |
| *Alcyone* (AKA-7) | 13 Nov-2 Dec 43 | Comdr. James Bertrand McVey, USN |
| *Bellatrix* (AKA-3) | 20 Nov-7 Dec 43 | Comdr. Charles A. Joans, USNR |
| *Thuban* (AKA-19) | 20-29 Nov 43 | Comdr. James C. Campbell, USNR |
| *Virgo* (AKA-20) | 20-29 Nov 43 | Comdr. Clayton H. McLaughlin, USNR |
| *Castor* (AKS-1) | 17 Nov-8 Dec 43 | Comdr. Harold Burton Herty, USN |

# 14

# Marshall Islands Occupation

*We were told that this entire operation involved considerably less than 200,000 Navy personnel and a very high percentage of the entire U.S. Fleet in the Pacific. From this it would seem that the entire Navy afloat cannot encompass much more than a half million officers and men. Where are the remaining one and one-half to two million the Navy is supposed to have? Certainly, somewhere many thousands can be gotten from shore billets or soft and now practically useless local defense jobs and relieve the pressure upon vexed and worried commanding officers. Let the shore-going commands suffer a bit and give the sea-going Navy a break. As far as special schools for transport personnel are concerned most of us say, "Give us the men and send the school personnel to sea to learn how the job is done."*

—Capt. Herbert B. Knowles, commander, Southern Transport Group, expressing frustration about manning levels, in his report on the capture and early occupation of Kwajalein. Attack cargo ships were part of the Transport Divisions within his group, and of other groups.[1]

Occupation of the Marshall Islands, which took place over the latter part of January and early February 1944, was a logical step forward in the advance of U.S. forces northward through the Gilbert and Ellice Islands. The probability of the Marshall Islands being the Allies next objective, must have been known to the Japanese, but uncertainty about the specific objectives, was likely fostered by the geography of the area. The Marshall Island group covered roughly an area some 180,000 miles square, consisting of many islands of various sizes. The largest atoll in the Marshalls and in the world is Kwajalein, which has a land area of only six square miles but surrounds a 655-square-mile lagoon.[2]

The general plan was for a Joint Expeditionary Force (Task Force 51) to capture three key points in the Marshalls, in order to establish air and naval bases from which to exercise dominance over the rest of the island group. The objectives were Roi and Kwajalein islands, about 45 miles apart on Kwajalein Atoll—which was believed to be the key

Japanese defense point of the Marshalls—and the lightly protected Majuro Atoll, 250 miles to the southeast.[3]

Map 14-1

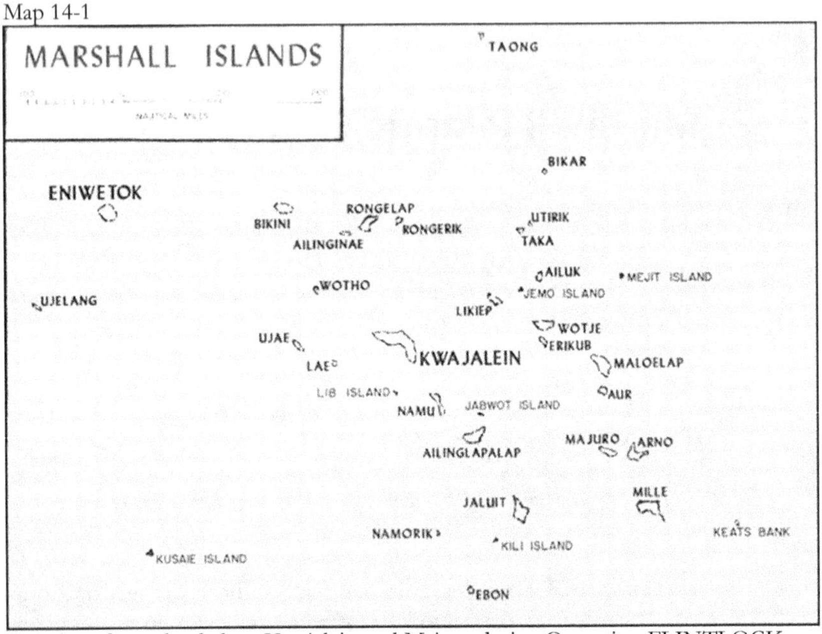

American forces landed on Kwajalein and Majuro during Operation FLINTLOCK, bypassing the other Japanese-held atolls in the Marshall Islands
http://www.ibiblio.org/hyperwar/USMC/III/maps/USMC-III-7.jpg

Kwajalein Atoll was a triangular grouping of ninety-three small islands. Because of the vast size of the atoll, Spruance's Expeditionary Force was split into Northern and Southern Landing Forces. The 4th Marine Division was tasked to take Roi-Namur (twin islands considered to be a single island) in the northeast quadrant of the atoll, while the Army's 7th Infantry Division seized the island of Kwajalein itself (44 nautical miles south of Roi-Namur) in the extreme southeast end of the atoll. Majuro Atoll on the eastern edge of the Marshalls, would also be seized in order to provide a fleet base and airfield for subsequent operations.[4]

This selection decision was Admiral Nimitz's alone. U.S. Navy commanders and planners were cautious after the bloodbath on Tarawa, in November 1943. Vice Adm. Raymond A. Spruance (commander, Fifth Fleet), Rear Adm. Richmond K. Turner (commander, Fifth Amphibious Force), and Marine Maj. Gen. Holland McTyeire "Howling Mad" Smith wanted to take the islands of Wotje and Maloelap in the southern Marshalls before attempting the heavily defended Kwajalein

Atoll. After Nimitz overruled them, Spruance and Turner continued to argue the point, until the Pacific Fleet commander, in his gentlemanly manner, offered to replace them if they didn't want to carry out the seizures, dubbed Operation FLINTLOCK.[5]

Vice Adm. Raymond A. Spruance was in overall command, with subordinate task force commanders assigned as shown in the following table. The actual attacks, landings, and seizures of all three objectives were carried out by Rear Adm. Richmond K. Turner's Expeditionary Force. This force comprised some 297 vessels with almost 84,415 troops embarked. Included were transports, troops, and cargo ships, as well as support naval vessels such as battleships, escort carriers, heavy and light cruisers, destroyers, etc.—needed for protection en route, at the objective, and for support of the landings—and a large number of miscellaneous craft. Of all these ships, only the cargo ships (the subject of this book) are identified in the table, a convention generally followed henceforth, so as to streamline material associated with the many operations in which they participated.

**U.S. Fifth Fleet (TF 50): Vice Adm. Raymond A. Spruance**
**Joint Expeditionary Force (TF 51): Rear Adm. Richmond K. Turner**
**Expeditionary Troops (TF 56): Maj. Gen. Holland McTyeire Smith**
**Carrier Force (TF 58): Rear Adm. Marc A. Mitscher**
**Neutralization Group (TG 50.15): Rear Adm. Ernest G. Small**
**Defense and Land Based Air (TF 57): Rear Adm. John H. Hoover**

| **Northern (Roi-Namur) TF 53: Rear Adm. Richard L. Conolly** | **Southern (Kwajalein) TF 52: Rear Adm. Richmond K. Turner** |
|---|---|
| Transports (53.2): | Advance Transport Unit (TU 52.5.1): |
| Capt. Pat Buchanan | Capt. John B. McGovern |
| TransDiv 24: Capt. Pat Buchanan | TransDiv 4: Capt. John B. McGovern |
| USS *Aquarius* (AKA-16) | USS *Virgo* (AKA-20) |
| TransDiv 26: Capt. A. D. Blackledge | Southern Transport Group (TG 52.5) |
| USS *Almaack* (AKA-10) | Capt. Herbert B. Knowles |
| TransDiv 28: Capt. Henry C. Flanagan | TransDiv 6: Capt. Thomas B. Brittain |
| USS *Alcyone* (AKA-7) | USS *Centaurus* (AKA-17) |
| | TransDiv 18: Capt. Herbert Knowles |
| | USS *Thuban* (AKA-19) |

| **Majuro TG 51.2: Rear Adm. Harry W. Hill** |
|---|
| USS *Electra* (AKA-4)[6] |

The landings and combat ashore by the Northern and Southern Forces at the several main and secondary objectives were complicated and difficult. The U.S. troops put ashore—Army and Marines—bore the brunt of the fighting and sustained the major part of the losses.

However, the assaults and seizures they carried out could not have been accomplished so expeditiously, and with relatively minor losses, without, as Adm. Chester Nimitz noted in his report on Operations in the Pacific Ocean Areas – February 1944, "a bombardment preparation exceeding in duration and intensity anything previously known to warfare, except possibly that at Verdun in World War I." Nimitz further related in amplification of this statement that:

> It has been estimated that 50% of the Japanese defenders were killed by the air, naval, and artillery bombardments prior to the assaults, not to mention the destruction of defenses and equipment and the stunning effect on the morale and fighting capacity of the survivors.[7]

The expeditionary force that simultaneously invaded both ends of Kwajalein Atoll was massive. It included 54,000 U.S. Marine and U.S. Army assault troops making the initial landings, with gunfire support provided by seven pre–World War II battleships, six escort carriers, and numerous cruisers and destroyers (about 300 ships total). The landings had been delayed from early January to the month's end in order to amass enough assault transports to execute them. Additional air support and cover (and suppression of Japanese bases on other atolls in the Marshall Islands) was provided by six fleet carriers and six light carriers with over 700 aircraft, accompanied by seven modern fast battleships, cruisers and destroyers. Japanese aircraft were swept from the skies before the landings took place and all enemy submarines in the area were sunk.[8]

The 4th Marine Division suffered 737 casualties, including 190 killed, and the 7th Infantry, 177 killed and 1,000 wounded, at Kwajalein. Estimated enemy losses totaled 3,472 dead, with 40 Korean laborers and 51 Japanese the only survivors. Prior to the invasion of Majuro Atoll, Army scouts had canvased Dalap and Uliga Islands, found no enemy present, and consequently Majuro was occupied without a fight.[9]

## INCREASING REQUIREMENTS FOR CARGO SHIPS

As the war in the Pacific expanded with combat on increasingly more fronts, there was an associated requirement for more and more cargo ships to support these operations, and to supply occupational forces ashore. Of the eleven USN cargo ships that earned battle stars for the occupation of Kwajalein and Majuro Atolls, six have already been mentioned in this book. For this operation, five, which had no previous reference in this book, were added by the Navy: *Alkes*, *Caelum*, *Electra*, *Mercury*, and *Rutilicus*.

Photo 14-1

USS *Almaack* (AK-27) at the Norfolk Navy Yard, 30 December 1941.
National Archives photograph #19-N-28324

**Cargo Ship Battle Stars for Occupation of Kwajalein and Majuro Atolls**

| Ship | Award Period | Ship Type | Comm |
|---|---|---|---|
| *Alkes* (AK-110) | 31 Jan-8 Feb 44 | EC2-S-C1 hull | 29 Oct 43 |
| *Caelum* (AK-106) | 3-8 Feb 44 | EC2-S-C1 hull | 22 Oct 43 |
| *Mercury* (AK-42) | 31 Jan-8 Feb 44 | C2 hull | 1 Jul 42 |
| *Rutilicus* (AK-113) | 3-8 Feb 44 | EC2-S-C1 hull | 30 Oct 43 |
| *Alcyone* (AKA-7) | 31 Jan-8 Feb 44 | C2 hull | 15 Jun 41 |
| *Almaack* (AKA-10) | 31 Jan-8 Feb 44 | C3E hull | 15 Jun 41 |
| *Aquarius* (AKA-16) | 31 Jan-8 Feb 44 | C2-S-B1 hull | 21 Aug 43 |
| *Centaurus* (AKA-17) | 31 Jan-5 Feb 44 | C2-S-B1 hull | 21 Oct 43 |
| *Electra* (AKA-4) | 31 Jan-8 Feb 44 | C2-T hull | 17 Mar 42 |
| *Thuban* (AKA-19) | 31 Jan-8 Feb 44 | C2-S-B1 hull | 10 Jun 43 |
| *Virgo* (AKA-20) | 31 Jan-4 Feb 44 | C2-S-B1 hull | 10 Jul 43 |

The *Alkes*, *Caelum*, and *Rutilicus* were Navy acquired and recently commissioned Liberty ships, the name given to the EC2 type ship designed for "Emergency" construction by the United States Maritime Commission, which President Franklin Delano Roosevelt nicknamed "ugly ducklings." The *Electra* and *Mercury* were older C2 type cargo ships, earning their first Pacific Theater battle stars. (*Electra* had an earlier one from the North Africa Occupation. As the war progressed, the Navy dispatched more and more cargo ships from other theaters to the Pacific.) Five of the remaining six cargo ships that took part in the assaults on Kwajalein and Majuro Atolls were also C2s. *Almaack*, the only C3E cargo ship of the group, was able to make 16.5 knots, laudable for a large, single-screw vessel.

| Hull | Production | Description |
|---|---|---|
| C2 | 19 ships, 9,758 DW tons | 459 feet, turbine or diesels, 15.5 knots |
| C2-S-B1 | 123 ships, 7,907 DW tons | 459 feet, turbine engines, 15.5 knots |
| C2-T | 3 ships, 8,656 DW tons | 459 feet, diesel engines, 15.5 knots |
| C3-E | 465 ships, 9,514 DW tons | 473 feet, turbine engines, 16.5 knots |
| EC2-S-C1 | 2,711 Liberty ships, 10,800 DW tons | Mass produced cargo ships, 441 feet, steam-reciprocating engines, 11 knots[10] |

DW: Dead weight

*Almaack*, and the other Navy C2 and C3 cargo ships were svelte, nimble vessels in comparison to the mass-produced, pedestrian 11-knot Liberty ships. However, the stout Liberties—shorter in length, narrower of beam, but deeper of draft—could carry more tonnage than the others. As hundreds of Liberties poured out of their builders' yards, many were dispatched to the Pacific Theater.

Photo 14-2

Liberty-type cargo ship USS *Caelum* (AK-106), circa 1945, location unknown. Naval History and Heritage Command photograph #NH 84657

Jumping ahead in the chronology of the war to make a point, an astounding eighty-seven USN cargo ships and attack cargo ships would earn battle stars for the assault and occupation of Okinawa. Many of these were Liberties, and some Victory ships (successors to the Liberty ships), as well as older, other type cargo ships.

## CAPTURE OF ENIWETOK (CATCHPOLE)

*Finally we killed them all. There was not much jubilation. We just sat and stared at the sand, and most of us thought of those who were gone—those whom I shall remember as always young, smiling and graceful, and I shall try to forget how they looked at the end, beyond all recognition.*

—Lt. Cord Meyer, USMC, describing the fighting on Parry Island during the assault of Eniwetok Atoll.[11]

With Kwajalein taken, the next objective was to capture Eniwetok Atoll. Located 326 miles west-northwest of Roi—and only 1,000 miles from the Mariana Islands—even the name of this western outpost of the Marshalls, which meant "land between the east and west," suggested the strategic importance of Eniwetok. Military commanders intended to use the atoll as an important staging point for U.S. Army, Navy, and Marine Corps forces in their progress from east to west toward Japan. The plan to capture Eniwetok and some lessor Marshall islands was code named Operation CATCHPOLE.[12]

The expeditious and successful operations against Kwajalein and Majuro made it possible to advance the planned date for the attack on Eniwetok by nearly three months. Rear Adm. Harry W. Hill, USN, commanded the Eniwetok Expeditionary Group, charged with the capture and occupation of Eniwetok. Brig. Gen. Thomas Eugene Watson, USMC, was in command of the Expeditionary Troops and the Eniwetok Landing Force.[13]

This force of eighty-nine vessels to seize Eniwetok was organized at Kwajalein, around the existing Expeditionary Reserve Force under Capt. Donald W. Loomis.

**Transport Group 51.14: Capt. Donald W. Loomis, USN**

Transports (Task Element 51.14.1): Capt. Donald W. Loomis
TransDiv 20 (Task Element 51.14.2): Capt. Donald W. Loomis
*Leonard Wood* (APA-12), *Heywood* (APA-6), *Arthur Middleton* (APA-25), *President Monroe* (AP-104), *Electra* (AKA-4)

TransDiv 30 (Task Element 51.14.3): Capt. Clinton A. Misson, USN
*Custer* (APA-40), *Neville* (APA-9), *Wharton* (AP-7), *Ashland* (LSD-1), *Mercury* (AK-42)[14]

The 22nd Marine Regiment, and the 1st and 3rd Battalions, 106th Infantry Regiment—5,760 Marines and 4,509 Army troops—sailed for Eniwetok Atoll, aboard transports that were largely veterans of the Guadalcanal and the Gilbert Island operations.[15]

On 17 February 1944, the 22nd Marine Regiment landed at Engebi Island on the north end of Eniwetok Atoll. The thousand or more Japanese defenders offered little resistance following prelanding shore bombardment, and Engebi was taken by late afternoon the next day, at the cost of eighty-five Marines killed.[16]

In southern operations against Eniwetok and Parry islands the opposing forces were much larger and, on Parry, enemy use of "hidey-hole" concealment made the combat more difficult. Having learned, via captured documents, that these islands were defended by troops of Maj. Gen. Yoshimi Nishida's 1st Amphibious Brigade, it was decided that both battalions of the 106th Army Infantry Regiment and a Marine Reserve battalion would assault Eniwetok Island instead of a lesser force. Fierce fighting ensued and, after much effort, the island was secured at 1630 on 21 February. The cost in American lives was 37 Marines killed and another 94 wounded. The Japanese garrison of about 800 men, except for 23 soldiers taken prisoner, was annihilated.[17]

At Parry Island, naval gunfire ships offshore had a lot of time to pound it before Marines began coming ashore at 0900 on 22 February. Assault troops pushed forward behind tanks, with demolition and flame-thrower parties directly behind to destroy each enemy nest. By that afternoon, the Marines advanced to the island's southern end where the remaining Japanese were in the open. Eniwetok Atoll including Eniwetok and Parry Islands was taken at a cost of 339 Americans killed and missing. Almost all the 3,500 Japanese defenders on the islands were killed and only about 100 taken prisoner.[18]

After these battles, many atolls in the Marshall Islands were still occupied by Japanese forces remained, but only four hosted airbases. The Fifth Fleet had bypassed Jaluit, Mili, Maloelap, and Wotje, which could be isolated until war's end through the use of Allied air power.[19]

**15**

# Admiralty Islands Landings and Hollandia Operation

*Within a week after the initial landing the First Cavalry Division had buried a Jap[anese] for each cavalryman landed on D-Day, with estimated total garrison of between 4000 and 5000 troops.*

—Commander Attack Group reporting on the Admiralty Islands operation, which was launched with a "reconnaissance in force" at Los Negros Island after an ineffective air reconnaissance had revealed no evidence of human activity. This misconception was discounted two days before the assault when Army scouts, who had gone ashore after being dropped off by a Catalina seaplane, reported that the area was "lousy with Japs."[1]

Photo 15-1

Manus Naval Base, Admiralty Islands, 19 May 1944.
National Archives photograph #80-G-254863

The capture and utilization of the powerful Japanese base at Rabaul, located on the northeast coast of New Britain, had been the main objective of Operation WATCHTOWER (which had opened with the landings at Guadalcanal, in August 1942). However, even after Allied movement up the New Guinea coast and finally into New Britain, the enemy stronghold still had close to 100,000 defenders. The garrison also had enough munitions, weapons, provisions, and supplies to withstand a long siege while inflicting large numbers of casualties on Allied invasion forces. In view of these circumstances, the decision was made to bypass Rabaul and to, instead, leap into the Admiralty Islands, which were ideally situated for development of facilities necessary for isolating Rabaul and to support the approach to the Philippines. Manus, the largest island in the group—spanning approximately 49 miles from east to west and 16 miles from north to south—offered ample space for military installations as well. Seeadler Harbor on its northeast coast, could accommodate a task force. Los Negros Island, which is located adjacent to Manus, forming the enclosure for the eastern half of the harbor, had sufficient flat land to construct a significant airfield that would enable the Allies to deny the Japanese access to the Bismarck Archipelago. Also, it would allow them to dominate a 1,000-mile square of neighboring ocean the corners of which were Bougainville, Solomon Islands; Truk in the Caroline Islands; the Palau Islands; and Biak Island off northwestern New Guinea.[2]

On the morning of 29 February 1944, the U.S. 1st Army Cavalry Division and detachments totaling 1,026 troops burst ashore at Hyane Harbor, on the eastern side of Los Negros Island. Planners had chosen the small, nearly landlocked Hyane, instead of the much larger and more accommodating Seeadler Harbor (some 15 miles long by 4 wide), or good beaches on the southeast coast of Los Negros, because the site was not an obvious choice and thus would likely not be as heavily defended. This assumption proved correct as the landing force encountered little enemy opposition during daylight on D-Day. However, it was heavily attacked that night and on the next. With the aid of Seventh Fleet gun batteries, abundant air power, and additional ground troops, the Americans advanced from Los Negros into Manus. By 3 April they were in control of Seeadler Harbor. A better base than Rabaul, and nearer Japan, the Admiralties became one of the most important staging points in the last fifteen months of the Pacific war.[3]

Map 15-1

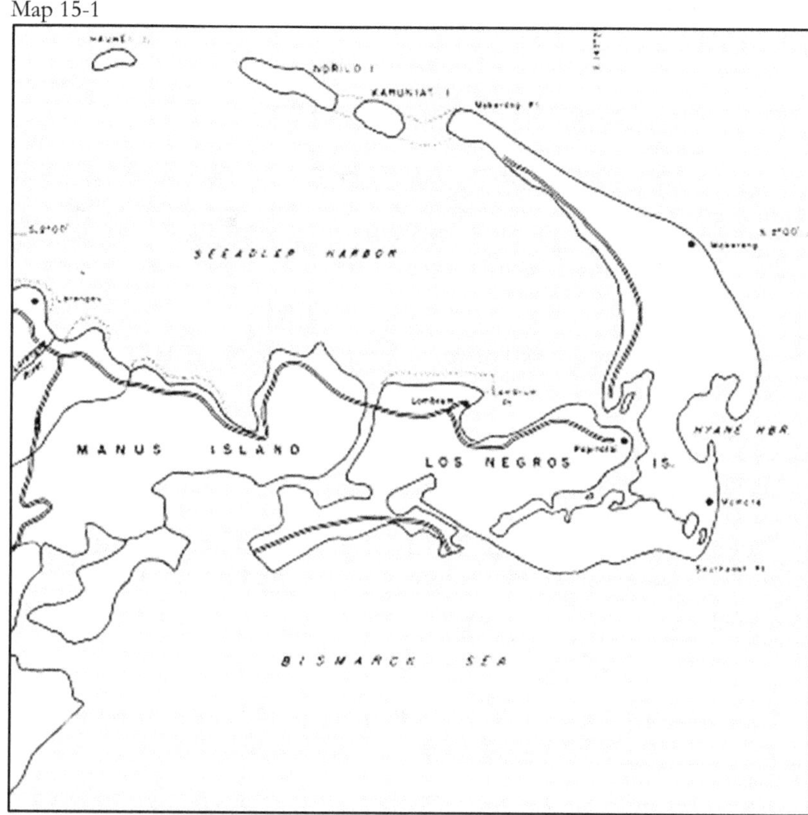


Manus and Los Negros, Admiralty Islands
http://www.ibiblio.org/hyperwar/USN/Building_Bases/maps/bases2-p297.jpg

## CARGO SHIP BATTLE STARS

Three cargo ships received battle stars for their participation in the Admiralty Island landings, during the periods identified in the table.

**Bismarck Archipelago Operation: Admiralty Island landings**

| Ship | Award Period | Ship Type |
|---|---|---|
| USS *Celeno* (AK-76) | 24-28 Mar 44 | EC2-S-C1 (Liberty) |
| USS *Aquarius* (AKA-16) | 11-12 Apr 44 | C2-S-B1 hull |
| USS *Centaurus* (AKA-17) | 5-9 Apr 44 | C2-S-B1 hull |

## HOLLANDIA OPERATION

> *Southwest Pacific Forces, covered by the Pacific Fleet will seize and occupy by simultaneous overwater operations, the AITAPE, HUMBOLDT BAY – TANAHMERAH BAY areas of New Guinea; isolate hostile forces to the eastward; establish at AITAPE minor air and naval facilities and in the HUMBOLDT - TANAHMERAH BAY areas a major air base, minor naval base, and an intermediate supply base for the purpose of supporting further operations to the westward.*
>
> —Commander, Task Force 77 Operation Plan No. 3-44, 3 April 1944.

All vessels shown in following tables are U.S. Navy ships except those specially noted as Australian (HMAS) ships.

**Western Attack Group (TG 77.1): Rear Adm. Daniel A. Barbey**
**Tanahmerah Bay landing force on Red Beaches**

Transports: Capt. Stevens
*Henry T. Allen* (APA-15), HMAS *Manoora* (F48), HMAS *Kanimbla* (F23), *Carter Hall* (LSD-3), *Triangulum* (AK-102)

**Central Attack Group (TG 77.2): Rear Adm. William M. Fechteler**
**Humboldt Bay landing force on White Beaches**

Transports: Comdr. Knight
HMAS *Westralia* (F95), *Humphreys* (APD-12), *Brooks* (APD-10), *Sands* (APD-13), *Gilmer* (APD-11), *Herbert* (APD-22), *Gunston Hall* (LSD-5), *Ganymede* (AK-104)

**Eastern Attack Group (TG 77.3): Capt. Alfred G. Noble**
**Aitape Bay landing force on Blue Beach**

Transports: Comdr. Mattie
*Kilty* (APD-15), *Ward* (APD-16), *Crosby* (APD-17), *Dickerson* (APD-21), *Talbot* (APD-7), *Schley* (APD-14), *Kane* (APD-18), *Dent* (APD-9), *Noa* (APD-24), *Belle Grove* (LSD-2), *Etamin* (AK-93)[4]

## RAN INFANTRY LANDING SHIPS

Assigned as part of the Western and Central Attack Groups were the Royal Australian Navy's three infantry landing ships—HMAS *Manoora*, HMAS *Kanimbla*, and HMAS *Westralia*—which had interesting pedigrees. The RAN, having decidedly fewer resources and ships than the USN, but needing to meet expansive defense requirements, acquired merchant vessels to help meet its original critical needs. Later, progress of the war brought new requirements, resulting in different usages of some vessels. Such was the case of these three ships, built as passenger vessels, and acquired by the RAN, then converted to armed merchant cruisers following the outbreak of war in 1939.

| HMAS *Manoora* | HMAS *Kanimbla* | HMAS *Westralia* |
|---|---|---|
| 12 Dec 39 – 6 Dec 47 | 6 Oct 39 – 25 Mar 1949 | 17 Jan 40 – 19 Sep 46 |
|   |   |  |
| Battle Honours | Battle Honours | Battle Honours |
| INDIAN OCEAN 1941-42 | | |
| PACIFIC 1942-45 | PACIFIC 1941-45 | PACIFIC 1941-45 |
| NEW GUINEA 1944 | NEW GUINEA 1942-44 | NEW GUINEA 1942-44 |
| LEYTE GULF 1944 | LEYTE GULF 1944 | LEYTE GULF 1944 |
| LINGAYEN GULF 1945 | LINGAYEN GULF 1945 | LINGAYEN GULF 1945 |
| BORNEO 1945 | BORNEO 1945 | BORNEO 1945 |

Photo 15-2

Painting by Frank Norton depicting HMAS *Manoora* intercepting the Italian merchant vessel *Romolo* in the Pacific Ocean, two days after Italy entered the war. *Romolo*'s crew set their ship aflame to prevent her capture, and took to lowered lifeboats.
Australian War Memorial photograph ART22328

The armed merchant cruisers were converted to infantry landing ships (LSIs), in 1943, for participation in United States and Australian amphibious assaults in the Southwest Pacific Area. In March 1942, the Australian Government, recognizing the importance of an amphibious capability to drive the Japanese out of the Pacific, had begun exploring, the requirements for Combined Operations training in Australia. ("Combined Operations" refer to operations conducted by military forces of two or more Allied nations acting together. "Joint Operations" are when military forces from the same nation work together, a term used elsewhere in the book.)[5]

Two training centres were established by the end of 1942: an Australian Army Combined Training Centre at Toorbul Point, Queensland, and a RAN Joint Overseas Operational Training School at Port Stephens (a small fishing village located to the north of Sydney). While construction work progressed ashore at Port Stephens, the school (titled HMAS Assault) commissioned aboard the armed merchant cruiser HMAS *Westralia* and began providing instruction for landing craft crews, beach parties (naval commandos), and operations signals teams. Assault transferred ashore, on 10 December 1942, and the two facilities were combined, in February 1943, as the Amphibious Training Centre, under the overall command of Rear Admiral Barbey.[6]

Photo 15-3

Vice Adm. Thomas C. Kinkaid, USN, commander, Seventh Fleet (right), and Rear Adm. Daniel A. Barbey, USN, commander Amphibious Force, Seventh Fleet, examine a map aboard Barbey's flagship, 5 January 1944. Naval History and Heritage Command photograph 80-G-214898

Photo 15-4

Near HMAS Assault, the shore-based Combined Operations School at Port Stephens, American 32nd Infantry Division troops storm ashore from RAN infantry landing craft *LCI-230* in the course of a practice landing.
Australian War Memorial photograph 304847

HMAS *Ping Wo*, an ex-Chinese steamer introduced to readers earlier in the book, was assigned to the training centre (HMAS Assault) for the transportation of water and stores for the LSIs based there, and also served as a training ship.[7]

Photo 15-5

Stores carrier HMAS *Ping Wo* plying between Sydney and Port Stephens while operating as a tender assigned to HMAS Assault, 28 September 1942.
Australian War Memorial photograph 301176

On the eve of the Hollandia operation, Comdr. Allan P. Cousin, RANR(S), was the commanding officer of HMAS *Manoora* and, as Senior Naval Officer, Australian Landing Ships (LSIs), he was responsible for *Kanimbla* and *Westralia* as well. Once converted to infantry landing ships, *Manoora, Kanimbla,* and *Westralia* each had a capacity of about 1,200 troops, landed from boats carried aboard the LSIs. The efficiency of the Australian LSIs in the subsequent operations drew widespread praise and Cousin, who held a Royal Australian Naval Reserve (Seagoing) commission, was promoted to Acting Captain in February 1945. He was also awarded a Distinguished Service Order Medal, for his "gallantry, fortitude and skill" during the amphibious assaults.[8]

## HOLLANDIA, NEW GUINEA OPERATION

Map 15-2

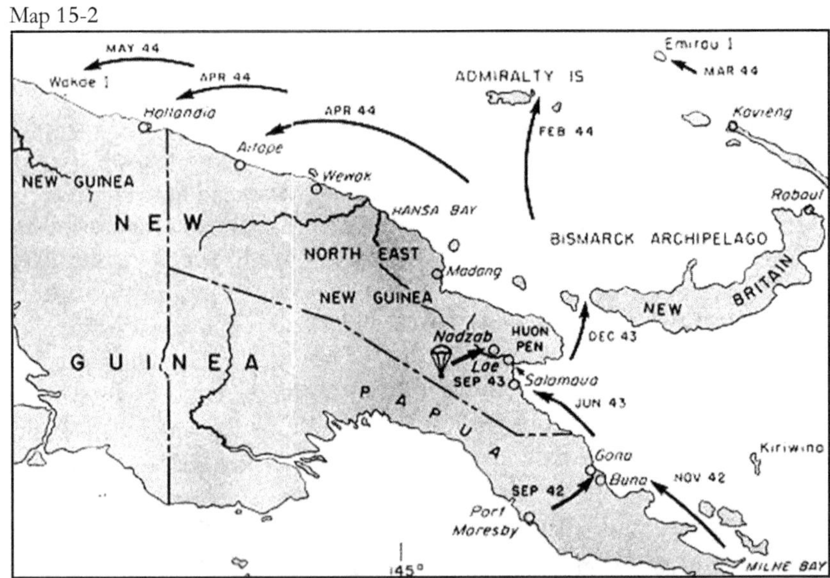


MacArthur's Australian and American forces advance up the east coast of Papua, New Guinea, along his road back to the Philippines (http://www.history.army.mil/books/AMH/Map23-43.jpg)

On 22 April 1944, the Allies launched the largest amphibious operation then carried out in the Southwest Pacific involving over two hundred ships. Intended to initiate the final stages of the isolation of Rabaul, it involved concurrent landings at three locations on the northwest coast of New Guinea. The western landing took place at Tanahmerah Bay, about thirty miles to the northwest of Hollandia, the site of an enemy army and air force base. The central landing was thirty miles to the east in Humboldt Bay, and the eastern landing about ninety miles farther eastward at Aitape. Hollandia, the name by which the entire area would become known to American forces, was a tiny settlement nestled at the head of Challenger Cove, an arm of Humboldt Bay. It was formerly the eastern-most Dutch outpost in the Netherlands East Indies.[9]

Map 15-3

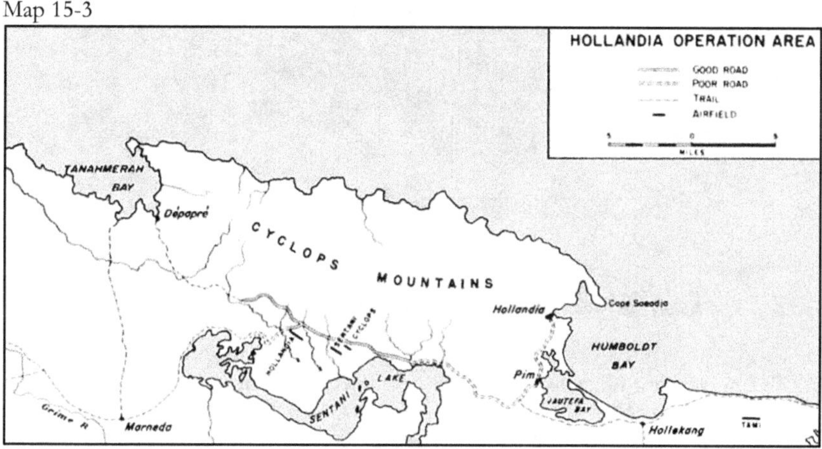


A closer view of the Tanahmerah – Humboldt Bay area
www.ibiblio.org/hyperwar/USA/USA-P-Approach/maps/USA-P-Approach-2.jpg

All the landings were a complete surprise. An intense pre-assault air and naval bombardment broke the back of any resistance, and little Japanese opposition resulted in practically no casualties during the landings. The enemy at Humboldt Bay retired into the hills and a party at Aitape left their defenses prior to the assault. Two Aitape airfields were captured, by 22 April, and fighter operations by the Allies commenced two days later. The airfields at Hollandia were taken, on 26 April 1944, and one airstrip was ready for use at month's end.[10]

Photo 15-6

Tank landing ships unloading on Beach Red Two, Tanahmerah Bay, Hollandia, New Guinea, on 22 April 1944.
National Archives photograph #80-G-251402

## CARGO SHIP BATTLE STARS

**Hollandia Operation: Aitape-Humboldt Bay-Tanahmerah Bay**

| Ship | Award Period | Ship Type |
|---|---|---|
| USS *Bootes* (AK-99) | 21 Apr-1 May 44 | EC2-S-C1 (Liberty) |
| USS *Ganymede* (AK-104) | 21-27 Apr 44 | EC2-S-C1 (Liberty) |
| USS *Triangulum* (AK-102) | 21-27 Apr 44 | EC2-S-C1 (Liberty) |
| USS *Centaurus* (AKA-17) | 21 Apr-5 May 44 | C2-S-B1 hull |
| USS *Virgo* (AKA-20) | 21 Apr-5 May 44 | C2-S-B1 hull |

16

# Capture of Saipan, Tinian, and Guam

*The presence of many transports in the waters off SAIPAN required vigilant and constant patrol activity for their protection. Destroyers, destroyer escorts, converted destroyers, LCI(G)s [gunboats], and SCs [sub-chasers] participated in these operations. They screened the transport area as the ships unloaded, accompanied them to sea on retirement, maintained anti-submarine watch, and patrolled close inshore against enemy amphibious counter-attacks, and evacuation attempts. The freedom of our forces from surface and submarine interference is evidence of the effectiveness of these operations.*

—Commander in Chief, U.S. Pacific Fleet and Pacific Ocean Areas describing in Operations in the Pacific Ocean Area – June 1944, dated 7 November 1944, actions taken to protect the transports (collective term for transports and cargo ships assigned to Transport Divisions) during the invasion of Saipan, Mariana Islands, from enemy attack.

The next Central Pacific Force objective was the southern Marianas, after newly won bases in the Marshalls were consolidated, and as soon as the necessary forces and supplies could be assembled. The Mariana Islands form part of an almost continuous chain of islands extending 1,350 miles southward from Tokyo. A series of mutually supporting airfields and bases, sited on islands of military value, afforded protected lines of air and sea communications from the home islands of the Japanese Empire through the Bonin and Volcano Islands and Marianas to Truk; then to the eastern Carolines and Marshalls, as well as to the western Carolines, the Philippines, and Japanese-held territory to the south and west. Allied occupation of the Marianas would effectively cut these protected lines of enemy communication, provide bases for control of sea areas farther west, and enable bombers to strike Tokyo and home islands of the Empire.[1]

The Marianas Operation, spanning a three-month period from early June until the end of August 1944, was divided into three phases: the capture of Saipan, followed by the capture of Guam, and then

Tinian. Admiral Spruance was in overall command, the amphibious forces were under Vice Admiral Turner, and the expeditionary forces were commanded by Marine Lieutenant General Smith. More than 600 vessels—ranging from battleships and aircraft carriers to cruisers, high-speed transports and tankers—greater than 2,000 aircraft, and some 300,000 Navy, Marine, and Army personnel took part in the capture of the Marianas.[2]

Map 16-1

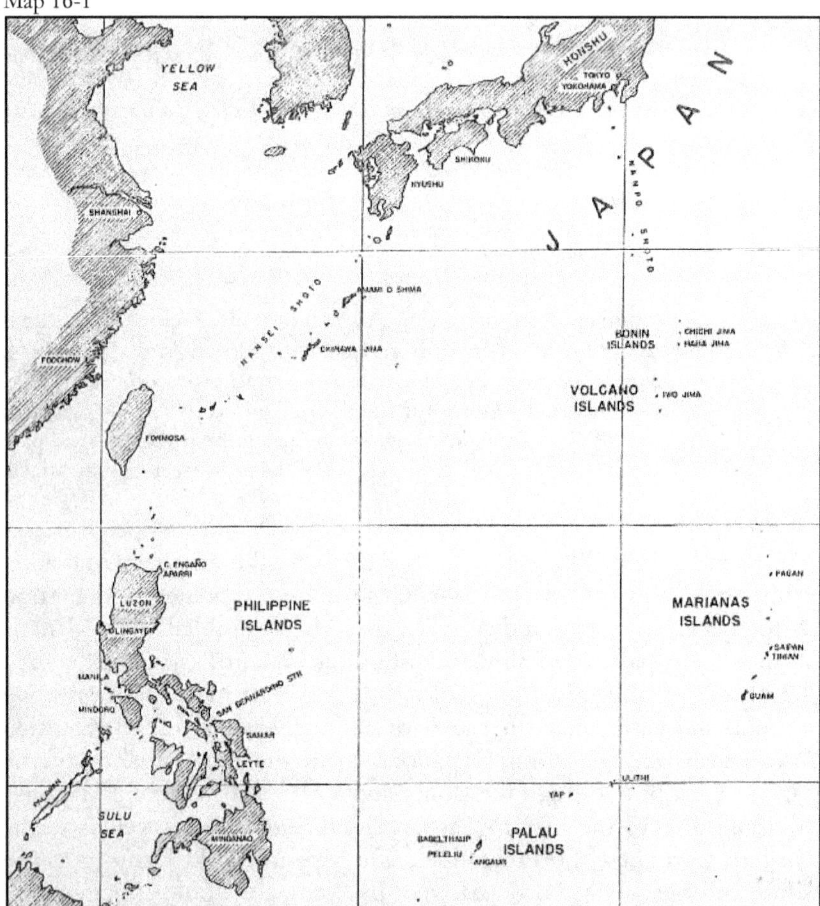


Southern Japanese islands and surrounding areas
*United States Navy at War, Second Official Report to the Secretary of the Navy Covering Combat Operation March 1, 1944, to March 1, 1945* by Fleet Admiral Ernest J. King

## PRE-ASSAULT LANDING PREPARATIONS

Map 16-2

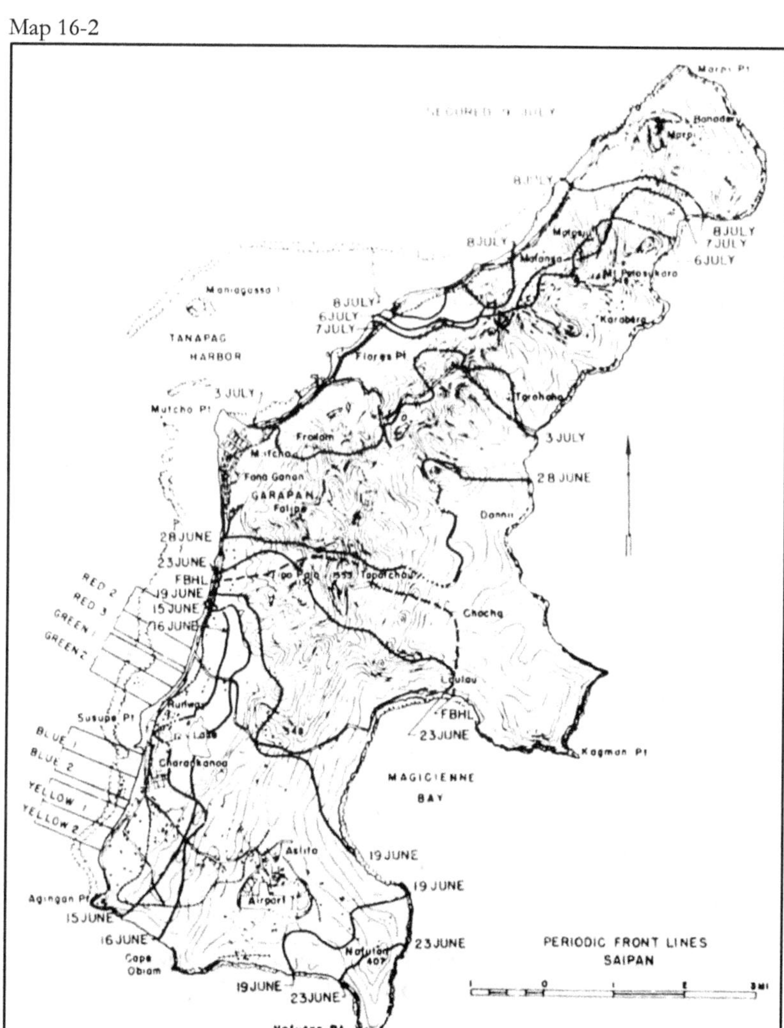


Saipan Island
Secret Information Bulletin No. 20: Battle Experience Supporting Operations for the Capture of the Marianas Islands (Saipan Guam and Tinian) June – August 1944, United States Fleet Headquarters of the Commander in Chief

Because enemy airbases on Marcus and Wake islands flanked the planned approach of Spruance's forces to the Marianas on the north, U.S. carrier aircraft struck these targets, on 19-20 May and 23 May, respectively. From about the beginning of June, land-based aircraft from the Admiralties, Green Island, Emirau Island, and Hollandia in New Guinea, neutralized enemy bases, particularly at Truk, Palau, and Yap. As the scheduled landings on Saipan drew nearer, U.S. Fifth Fleet carriers and battleships opened the way for the amphibious assault by striking targets on that island. After achieving air control over the Marianas, on 11 June, the fleet and its aircraft attacked enemy air facilities, defense installations, and shipping in the vicinity.[3]

## INITIAL MARIANA ISLANDS ASSAULT TIMELINE

June 15th was designated as D-Day, the day on which the initial landings on Saipan would take place. The opening day of assault on Guam and Tinian were denoted W-Day and J-Day. For planning purposes, these dates were considered to be D+3 (18 June) for Guam, and D+20 (5 July) for Tinian. However, because strong enemy resistance encountered by U.S. forces on Saipan necessitated using the entire floating reserve for that attack, W and J Days were considerably delayed.[4]

## CARGO SHIPS / ATTACK CARGO SHIPS ASSIGNED

The enormity of the planned assaults on the southern Mariana Islands, were evidenced by the sizable number of transports and cargo ships, assigned to deliver large numbers of Marine and Army combat troops, equipment, and supplies to their landing beaches.

Because of the theme of this book, only the cargo ships (AKs and AKAs) are identified in the below abbreviated task force organization—and not the transports which comprised the bulk of the Transport Divisions to which the cargo ships were assigned. The award periods associated with battle stars earned by the cargo ships are also cited, as are the hull types of the AKs and AKAs (only one was a Liberty ship).

**Task Force 51 (Joint Expeditionary Force):**
**Vice Adm. Richmond K. Turner**

**Task Force 52 (Northern Attack Force):**
**Vice Adm. Richmond K. Turner**
**Demonstration Group (TG 52.9): Capt. G. D. Morrison**

| Cargo Ship | Award Period | Hull Type |
|---|---|---|
| USS *Alhena* (AKA-9) | 15-22 Jun 44 | C2-S hull |
| USS *Jupiter* (AK-43) | 15-26 Jun 44 | C2 hull |
| USS *Hercules* (AK-41) | 15-26 Jun 44 | C3-E hull |
| **Transport Division 30 (TU 52.9.2): Capt. Clinton A. Misson** | | |
| USS *Bellatrix* (AKA-4) | 16-22 Jun 44 | C2 hull |

**Western Landing Group (TG 52.2): Rear Adm. Harry W. Hill**
**Transport Group Able (TG 52.3): Capt. Herbert B. Knowles**

| TransDiv 10 (TU 52.3.1): Capt. G. D. Morrison | | |
|---|---|---|
| USS *Alhena* (AKA-9) | 15-22 Jun 44 | C2-S hull |
| USS *Jupiter* (AK-43) | 15-26 Jun 44 | C2 hull |
| USS *Hercules* (AK-41) | 15-26 Jun 44 | C3-E hull |
| **TransDiv 18 (TU 52.3.2): Capt. Herbert B. Knowles** | | |
| USS *Alcyone* (AKA-7) | 15-22 Jun 44 | C2 hull |
| **TransDiv 28 (TU 52.3.3): Capt. Henry C. Flanagan** | | |
| USS *Electra* (AKA-4) | 15-26 Jun 44 | C2-T hull |
| **Transport Group Baker (TG 52.4): Capt. Donald W. Loomis** | | |
| **TransDiv 20 (TU 52.4.1): Capt. Donald W. Loomis** | | |
| USS *Thuban* (AKA-19) | 15 Jun-28 Jul 44 | C2-S-B1 hull |
| **TransDiv 26 (TU 52.4.2): Capt. R. E. Hanson** | | |
| USS *Almaack* (AKA-10) | 15-24 Jun 44 | C3-E hull |
| **TransDiv 30 (TU 52.4.3): Capt. Clinton A. Misson** | | |
| USS *Bellatrix* (AKA-3) | 16-22 Jun 44 | C2 hull |

**Joint Expeditionary Force Reserve (TG 51.1):**
**Rear Adm. William H. P. Blandy**
**Reserve Group One (TG 51.18): Rear Adm. William H. P. Blandy**

| TransDiv 7 (Temporary) (TU 51.18.1): Capt. C. G. Richardson | | |
|---|---|---|
| USS *Fomalhaut* (AKA-5) | 16-24 Jun 44 | C1-A hull |
| **TransDiv 32 (Temporary) (TU 51.18.2): Capt. M. O. Carlson** | | |
| USS *Auriga* (AK-98) | 16-28 Jun 44 | C1-B hull |
| **Reserve Group Two (TG 51.19): Capt. Charles Allen** | | |
| **TransDiv 14 (Temporary) (TU 51.19.1): Capt. Charles Allen** | | |
| USS *Leonis* (AK-128) | 20 Jun-3 Jul 44 | EC2-S-C1 hull (Liberty) |

**Southern Attack Force (TF 53): Rear Adm. Richard L. Conolly**
**Northern Attack Group (TG 53.1)**
**Northern Transport Group (TG 53.3): Capt. Pat Buchanan**

| TransDiv 2 (TU 53.3.1): Capt. H. D. Baker | | |
|---|---|---|
| USS *Titania* (AKA-13) | 21-26 Jul 45 | C2-F hull |
| **TransDiv 8 (TU 53.3.2): Capt. Frank R. Talbot** | | |
| USS *Libra* (AKA-12) | No BS for Saipan | C2-F hull |
| **TransDiv 12 (TU 53.3.3): Capt. Pat Buchanan** | | |
| USS *Aquarius* (AKA-16) | 15-25 Jun 44 | C2-S-B1 hull |
| **Southern Attack Group (TG 53.2)** | | |
| **Southern Transport Group (TG 53.4): Capt. John B. McGovern** | | |
| **TransDiv 4 (TU 53.4.1): Capt. John B. McGovern** | | |
| USS *Virgo* (AKA-20) | No BS for Saipan | C2-S-B1 hull |
| **TransDiv 6 (TU 53.4.2): Capt. Thomas B. Brittain** | | |
| USS *Centaurus* (AKA-17) | No BS for Saipan | C2-S-B1 hull[5] |

The assault landings at Saipan were carried out by troops of the Western Landing Group, and the Joint Expeditionary Force Reserve. The Northern Attack Force conducted an amphibious feint, designed to draw Japanese defense forces away from the main landings. The Southern Attack Force (intended to assault Guam three days after the landings began on Saipan, before fierce fighting ashore delayed this action) stood by off Saipan per orders of commander, Fifth Fleet. On 25 June, after "marking time" for several days, the force departed for Eniwetok to await orders to proceed to Guam.[6]

Of course, the mere presence of a ship at, or near Saipan could bring danger, as witnessed by the officers and men of the USS *Mercury*, which didn't arrive at Saipan until 26 June, and earned a battle star for the period, 27 June to 2 July 1944. (It appears that a mistake was made, and the commencement date for the award should actually be 26 June.)

## USS *MERCURY* "DOWNS" JAPANESE AIRCRAFT

Photo 16-1

Cargo ship USS *Mercury* (AKS-20/formerly AK-42) under way, 24 November 1951. National Archives photograph #USN 429250

USS *Mercury* (AK-42) arrived at Saipan, on 26 June, after a 3,500-mile voyage from Honolulu, loaded with cargo and embarked troops of the 804th and 805th Army Aviation Engineers. At 2052 that night, as she lay anchored off the island, a flash red warning of an impending enemy air raid sent her crew to General Quarters, and a LCVP (Higgins boat) from the ship left her to lay a protective smoke screen. Later at 2131, after securing from battle stations, the roar of a plane's engine was heard coming in from off her port bow. A twin-engine medium bomber—believed to be a Betty—then broke through the smoke, about 200 yards distant and 90 feet off the water, headed directly for the ship's superstructure.[7]

The plane released a torpedo, which falling, struck *Mercury* and entered the amidships living spaces on the 01 level (first deck above the main deck), port side. With no chance to arm, it broke in two. The pilot, trying to gain altitude, cleared *Mercury*'s stack, but struck her starboard boom at No. 4 hold, causing the plane to spin into the water about 500 yards off the ship's starboard quarter. The body of the demolished torpedo was found in the First Lieutenant's stateroom. Upon impacting the ship, its unexploded warhead had burst open, spreading TNT in powder form over the hatch cover, the port boat deck, and the bridge deck.[8]

Photo 16-2

Entry hole from a Japanese aircraft-dropped torpedo striking USS *Mercury* (AK-42), on the night of 26 June 1944, as she lay at anchor off Saipan. National Archives photograph 80-G-270918

Chief Commissary Steward Clyde E. Saylor, USN, was killed (and buried ashore the following day); two others were wounded in action; and 67 men subsequently required medical care for dermatitis caused by exposure to the powder. Several days later, on 6 July, fifteen crewmen were placed on the ship's "binnacle list" (sick list), for dermatitis venenata caused by TNT poisoning.[9]

Activities aboard *Mercury* in the next few days after the failed torpedo plane attack, were typical of ships discharging cargo off hostile shores—ceaseless work interrupted by calls to "battle stations" for enemy air raids, or warnings of aircraft active in the area:

- 26 June: 2140 general quarters, flash red on orders of CTF 52
  2318 secured from general quarters
- 27 June: 1845 flash red, general quarters on orders of CTF 52
  1902 secured from general quarters
  2056 flash red, general quarters on orders of CTF 52
  2205 secured from general quarters
  2222 flash red, general quarters on orders of CTF 52
  2232 eight BOMBS hit water 500 yards off starboard bow between *Mercury* and *Cambria* (APA-36)
- 28 June: 0012 four BOMBs hit water approximately 3,000 yards northeast of *Mercury*
  0035 secured from general quarters
  2039 flash red, general quarters on orders of CTF 52
  2137 secured from general quarters
- 30 June: 1951 flash red, general quarters on orders of CTF 52
  2020 secured from general quarters
  2046 flash red, general quarters on orders of CTF 52
  2149 secured from general quarters
  2257 flash red, general quarters on orders of CTF 52
  2327 secured from general quarters[10]

Finally, on 1 and 2 July, each day there was only one Flash Red call to General Quarters received aboard *Mercury*, as the final cargo and equipment aboard her were unloaded, and remaining troops were disembarked. At a little past noon on the 2nd, *Mercury* put to sea, and joined Task Group 51.4 for passage to Eniwetok, Marshall Islands. She departed there, on 8 July, with the task group, bound for Pearl Harbor.[11]

All organized resistance at Saipan ended on 9 July.[12]

## CAPTURE OF GUAM

Following heavy, and long-continued surface and air bombardments, U.S. troops landed on beaches on both sides of Apra Harbor, on 21 July. The 3rd Marine Division went ashore on Blue, Green, and Red beaches at Asan, north of Apra Harbor. Beaches Yellow and White at Agat were designated for the 1st Provisional Marine Brigade, and the

U.S. Army 77th Infantry Division. Enemy resistance was determined, but did not, at any time, place in doubt the outcome of the operation. By 10 August, all organized resistance came to an end.[13]

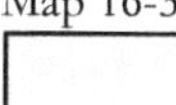

Map 16-3

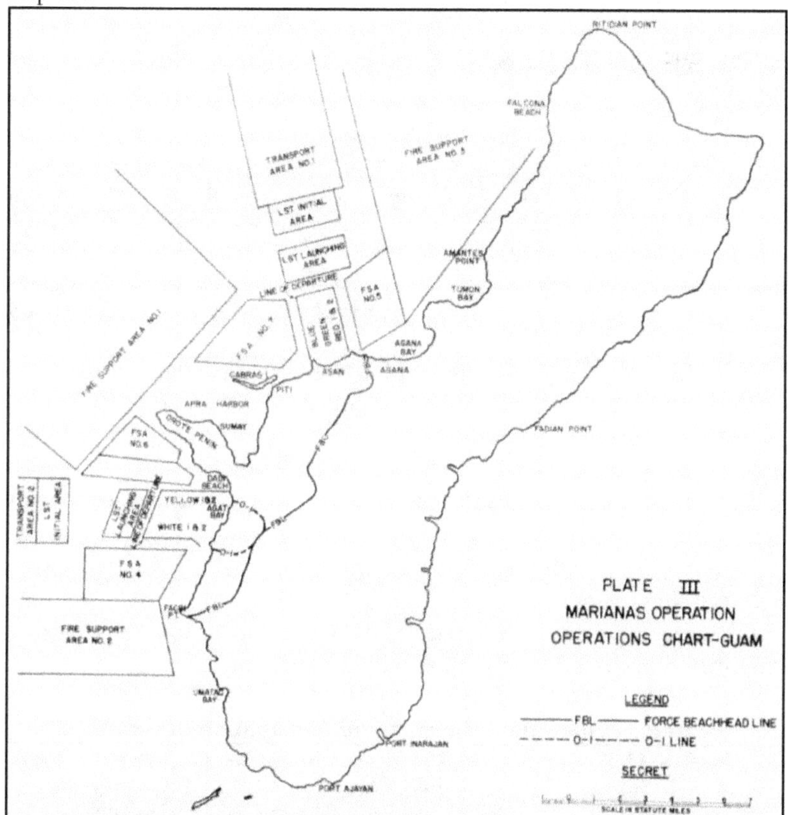


Disposition of U.S. forces for the invasion of Guam
Commander in Chief, U.S. Pacific Fleet and Pacific Ocean Areas, Operations in the Pacific Ocean Area – July 1944, 22 December 1944

Fortunately, there was no fire from the beach by enemy emplaced guns or mobile artillery, on the transport area or initial assault craft wave. Only after the first wave had landed, did mortar fire open on the beaches and boat lanes. The assault transports were unloaded rapidly, unhampered by air attack or forced nightly retirements. During the first four days of the operation, the average material landed daily was 10,000 tons. This was achieved despite poor conditions: there being no passages through a reef edge located 200-500 yards offshore the landing beaches.[14]

Troops and supplies were brought to the edge of the reef in landing craft for transfer to pontoon barges and causeways secured to moorings off the reef. Twenty-five light cranes, mounted on the barges, served admirably in transferring supplies from the landing craft to amphibious vehicles, for movement ashore.[15]

## CAPTURE OF TINIAN

*During the entire operation, we were impressed with the ability of the Japanese to conceal their installations and personnel. In spite of the large number of people on SAIPAN, TINIAN, and GUAM, few were seen from the ship. On the other hand, our troops, tanks, and motor transport, could frequently be seen. Exposure must more often be necessary for troops on the offensive than those on the defensive, and the defensive tactics of the Japanese probably partly explains their concealment.*

—Observation of the commanding officer of the light cruiser USS *Cleveland* (CL-55).[16]

The 2nd and 4th Marine Divisions were used for the assault of Tinian. American casualties between 24 July-1 August 1944 (assault through end of resistance) were light: 1,820 total, with 290 killed, 1,515 wounded, and 24 missing in action.[17]

The cargo ships that took part in the Guam and Tinian operations, are identified in the table; not all received battle stars.

| **Marianas Operation: Capture and Occupation of Guam** | | |
|---|---|---|
| USS *Alkes* (AK-110) | 8-10 Aug 44 | EC2-S-C1 (Liberty) |
| USS *Ara* (AK-136) | 27 Jul-15 Aug 44 | EC2-S-C1 (Liberty) |
| USS *Cor Caroli* (AK-91) | 27 Jul-15 Aug 44 | EC2-S-C1 (Liberty) |
| USS *Draco* (AK-79) | 27 Jul-9 Aug 44 | EC2-S-C1 (Liberty) |
| USS *Rutilicus* (AK-113) | 10 Aug 44 | EC2-S-C1 (Liberty) |
| USS *Sterope* (AK-96) | 21 Jul-9 Aug 44 | EC2-S-C1 (Liberty) |
| USS *Vega* (AK-17) no BS | 1-15 Aug 44 | "Hog Island" ship |
| USS *Alcyone* (AKA-7) no BS | 22-29 Jul 44 | C2 hull |
| USS *Almaack* (AKA-10) no BS | 22-29 Jul 44 | C3-E hull |
| USS *Alshain* (AKA-55) | 21 Jul-3 Aug 44 | C2-S-B1 hull |
| USS *Aquarius* (AKA-16) no BS | 21-26 Jul 44 | C2-S-B1 hull |
| USS *Centaurus* (AKA-17) | 21-27 Jul 44 | C2-S-B1 hull |
| USS *Libra* (AKA-12) | 21-25 Jul 44 | C2-F hull |
| USS *Titania* (AKA-13) no BS | 26 Jul 44 | C2-F hull |
| USS *Virgo* (AKA-20) | 21-27 Jul 44 | C2-S-B1 hull |
| **Tinian Capture and Occupation** | | |
| USS *Thuban* (AKA-19) | 24-28 Jul 44 | C2-S-B1 hull |

**17**

# Capture of South Palau Islands

*Enemy rifle and machine gun sniper fire appeared to be inaccurate. It is believed that casualties suffered by Navy personnel from enemy fire were made by lucky enemy shots. Very little data is available to this command on the moored mines in the area. They were estimated by Demolition Team Observers to range in size from 450 to 2000 pounds. They were of the contact-horn type and were non-sensitive to counter-mining or water disturbances in their vicinity. They were moored at depths from six to eleven feet at low water.*

—Capt. R. E. Hanson, commander Transport Division 26, in his action report on the assault landing on Angaur.[1]

In September 1944, the final barriers to Allied movement into the Philippines were eliminated by simultaneous landings of Central Pacific forces in the southern Palau group, and of Southwest Pacific forces on Morotai Island. Morotai was needed for establishment of an air and naval support base roughly halfway between the existing Allied bases in New Guinea and the location of the first invasion in the Philippines, then planned to be Mindanao, about 650 miles north of New Guinea.[2]

Acquisition of the Palau Islands and Morotai (not taken up in this book) brought the Philippines within range of Allied land-based aircraft, and pushed the Japanese back on their inner defenses in the Pacific. The landings were preceded in early September, by heavy attacks on enemy bases which might otherwise interfere with operations. The Fast Carrier Forces of the Central Pacific ranged through the northern string of Japanese-held bases, striking airfields and harbors of the Bonin and Volcano islands; Yap Island; Ulithi Atoll; and Mindanao and the Visayas in the Philippines (see map on next page). Concurrently, land-based aircraft of the Southwest Pacific were delivering strikes against the more southerly enemy bases. As a result of these attacks, air and sea opposition were almost absent during the subsequent landings on these islands.[3]

Map 17-1

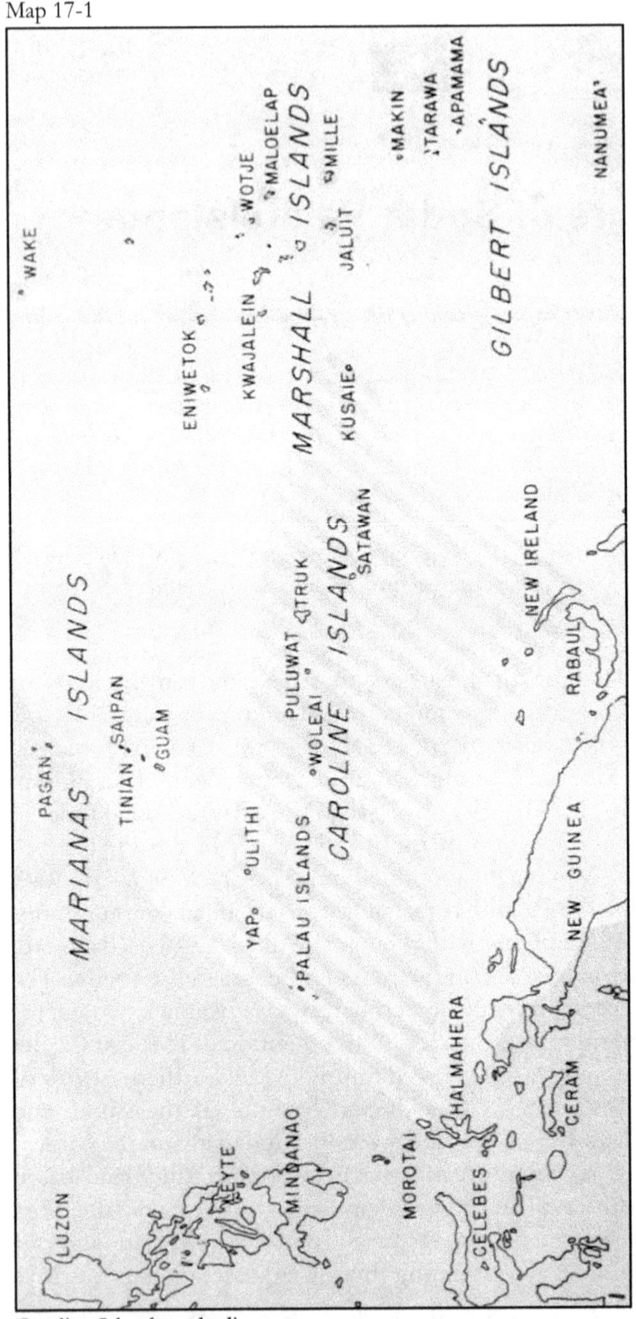


Caroline Islands and adjacent areas
Commander in Chief, U.S. Pacific Fleet and Pacific Ocean Areas, Operations in the Pacific Ocean Areas – September 1944, 7 March 1945

Map 17-2

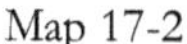

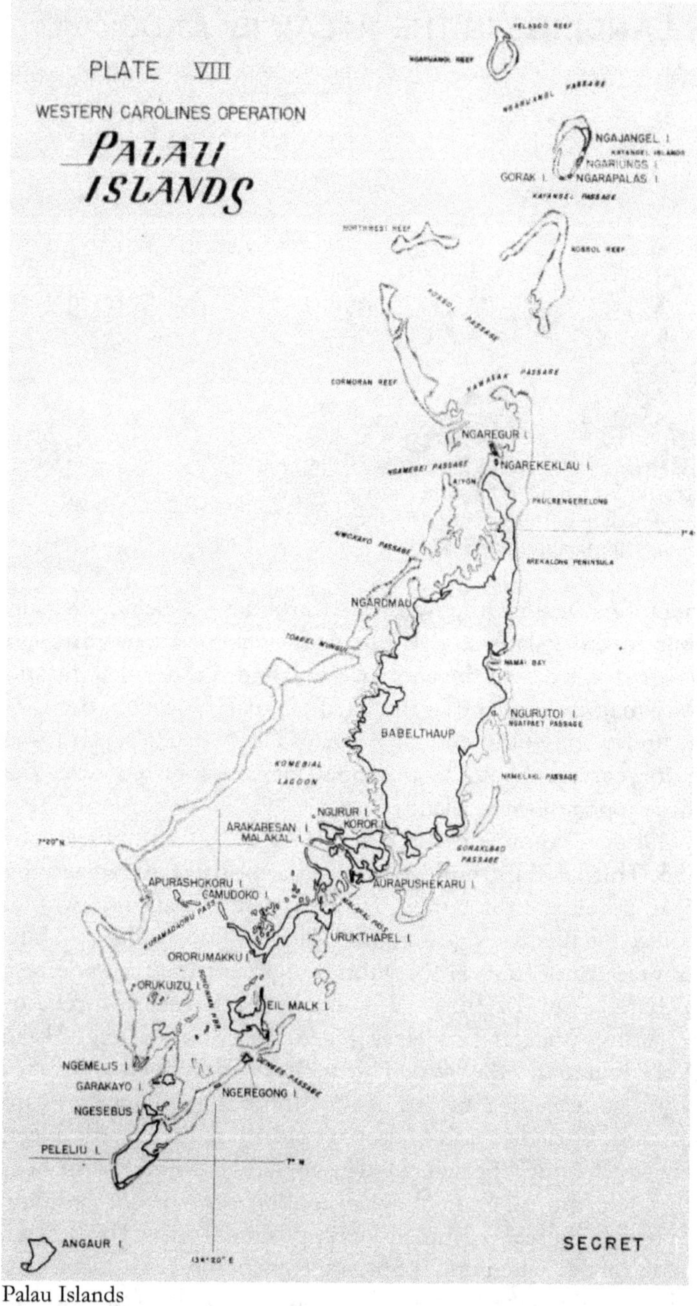

Palau Islands
Commander in Chief, U.S. Pacific Fleet and Pacific Ocean Areas,
Operations in the Pacific Ocean Areas – September 1944, 7 March 1945

## THIRD FLEET RESPONSIBLE FOR PELELIU AND ANGAUR LANDINGS IN THE PALAU ISLANDS

Photo 17-1

Smoke and dust rise from the Peleliu shore, during the final stages of the pre-invasion bombardment, 15 September 1944. National Archives photograph #80-G-59497

Planned operations against the Western Carolines—Peleliu, Angaur, and Ngesebus in the Palaus, and Ulithi Atoll—from August through October 1944, extended over the entire Central and Western Pacific and involved every major command in the Pacific Area. However, the U.S. Third Fleet, under the command of Adm. William F. Halsey Jr., was responsible for carrying out the principal tasks associated with the assault of these Japanese-held islands.[4]

Admiral Nimitz, commander in chief of the U.S. Pacific Fleet—of which the U.S. Third and Fifth Fleets were a part—had reorganized the Pacific Fleet to accelerate the tempo/frequency of operations. At that time, the South Pacific Area command (also previously titled, Third Fleet) under Vice Adm. John H. Newton, was maintained at Noumea, then a rear base. On 15 June 1944, he had relieved its previous commander, Adm. William F. Halsey Jr., so that Halsey might take operational command of the new Third Fleet, comprised of forces previously only commanded by Adm. Raymond A. Spruance. From that time forward, Spruance and Halsey would share the Pacific Fleet, with one fleet commander (Spruance) planning and training for the next operation(s), while the other (Halsey) executing the current fighting. Naval forces under Halsey's command were designated the Third Fleet, and these same forces, when under Spruance, were the Fifth Fleet.[5]

Admiral Halsey assigned responsibility for the Palau operation to Vice Adm. Theodore S. Wilkinson, USN, whose Joint Expeditionary

Force (Task Force 31) consisted of the Western Attack Force (Task Force 32) under Rear Adm. George H. Fort, and the Expeditionary Troops (Task Force 36) commanded by Maj. Gen. Holland M. Smith, USMC. Fort's Attack Force was split into a Peleliu Attack Group (Task Group 32.1) under Rear Admiral Fort, and the Angaur Attack Group of Rear Adm. William H. P. Blandy (Task Group 32.2). The cargo ships assigned to these task groups, and to the Western Garrison Group, are identified below, as well as their hull types, and award periods for battle stars received.

**Peleliu Attack Group (TG 32.1): Rear Adm. George H. Fort**

| Ship | Award Period | Hull Type |
|---|---|---|
| **TransDiv 6 (Temporary): Capt. John C. Lester** | | |
| USS *Centaurus* (AKA-17) | 6 Sep-14 Oct 44 | C2-S-B1 hull |
| USS *Hercules* (AK-41) | 6 Sep-14 Oct 44 | C3-E hull |
| **TransDiv 8 (Temporary): Capt. Frank R. Talbot** | | |
| USS *Virgo* (AKA-20) | 9-24 Sep 44 | C2-S-B1 hull |
| **TransDiv 24 (Temporary): Capt. Thomas B. Brittain** | | |
| USS *Aquarius* (AKA-16) | 6 Sep-14 Oct 44 | C2-S-B1 hull |

**Angaur Attack Group (TG 32.2): Rear Adm. William H. P. Blandy**

| Ship | Award Period | Hull Type |
|---|---|---|
| **TransDiv 20 (Temporary): Capt. Donald W. Loomis** | | |
| USS *Electra* (AKA-4) | 6 Sep-14 Oct 44 | C2-T hull |
| **TransDiv 26 (Temporary): Capt. R. E. Hanson** | | |
| USS *Jupiter* (AK-43) | 6 Sep-14 Oct 44 | C2 hull |
| **TransDiv 32 (Temporary): Capt. M. O. Carlson** | | |
| USS *Arneb* (AKA-56) | 6 Sep-14 Oct 44 | C2-S-B1 hull |

**Western Garrison Group (TG 31.4):**
**Comdr. Charles A. MacGowan, USN (Ret.)**

| Ship | Award Period | Hull Type |
|---|---|---|
| USS *Matar* (AK-119) | 6 Sep-14 Oct 44 | EC2-S-C1 hull (Liberty)[6] |

## ASSAULTS ON PELELIU AND ANGAUR ISLANDS

Prior to dawn at 0515 on 15 September, the transports of the Peleliu Attack Group arrived in the transport area. Fire support ships started their preliminary bombardment fifteen minutes later, as tank landing ship (LST) groups were approaching their assigned areas. Following their arrival, the transports and LSTs commenced launching boats and LVTs (tracked landing vehicles), while the two dock landing ships assigned to Transport Divisions launched their tank landing craft, carrying medium-size tanks.[7]

Air support strikes and gunfire support preceded a heavy rocket barrage laid down by eighteen LCI(G) infantry landing craft, each specially fitted with 42 rocket launchers. This was followed by a final strafing rocket and dive-bombing attack on the beach area by U.S. aircraft. At 0832, the assault troops of the First Marine Division landed on Peleliu. Shortly after landing, the beach area came under enemy fire from mortars and light artillery defiladed (protected from observation and frontal fire by obstacles) on the high ground to the north of the beach, and on an adjacent small island to the south. These defense positions were well camouflaged and difficult to locate.[8]

The transports did not receive enemy fire from the island, and started moving in to the unloading area at about 0930. Only emergency supplies were landed on D-Day, to support the Marines who succeeded in gaining a narrow beachhead by evening. Moving forward the following day, they succeeded in capturing the airfield, and steady, slow advances continued in other areas in spite of stubborn enemy resistance.[9]

At 0836 on 17 September, the U.S. Army 81st Infantry Division, less one Regimental Combat Team (RCT), was landed on Angaur. Combat operations there progressed rapidly, and organized resistance ceased in late morning on the 20th, except for a small pocket in caves in the northwest corner of the island. American casualties were 260 killed, 1,354 wounded. The RCT (which was to have been employed at Angaur) was shifted from Angaur to Peleliu, on 22 September, to reinforce the First Marines and later to garrison the island.[10]

## LARGE AMERICAN CASUALTIES ON PELELIU

On Peleliu, surviving Japanese troops under the command of Col. Kunio Nakagawa mounted a fanatical resistance from numerous interconnected caves, and emplacements, in the very rugged hills in the northern section of the island. This fighting resulted in many American casualties, until the bulk of this enemy opposition was overcome, on 13 October 1944. Organized resistance did not end until the night of 24 November, when the last of Nakagawa's garrison was killed, following a two and one-half month-long defense of the island, at a cost of 1,950 U.S. Marines and soldiers killed.[11]

The biggest threat posed to cargo ships in the Palau Islands were mines; four USN units were sunk—minesweepers *Perry* and *YMS-19*, and amphibious landing craft *LCI-459* and *LCT-579*—and numerous other minesweepers and craft damaged by mines.[12]

| Date | Ship/Craft/Location | Date | Ship/Craft/Location |
|---|---|---|---|
| 13 Sep 44 | USS *Perry* (DMS-17) Angaur | 24 Sep 44 | USS *YMS-19* Angaur |
| 19 Sep 44 | USS *LCI-459* Peleliu | 4 Oct 44 | USS *LCT-579* Angaur |

18

# Liberation of the Philippines

*Should we lose in the Philippines operations, even though the fleet should be left, the shipping lane to the south would be completely cut off so that the fleet, if it should come back to Japanese waters, could not obtain its fuel supply. If it should remain in southern waters, it could not receive supplies of ammunition and arms. There would be no sense in saving the fleet at the expense of the loss of the Philippines.*

—Adm. Soemu Toyoda, Imperial Japanese Navy, discussing Vice Adm. Takeo Kurita's mission to destroy completely the transports in Leyte Bay following the American invasion of the Philippines, and why there were no restrictions as to the damage that his force might take.[1]

The U.S. Sixth Army went ashore at Leyte Island, on 20 October 1944, two months and two years after the first landings were conducted in the Guadalcanal area of the Solomon Islands. This significant operation was predicated and made possible by the ones that opened the way. Amphibious landings in the Marianas, in June and July 1944, breached Japan's strategic inner defense ring and gave the Americans a base from which B29 bombers could attack the Japanese home islands. Following the invasion of Saipan, the Japanese counterattacked in the Battle of the Philippine Sea, fought between the First Mobile Fleet and American Fifth Fleet, from 19 to 21 June. In what one American aviator termed "The Great Marianas Turkey Shoot," the U.S. Navy destroyed three enemy aircraft carriers—the *Hijo*, *Shokaku*, and *Taiho*—some 480 planes, and nearly as many aviators. The devastating loss left the Japanese with virtually no carrier-based aircraft or experienced pilots for the forthcoming Battle of Leyte Gulf.[2]

## MACARTHUR AND HALSEY JOIN FORCES

The naval force (comprised of units of the American Third and Seventh Fleets) assembled for the invasion of Leyte was not quite as large as the one that had taken part in June in the invasion of Normandy, but it had more striking power. Embarked aboard the assault vessels were the U.S. Sixth Army's Tenth and Fourteenth Corps commanded by Lt. Gen. Walter Krueger.[3]

The operations in the Central Philippines brought together in a significant way, for the first time, forces of Admiral Nimitz's Pacific Ocean Areas command, and those of General MacArthur's South West Pacific Area command. Under Nimitz, Halsey's South Pacific Forces had evolved into the U.S. Third Fleet, and Spruance's Central Pacific Forces the U.S. Fifth Fleet. U.S. Southwest Pacific naval forces, which operated directly under MacArthur's command had similarly upgraded to the U.S. Seventh Fleet under Vice Adm. Thomas C. Kinkaid, USN.

As previously explained, per direction by Nimitz, Halsey's and Spruance's previous forces were essentially combined. When Halsey commanded them, they were designated the Third Fleet, and when Spruance, alternating with Halsey, took the reins, the Fifth Fleet. Following the landings by Central Pacific (Third Fleet) forces at Palau, and those by Southwest Pacific Forces on Morotai, planned landings on Yap (phase II of the Palau operation) by Third Fleet forces were cancelled, and the forces that were to carry them out were assigned to the Seventh Fleet for the landings at Leyte.[4]

General MacArthur was overall in charge of the Philippines operation and, under him, Vice Admiral Kinkaid the naval operations. The three major task organizations of the Seventh Fleet were:

- Task Force 77 (Covering and Support) under Vice Adm. Thomas C. Kinkaid, USN (as well as overall naval operations)
- Task Force 78 (Northern Attack Force) under Rear Adm. Daniel E. Barbey, USN
- Task Force 79 (Southern Attack Force) under Vice Adm. Theodore S. Wilkinson, USN[5]

Covering the landing operations, though remaining under the control of commander in chief, Pacific Fleet and Pacific Ocean Areas (Admiral Nimitz), were the fast battleships and carriers of the Third Fleet, commanded by Admiral Halsey.[6]

The weather at the entrance to Leyte Gulf at daybreak on 20 October was cloudy with altostratus and partial swelling cumulus, a visibility to seaward of twelve miles, and light winds from the southeast. Planners had been concerned that a typhoon might pass through the area and cause retirement or diversion of the forces en route from New Guinea. However, the conditions on "A-Day" (Assault Day) were perfect as described by commander, Third Amphibious Force:

The assault proceeded on schedule following the preliminary bombardment by ships' gunfire and aircraft, a slight onshore tendency of the almost imperceptible wind conveniently drifting the smoke and dust of the bombardment off the beaches and into the interior. The airborne beach observer had made his required report earlier, but the report was unnecessary in this case due to the almost complete absence of surf.[7]

Map 18-1

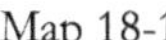

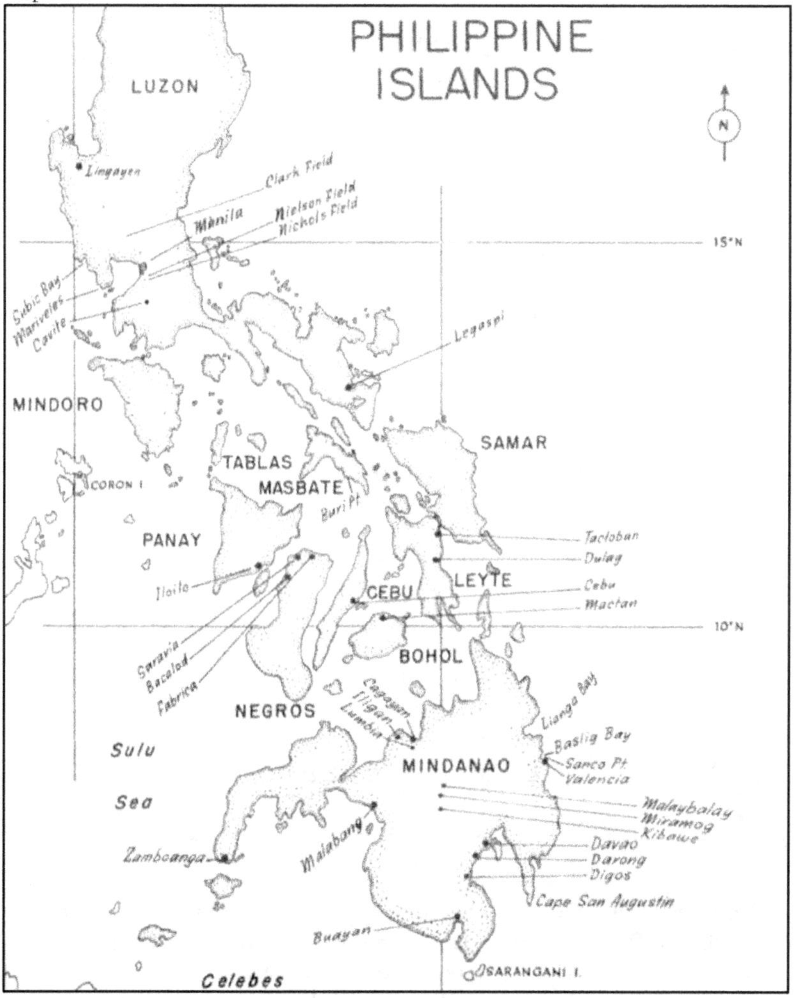

Philippine Islands
Commander in Chief, U.S. Pacific Fleet and Pacific Ocean Areas, Operations in the Pacific Ocean Areas - September 1944, 7 March 1945

The landings at Tacloban, located in northeast Leyte on an inlet of the Leyte Gulf, and at Dulag, twenty-five miles to the southward, were made against little opposition. Naval historian Samuel Eliot Morison noted about the operation: "The Leyte landings were easy, compared with most amphibious operations in World War II—perfect weather, no surf, no mines or underwater obstacles, slight enemy resistance, mostly mortar fire." With this beginning, the liberation of the Philippines was off to a good start.[8]

## USN CARGO SHIPS/RAN INFANTRY LANDING SHIPS

The USN cargo ships assigned to the Northern and Southern Attack Forces are identified in the tables, as well as the three RAN infantry landing ships comprising Transport Division 20. The LSIs were a part of the Panaon Attack Group within the Northern Attack Force.

**NORTHERN ATTACK FORCE (TASK FORCE 78): Rear Adm. Daniel E. Barbey (landing of X Corps in the Tacloban area)**

**Palo Attack Group (TG 78.1): Rear Adm. Daniel E. Barbey**
**U.S. Army 24th Infantry Division (Left beaches)**
**Transport Unit: Capt. Thomas B. Brittain**
**TransDiv 6 (temporary): Capt. H. D. Baker**

| Ship | Award Period | Hull Type |
|---|---|---|
| USS *Hercules* (AK-41) | 13 Oct-29 Nov 44 | C3-E hull |
| USS *Titania* (AKA-13) | 13-20 Nov 44 | C2-F hull |
| **TransDiv 24 (temporary): Capt. Thomas B. Brittain** | | |
| USS *Aquarius* (AKA-16) | 13 Oct-20 Nov 44 | C2-S-B1 hull |

**San Ricardo Attack Group (TG 78.2): Rear Adm. William M. Fechteler**
**U.S. Army 1st Cavalry Division (Right beaches)**
**Transport Unit: Capt. M. O. Carlson**

| **TransDiv 20 (temporary): Capt. Donald W. Loomis** | | |
|---|---|---|
| USS *Electra* (AKA-4) | 13-30 Oct 44 | C2-T hull |
| **TransDiv 32 (temporary): Capt. M. O. Carlson** | | |
| USS *Arneb* (AKA-56) | 13 Oct-29 Nov 44 | C2-S-B1 hull |

**Panaon Attack Group (TG 78.3): Rear Adm. Arthur D. Struble**
**U.S. 21st Regimental Combat Team / 24th Infantry Division**
**Transports: Comdr. Allan Paterson Cousin, RANR(S)**

| **TransDiv 20 (temporary): Capt. Donald W. Loomis** | | |
|---|---|---|
| HMAS *Manoora* | HMAS *Kanimbla* | HMAS *Westralia* |

**Reinforcement Group One (TG 78.6): Capt. Frank R. Talbot**

**TransDiv 26 (temporary): Capt. R. E. Hanson**

| | | |
|---|---|---|
| USS *Jupiter* (AK-43) | 16 Oct-29 Nov 44 | C2 hull[9] |

**SOUTHERN ATTACK FORCE (TASK FORCE 79): Vice Adm. Thomas C. Kinkaid (landings of XXIV Corps in the Dulag area)**

**Attack Group Able (TG 79.1): Rear Adm. Richard L. Conolly**
**U.S. 7th Infantry Division (left beaches)**

**Transport Group Able (TG 79.3): Capt. C. G. Richardson**

**TransDiv 7 (temporary): Capt. C. G. Richardson**

| | | |
|---|---|---|
| USS *Thuban* (AKA-19) | 20 Oct-18 Nov 44 | C2-S-B1 hull |

**TransDiv 30 (temporary): Capt. Clinton A. Misson**

| | | |
|---|---|---|
| USS *Chara* (AKA-58) | 20 Oct-18 Nov 44 | C2-S-B1 hull |

**TransDiv 38 (temporary): Capt. Charles Allen**

| | | |
|---|---|---|
| USS *Alshain* (AKA-55) | 20 Oct-18 Nov 44 | C2-S-B1 hull |

**TransDiv "X-Ray" (temporary): Capt. J. A. Snackenberg**

| | | |
|---|---|---|
| USS *Mercury* (AK-42) | 20 Oct-18 Nov 44 | C2 hull[10] |

**Attack Group Baker (TG 79.2): Rear Adm. Forrest B. Royal**
**U.S. 96th Infantry Division (right beaches)**

**Transport Group Baker (Task Group 79.4): Capt. Herbert B. Knowles**

**TransDiv 10 (temporary): Capt. G. D. Morrison**

| | | |
|---|---|---|
| USS *Capricornus* (AKA-57) | 20 Oct-14 Nov 44 | C2-S-B1 hull |

**TransDiv 18 (temporary): Capt. Herbert B. Knowles**

| | | |
|---|---|---|
| USS *Alcyone* (AKA-7) | 20 Oct-18 Nov 44 | C2 hull |

**TransDiv 28 (temporary): Capt. Henry C. Flanagan**

| | | |
|---|---|---|
| USS *Almaack* (AKA-10) | 20 Oct-14 Nov 44 | C3-E hull |
| USS *Auriga* (AK-98) | 20 Oct-18 Nov 44 | C1-B hull[11] |

## LANDINGS IN THE PANAON AREA

*The Task Group Commander desires to commend the Commanding Officers of the* MANOORA, KANIMBLA, *and* WESTRALIA *for the excellent and smooth manner in which the unloading was carried out.*

—Rear Adm. Arthur D. Struble, USN, commander Panaon Attack Group (TG 78.3) recognizing the important contributions of the Royal Australian Navy's three infantry landing ships to the Allied landings at Panaon Island, Philippines.[12]

On 20 October, as the Northern and Southern Attack Forces landed their troops on the east coast of Leyte, the 21st Regimental Combat

Team of the Panaon Attack Group came ashore some seventy miles south of the main landing beaches to secure the strait between Leyte and Panaon Islands. As previously noted, it was generally much easier to mount an amphibious assault against a large island with an expansive coastline, offering the possibility of finding lightly defended or undefended beaches, than a small, well-defended one with coastal guns, and other fortifications and barriers guarding a finite number of accesses from the sea.[13]

The Panaon Attack Force found its designated landing beaches undefended by enemy ground forces, and the local population excited about finally throwing off the yoke of their Japanese occupiers. Comdr. Allan P. Cousin—commanding the infantry landing ships HMAS *Manoora*, *Kanimbla*, and *Westralia*—described the atmosphere that existed during landing operations:

> Great numbers of Filipinos in canoes thronged around the ships throughout the whole day. Food, cigarettes and clothing were handed out to these people, whose gratitude at deliverance from the Japanese was most sincere and moving. This day was a real "red letter day" in their lives. There can be no doubt whatsoever of the really sincere and friendly attitude of all the Filipino people to the Forces of the United States of America.[14]

## ENEMY AIR ATTACK DEVELOPS LATER THAT DAY

Pre-landing bombardment scheduled on the morning of 20 October for the Panaon landing was cancelled when information was received that it was almost certain that no Japanese were on Panaon Island, or on the extreme south end of Leyte Island. At 0845, the infantry landing ships arrived in the inner transport area, and after lowering boats, began disembarking troops. All troops were off the LSIs by 0926, and the landing ships, finding the situation secure moved in toward the beach to discharge cargo.[15]

Protective aircraft from USN escort carriers remained overhead throughout the day, and no enemy interference was experienced while discharging troops and cargo. At 1751, HMAS *Manoora*, *Westralia*, and *Kanimbla* weighed anchor, and departed in formation, screened by destroyers. Minutes later, two, possibly more, Japanese aircraft, attacked the formation one at a time, as described by Cousin:

> At 1804 one enemy plane (EMILY) was engaged by destroyer escort and driven off. At 1826 another enemy plane (probably DINAH) was engaged by escort and driven off. At 1841, possibly the same plane endeavored to attack from the starboard side, but was

unsuccessful. At 1842 *MANOORA* opened fire on the plane. At 1846, an enemy plane endeavored to attack from the port side, possibly the same plane which had gone round ahead of the convoy. All ships threw up an intense barrage which was successful in driving the plane off. At 1852 destroyers astern opened fire, probably on the same plane, which withdrew to the South Westward. No further attacks developed.[16]

Photo 18-1

Drawing by Frank Norton of the Australian infantry landing ships HMAS *Manoora* (ahead), *Westralia*, and *Kanimbla*, with support from U.S destroyers, in action against Japanese torpedo bombers in the Leyte Gulf.
Australian War Memorial photograph ART21172

## JAPANESE BEGIN EMPLOYING SUICIDE AIRCRAFT

> *Carrier air strikes had so well neutralized enemy airfields that there were no attacks on LEYTE Gulf by Japanese planes in large force during the unloading phase of the operation. Troops and cargo were discharged rapidly and according to plan. By the 24th, most of the ships of the first echelon had completed unloading, and were on their way to rear areas. Small numbers of enemy planes persistently attacked the transport area, however, and suicide-dive tactics developed here to serious proportions.*
>
> —Commander in Chief, U.S. Pacific Fleet and Pacific Ocean Areas, Operations in the Pacific Ocean Areas – October 1944, 31 May 1945.

Beginning at Leyte Gulf, and continuing throughout the remainder of the war, the Japanese employed suicide as well as conventional aircraft in attacks against Allied forces. Naval vessels which suffered the most damage from these enemy attacks in the Leyte Gulf area were as follows by date:

- 20 October: Light cruiser USS *Honolulu* (CL-48) was hit by an aircraft torpedo, but able to retire under her own power
- 21 October: Heavy cruiser HMAS *Australia* (D84) was struck on her bridge by a suicide plane
- 24 October: Old fleet tug USS *Sonoma* (ATO-12) was crashed by a flaming bomber on her starboard side amidships; the Infantry landing craft *LCI-1065* was hit by a plane, burned, and was lost; and the attack transport USS *Fremont* (APA-44) was strafed
- 28 October: A suicide plane crashed close aboard the light cruiser USS *Denver* (CL-58), doing some damage[17]

The loss of life from these attacks was particularly grievous aboard HMAS *Australia.* Her commanding officer, Capt. Emile F. V. Dechaineux, and 29 other officers and ratings were killed or died of wounds, and Commodore John A. Collins and a further 64 were injured. *Australia* was escorted by HMAS *Warramunga* to Manus Island and then to Espiritu Santo in the New Hebrides for repairs.[18]

## USN CARGO SHIPS STRAFED AND BOMBED

Photo 18-2

Cargo ship USS *Auriga* (AK-98), a part of the Leyte assault forces, was strafed by two Japanese aircraft, on 25 October 1944, severely wounding Comdr. John G. Hart, D-M, USNR (her commanding officer), and four crewmembers who also suffered shrapnel wounds while at their battle stations.
Naval History and Heritage Command photograph #NH 91135

Although not crashed by suicide planes, or victims of air-dropped torpedoes, other Allied ships suffered loss of life to enemy aircraft in Leyte Gulf, including the cargo ship USS *Auriga* (AK-98). A unit of the Southern Attack Force's Attack Group Baker, *Auriga* witnessed, while unloading cargo in the Dulag area, on 20 October, a Japanese torpedo plane approaching from over the top of Catmon Hill, Leyte Island, and the subsequent drop of the torpedo which struck the *Honolulu*. The aircraft then escaped through heavy anti-aircraft fire from the battleship *Tennessee* (BB-43) and an accompanying destroyer, by flying very close to the water (cleverly, below the depressions angle of their guns).[19]

The crew of *Auriga* was called to General Quarters (GQ) twice on D-Day, for enemy aircraft in the area, but none approached the ship close enough for engagement by her gun crews. A similar story was experienced the following day, when GQ was set four times for planes in the area. On each occasion, none were sighted. The same situation persisted for most of 22 October. General Quarters was set whenever planes were reported in the area by the radar guard (ship assigned to search with radar and report any air contacts made). However, when no enemy planes were sighted each time, *Auriga* secured from GQ.[20]

On the evening of 22 October, General Quarters was again sounded, this time at 1845. Fifteen minutes later, *Auriga*'s guns opened fire on an enemy plane observed flying seaward over Leyte Island, approaching her broad on her port bow. Because of a thick protective smoke screen, it was impossible to identify the aircraft, but it was believed to be a torpedo bomber owing to its low altitude. No hits were observed, and *Auriga* ceased firing when the plane disappeared into the smoke screen.[21]

Smoke screens proved to be very effective in the evening, at night, and especially during morning twilight. However, during the day, smoke hampered the ability of lookouts, and served as cover for enemy planes, enabling them to get in close to shipping before visual detection.[22]

As *Auriga* continued to discharge cargo, on 23 October, this work was disrupted only twice by a requirement to man battle stations. Although General Quarters was set upon radar guard reports of enemy aircraft in the area, each time, no aircraft were encountered.[23]

The following day, the 24th, brough gun action. There were two calls to battle stations that morning, each lasting about two hours, but no encounters with planes. However, this changed after the General Quarters alarm sounded at 1715, again calling her crew to their battle stations. As two "Betty" bombers closed the ship, her guns opened fire on the nearest one. It was flying on a course parallel to the ship's heading, about 1,000 yards distant at an altitude of 800 feet. Several

20mm tracer rounds were seen to hit the plane before it disappeared in the smoke screen, while maintaining the same altitude. After securing from General Quarters (following the departure of Japanese aircraft from the area), *Auriga*'s crew was called to battle stations once more, then secured after report of the area clear of planes.[24]

In the early morning hours of 25 October, *Auriga* was discharging artillery, ammunition, and other combat equipment, and disembarking Marine and Army personnel. At 0309, General Quarters was set for enemy planes in the area, and enemy surface vessels engaging Allied forces approximately thirty miles distant. (This was part of the Battle of Leyte Gulf, 23-26 October 1944—a decisive air and sea battle in progress, which crippled the Japanese Combined Fleet, desperately trying to prevent the U.S. invasion of the Philippines.) Aboard the cargo ship, General Quarters was secured at 0645, there having been no encounters with enemy planes.[25]

## *AURIGA* STRAFED

> *This ship underwent a strafing attack by two... [of three] enemy planes approaching from the stern out of the sun. The ship had a heading of 280° True and the planes approached on courses parallel to this heading. One, identified as a Japanese BETTY bomber turned away over the ship to starboard. It approached to within 800 yards distant of the ship at an altitude estimated at 900 feet. The waist gunner was observed to be firing at this ship.*
>
> —USS *Auriga* War Diary, October 1944.

After only a brief reprise, at 0707 the General Quarters was sounded aboard *Auriga* once again, calling her crew to battle stations. A few minutes later, three enemy planes were sighted dead astern. Two approaching the cargo ship were taken under fire. The first one, a Betty bomber, strafed *Auriga* with machine gun fire, then turned away to starboard, trailing smoke. Hit in the tail and rear fuselage by 20mm tracer rounds, the aircraft suffered no loss of altitude and disappeared in the vicinity of Vigia Point on Leyte.[26]

The second plane, a Frances (Yokosuka P1Y1 Navy land-based bomber), came in over *Auriga*'s port quarter and dived at the ship. No bombs were released and the attacker was hit by 20mm fire the entire length of its fuselage, when about 800 yards distant. After drawing abeam of *Auriga*, the aircraft began smoking badly, and crashed in flames, just at the end of a beach, south of Libernah Head, Leyte—shot down by the cargo ship.[27]

*Auriga*'s commanding officer, Comdr. John Glenn Hart, D-M, USNR, on the bridge during these attacks, was severely wounded by shrapnel, which passed through his back and penetrated his right lung and chest. Four members of the crew were also wounded. Hart was a U.S. Merchant Marine Reserve officer; the "D-M" following his rank, denoted "Deck officers, commissioned and warrant, including boatswains, qualified for deck or appropriate administrative duties."[28]

At 0730, *Auriga*'s gun crews ceased firing and, following no other ensuing enemy activity, the ship secured from General Quarters at 0921. The alarm was sounded once again at 1136, when enemy planes were again reported in the area. An hour or so later, a Zeke (Japanese Zero fighter) dove out of the clouds and smoke off the ship's starboard bow. *Auriga*'s gun crews immediately opened on it, but only about 300 yards away, flying low at an altitude of 600 feet, it climbed out of sight rapidly, unharmed by the ship's gunfire.[29]

Three minutes after securing from General Quarters at 1520, *Auriga* got under way from the transport area. During her transit to the northern anchorage in San Pedro Bay, at the northwest end of Leyte Gulf, battle stations were manned upon a sighting of a formation of six Val dive bombers, broad on her port bow, diving on other Allied shipping. Her 3-inch/50-caliber gun mount opened fire on one of the enemy aircraft, 1,500 yards distance. No hits were observed, and "cease fire" was ordered as the planes passed low over the shipping.[30]

Commander Hart was transferred that evening to the amphibious force command ship USS *Mount Olympus* (AGC-8) for medical care and, the following day, on to the hospital ship USS *Mercy* (AH-8) for further treatment and disposition. The injured sailors apparently were not wounded sufficiently to warrant transfer off *Auriga* as casualties. The Senior Medical Officer on board *Mount Olympus* listed Hart in a Casualty and Casualty Evacuation Report dated 31 October 1944, but not the other men.[31]

Upon Hart's departure from *Auriga*, Lt. Albion B. Woodside, D-V(G), USNR, assumed duties as commanding officer. Woodside was a Volunteer Reserve (General Service) officer, whose D-V(G) meant, "Deck officers, commissioned and warrant, including boatswains and gunners, qualified for general detail afloat or ashore."

## *HYPERION* BOMBED

*During our fifteen days at Leyte we discharged our cargo, went to general quarters eighty seven times, stayed at general quarters a total of seventy three hours and seventeen minutes, were subject to thirty seven Jap[anese] plane attacks, were hit by a phosphorus bomb, and were officially credited with shooting down two Jap[anese] planes.*

—Experiences of the cargo ship USS *Hyperion* (AK-107) which, on 29 October 1944, nine days after the assault landings, arrived in Leyte Gulf in convoy with thirty-three ships.[33]

USS *Hyperion* was one of three cargo ships, not part of the initial assault forces, but which arrived in the Leyte Gulf with echelons of other ships in support of combat forces ashore. As Japanese aircraft continued to mount attacks on Allied shipping, they earned battle stars for the periods indicated in the table. All three were former Liberty ships delivered by their builders to the Navy, commissioned, converted, fitted out, and put through requisite trials in preparation for subsequent naval service.

| Ship | Award Period | Hull Type |
|---|---|---|
| USS *Hyperion* (AK-107) | 29 Oct-12 Nov 44 | EC2-S-C1 hull (Liberty) |
| USS *Murzim* (AK-95) | 23 Oct-29 Nov 44 | EC2-S-C1 hull (Liberty) |
| USS *Triangulum* (AK-102) | 23-28 Nov 44 | EC2-S-C1 hull (Liberty) |

Photo 18-3

Cargo ship USS *Murzim* (AK-95) headed seaward, location and date unknown. Naval History and Heritage Command photograph #NH 78577

# 19

# Lingayen Gulf Landings

*Heavy suicide attacks on our shipping by "Kamikaze" pilots…sank or badly damaged nearly 30 combatant vessels participating in the invasion, and which for the first time became a really serious menace to the success of an operation.*

—Commander in Chief, U.S. Pacific Fleet and Pacific Ocean Areas, Operations in Pacific Ocean Areas – January 1945, 31 July 1945.

*Considering the Japanese philosophy of life and war it is no surprise that there is now a wave of suicide attacks; that they did not come sooner is more of a surprise.*

*The Japs unquestionably have given considerable study to this method of attack. It is not a haphazard procedure. It has been suggested that in attacks to date the planes came in from the same direction over land fade [shielding] under the CAP [protective combat air patrol] and dive from approximately the same altitude.*

*The real defense to these attacks would seem to be accurate gunnery…. There is indicated and immediate need for more intense training of all gunnery personnel, supplemented with special training for all lookouts on the probable methods of approach and attack.*

—Commander Task Unit 79.3.2 (ComTransDivEight), Operation and Action Report of Amphibious Assault on Lingayen Area, Luzon, Philippines, 12 January 1945.

Photo 19-1

Amphibious command ship USS *Blue Ridge* (AGC-2), left, and USS *Thuban* (AKA-19) off the transport area, white beach, during Lingayen Gulf landings, on 9 January 1945. National Archives photograph #80-G-300633

As the beginning of January 1945, Leyte was firmly in the control of U.S. Army ground forces, with only mopping up of isolated enemy remnants remaining, and Central Philippines waters largely having been cleared of all enemy combatant ships, except for a few submarines. The stage was set for the next major phase of the reconquest of the Philippines, amphibious landings at Lingayen Gulf, Luzon. The several interrelated objectives of the Lingayen operation were as follows: prompt seizure of the Central Luzon area; destruction of the principal defense forces; denial to the enemy of the northern entrance to the South China Sea; and obtainment of bases for the support of further operations against the Japanese.[1]

Map 19-1

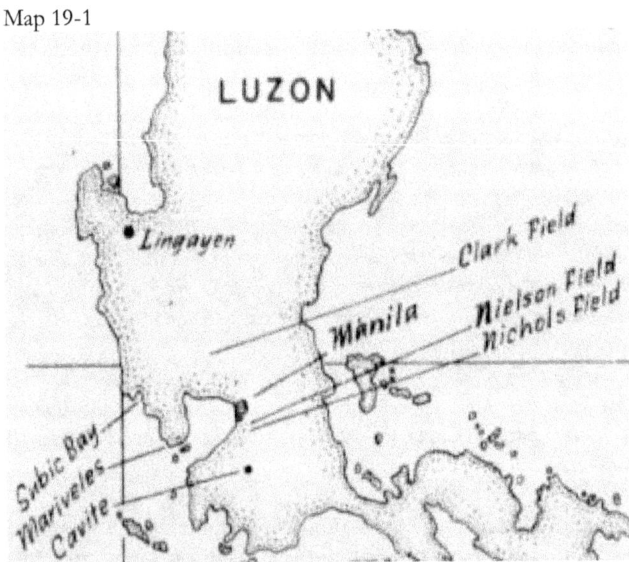


Luzon Island in the northern Philippines

The invasion of northwestern Luzon was larger than any previous one in the Pacific, including that at Leyte Gulf. Accordingly, naval forces of the Pacific Fleet had to be heavily committed, either in the form of combatant and transport shipping temporarily placed under Seventh Fleet command, or as covering forces. The latter function involved Halsey's Fast Carrier Task Force neutralizing enemy air strength in Formosa (today Taiwan), northern Luzon, and the Ryukyu Islands, which stretch southwest from Kyushu to Taiwan. The Japanese, having fewer planes available to oppose the landings, employed Kamikaze as well as conventional aircraft attacks.[2]

## AKS / AKAS ASSIGNED TO LINGAYEN OPERATION

The Luzon landings involved two attack forces: The San Fabian Attack Force landed its troops in the easternmost part of Lingayen Gulf, on both sides of the town of San Fabian, whereas those of the Lingayen Attack Force went ashore opposite the town of Lingayen in the southern gulf. A measure of the quantity of naval forces committed to Luzon, was the large assignment of cargo ships to the operation, eighteen in total. These included seven to the San Fabian Attack Force, six with the Lingayen Attack Force, and the remaining five to a reinforcement group scheduled to arrive after the initial landings.

With the exception of Transport Division Eight, only the cargo ships, and not the more numerous troop transport ships assigned to each division, are identified in the tables. All of the units of TransDiv 8 are listed because it counted among its members, HMAS *Manoora*, HMAS *Kanimbla*, and HMAS *Westralia*, part of the RAN's contribution to the operations. The latter infantry landing ship would claim one of the Japanese planes shot down by Allied ships.

Photo 19-2

Painting by James Turnbull of LCVPs (Higgins boats) laying a smoke screen around the ships anchored in Lingayen Gulf. Bomber and Kamikaze attacks usually came in the few minutes of twilight either before sunrise or after sunset (especially suicide attacks). Naval History and Heritage Command photograph #88-159-KM

While the Luzon Attack Force suffered terrible ship and personnel losses in the Lingayen operation, its cargo ships fared much better as a result of fewer enemy aircraft attacks during transit to the Lingayen Gulf, and as a result of defensive smoke screens laid in the transport areas by USN vessels.

Of the eight AKs/AKAs that garnered battle stars, all were earned on S-Day (9 January, the day of the amphibious landings), except for those of *Aquarius* (AKA-16) and *Libra* (AKA-12). Explanation follows in the next section.

**Luzon Attack Force (TF 77): Vice Adm. Thomas C. Kinkaid**
**Reinforcement Group (TG 77.9): Rear Adm. Richard L. Conolly**

| **TransDiv 2: Capt. W. S. Popham** | | |
|---|---|---|
| USS *Algol* (AKA-54) | No battle star | C2-S-B1 hull |
| **Bougainville Transport Unit: Captain Ballreich** | | |
| USS *Libra* (AKA-12) | 11 Jan 45 | C2-F hull |
| **Oro Bay Transport Unit** | | |
| USS *Uvalde* (AKA-88) | No battle star | C2-S-B1 hull |
| **Finschhafen Transport Unit** | | |
| USS *Warrick* (AKA-89) | No battle star | C2-S-B1 hull |
| **Noemfoor Transport Unit: Captain Davis** | | |
| USS *Diphda* (AKA-59) | No battle star | C2-S-B1 hull |

**San Fabian Attack Force (TF 78): Vice Adm. Daniel E. Barbey (landing of U.S. 6th and 43th Infantry Divisions of I Corps, on the easternmost beaches of Lingayen Gulf)**

**White Beach Attack Group (TG 78.1): Vice Adm. Daniel E. Barbey**
**Transport Group White Beach: Commodore C. G. Richardson**

| **TransDiv 7 (temporary): Commodore C. G. Richardson** | | |
|---|---|---|
| USS *Thuban* (AKA-19) | No battle star | C2-S-B1 hull |
| **TransDiv 6 (temporary): Capt. D. L. Ryan** | | |
| USS *Hercules* (AK-41) | No battle star | C3-E hull |
| **TransDiv 24 (temporary): Captain Welsh** | | |
| USS *Aquarius* (AKA-16) | 4-10 Jan 45 | C2-S-B1 hull |

**Blue Beach Attack Group (TG 78.5): Rear Adm. William M. Fechteler**
**Transport Group Blue Beach: Commodore Donald W. Loomis**

| **TransDiv 20 (temporary): Commodore Donald W. Loomis** | | |
|---|---|---|
| USS *Electra* (AKA-4) | No battle star | C2-T hull |
| USS *Auriga* (AK-98) | No battle star | C1-B hull |
| **TransDiv 26 (temporary): Capt. Harold J. Wright** | | |
| USS *Jupiter* (AK-43) | No battle star | C2 hull |
| **TransDiv 32 (temporary): Capt. Charles Allen** | | |
| USS *Mercury* (AK-42) | No battle star | C2 hull |

**Lingayen Attack Force (TF 79): Vice Adm. Theodore S. Wilkinson (landing of U.S. 37th and 40th Infantry Divisions of XIV Corps, across the middle of Lingayen Gulf's southern shores)**

**Attack Group Able (TG 79.1): Rear Adm. Ingolf N. Kiland**
**Transport Group Able: Commodore M. O. Carlson**
**TransDiv 28 (temporary): Commodore M. O. Carlson**

| | | |
|---|---|---|
| USS *Almaak* (AKA-10) | 9 Jan 45 | C3-E hull |

**TransDiv 8 (temporary): Capt. Samuel P. Jenkins**

| | | |
|---|---|---|
| USS *Titania* (AKA-13) | 9 Jan 45 | C2-F hull |
| USS *Sarasota* (APA-204) | 9 Jan 45 | |
| HMAS *Manoora* | | |
| HMAS *Kanimbla* | | |
| HMAS *Westralia* | | |

**TransDiv 38 (temporary): Capt. G. W. Johnson**

| | | |
|---|---|---|
| USS *Alshain* (AKA-55) | 9 Jan 45 | C2-S-B1 hull |

**Attack Group Baker (TG 79.2): Rear Adm. Forrest B. Royal**
**Transport Group Baker: Commodore Herbert B. Knowles**
**TransDiv 10 (temporary): Capt. S. M. Haight**

| | | |
|---|---|---|
| USS *Capricornus* (AKA-57) | No battle star | C2-S-B1 hull |

**TransDiv 18 (temporary): Commodore Herbert B. Knowles**

| | | |
|---|---|---|
| USS *Alcyone* (AKA-7) | 9 Jan 45 | C2 hull |

**TransDiv 30 (temporary): Captain Short**

| | | |
|---|---|---|
| USS *Chara* (AKA-58) | 9 Jan 45 | C2-S-B1 hull[3] |

## MOVEMENT OF FORCES TO THE OBJECTIVE

*The actual landings in the LINGAYEN area were relatively lightly opposed, as was the initial progress overland through the central plains of LUZON toward MANILA. By contrast, the Bombardment and Fire Support Groups, and the Escort Carrier and Minesweeping Groups, which preceded by three days the passage of the Amphibious Forces, were treated at LINGAYEN Gulf to a series of suicide air attacks of unparalleled ferocity, thus absorbing the efforts of the bulk of remaining Japanese air strength in, or capable of being staged to, the area.*

—Commander in Chief, U.S. Pacific Fleet and Pacific Ocean Areas, Operations in Pacific Ocean Areas – January 1945, 31 July 1945.

Movement of the San Fabian Attack Force to Lingayen Gulf and its landing operations, on 9 January, were relatively benign. Only one of the cargo ships assigned to this force, USS *Aquarius*, earned a battle star, and it appears there is an error associated with her award period (4-10 January). On 3 January, she was proceeding in company with units of

Task Group 78.1, when the combat air patrol shot down a "Frances" (Yokosuka P1Y1 Navy land-based bomber). Nothing of note occurred the following day.[4]

A week later, on 10 January, having completed unloading as part the San Fabian Attack Force (TF 78), *Aquarius* got under way, and joined Task Unit 79.14.2 en route to the Leyte Gulf. That evening, an enemy plane made a suicide attack on the convoy, and crashed into the superstructure of USS *DuPage* (APA-14). The transport lost 35 killed and 136 wounded; 5 men who were blown over the side were picked up by destroyers. Despite damage suffered, *DuPage* was able to proceed, and arrived safely at Leyte three days later to transfer her casualties and undergo emergency repairs.[5]

Bound for Lingayen Gulf, Task Groups 79.1 and 79.2, of the Lingayen Attack Force had departed Manus, on 31 December. Their route led through Surigao Strait into the Mindanao Sea, through the Sulu Sea, and into the China Sea. On 6 January, the convoy encountered its first enemy aircraft. A "Zeke" (Zero fighter) streaked across the convoy with four F4U Corsair fighter aircraft in hot pursuit, before the American planes opened fire and sent it crashing into the sea.[6]

On 8 January, soon after entering the South China Sea, the convoy came under several attacks. The first occurred at 0910, when two twin-engine planes dove out of the sun. As ships' guns opened fire, they dropped bombs which splashed into the water a few yards from the cargo ship USS *Almaak* (AKA-10). No damage was done because the bombs did not detonate. One fell about 1,000 yards ahead of the amphibious force command ship USS *Mt. McKinley* (AGC-7), but she also sustained no damage. The bombers fled upon the approach of friendly aircraft.[7]

That evening, four enemy planes were shot down by the combat air patrol as they approached the convoy. One plane broke through and, at 1855, crashed into the escort carrier USS *Kitkun Bay* (CVE-71), stationed between the two transport groups. The Kamikaze hit the vessel on her port side, amidships at the waterline. *Kitkun Bay* dropped out of the formation, smoking and listing heavily. Sea water poured in through the breech in her hull; it caused a 13-degree list, and she was settling by the stern. Her crew, after heroic effort, were ultimately able to successfully combat fires and flooding, and save their ship, but the attack killed sixteen men and wounded another thirty-seven.[8]

Other attacks followed those made earlier, in which enemy pilots had focused their efforts on large, "high-value" units: command ships, carriers, and those service force ships loaded with war materiel. At 1857, several "bogies" (unknown aircraft) closed the task group and dropped two bombs that fell in the water near the cargo ship USS *Titania* (AKA-13), fortunately causing no harm.[9]

In the final attack that day, at 1903 an enemy plane approached the convoy from the port quarter, flying high, and commenced a suicide dive aimed at HMAS *Westralia*, positioned in the fourth column of the formation. Along with other ships, the cargo ship USS *Alshain* (AKA-55) opened fire with her 5"/38 and twin 40mm guns, and later with her port 20mm guns. The fire was accurate, as hits were observed on the plane. The Kamikaze impacted the water, on the starboard side of *Westralia*, near her stern, and exploded. This affected steering which caused the landing ship to leave the formation. She was able to maintain speed and, after effecting repairs, resumed station in her column. *Westralia* claimed credit for shooting down her attacker, but apparently it was also hit by other ships shooting at the same target.[10]

## S-DAY (9 JANUARY 1945) IN LINGAYEN GULF

Photo 19-3

Anti-aircraft fire from ships of the U.S. Navy task force in Lingayen Gulf, Luzon, 10 January 1945.
Naval History and Heritage Command photograph #80-G-304355

The Luzon Attack Force reached Lingayen Gulf in the early morning on 9 January, and soon after its attack groups proceeded to their respective beaches near San Fabian and Lingayen. Since none of the cargo ships assigned to the San Fabian Attack Force earned battle stars for the landings there, the remaining portion of this account will focus on enemy action against the Lingayen Attack Force. The cargo ship USS *Chara* (AKA-58) and the other units of Transport Group Baker (TG 79.2) arrived in the transport area off Lingayen at 0700, and the ships separated preparatory to anchoring, and lowering boats.[11]

While boats were being launched, enemy planes appeared. At 0745, *Chara* opened fire on a single enemy aircraft with short bursts from her 40mm and 20mm guns. No hits were observed. As there were other ships and aircraft in the vicinity, the collective AA fire from ships was heavy. Despite this defense, one plane flew directly into USS *Columbia* (CL-56), resulting in the light cruiser losing 17 killed, 97 wounded, and 6 crew missing in action. Another was shot down by the fighter patrol. Three boat crew members were wounded, one fatally, by falling shrapnel while in an LCVP lying alongside *Chara*.[12]

## ATTACKS BY JAPANESE SUICIDE MOTORBOATS

> *At 0310 on 10 January, several enemy assault demolition boats carrying depth-charges made suicide attacks in the transport area, resulting in the sinking of* LCI(M) 974*; abandonment of* LCI(G) 365*; serious damage to* WAR HAWK *(AP) and* LST 1026*; and minor damage to* ROBINSON *(DD),* PHILIP *(DD), and LSTs* 610 *and* 925. *An expedition of LCT(G)s and LCI(M)s were organized to search out enemy suicide craft along the coast during the day, but no more were found.*
>
> —Commander in Chief, U.S. Pacific Fleet and Pacific Ocean Areas, Operations in Pacific Ocean Areas – January 1945, 31 July 1945.

As the first wave of landing craft hit the beaches as scheduled, unloading progressed favorably despite the crews of the cargo ships spending considerable time at their battle stations. At 0400, after *Chara* received a report of the presence of Japanese suicide boats in the Gulf, transfer of cargo ashore was curtailed. Between midnight and dawn of the first night, several ships were attacked by Japanese suicide boats, and suicide barges as well dynamite-laden swimmers were reported in the area. These threats made it necessary at night to darken ship, and suspend unloading.[13]

That night, a smoke plan was put into effect, but the screen proved equally advantageous to the enemy. The cover of the smoke, enabled Japanese to sneak among the ships with their suicide boats and their human bomb swimmers. Ceasing unloading at night helped to prevent the mistaking of landing craft for enemy boats. The cargo ship *Almaack* (AKA-12) stationed armed picket boats circling her to ward off any would-be attacker. Sporadically throughout the night, the chatter of machine-gun fire was heard as some alert boat crewman fired at a floating object—a suicide swimmer perhaps.[14]

Photo 19-4

Japanese motorboat on the beach at Lingayen Gulf. PT boats destroyed many of these *Shinyo* (suicide boats laden with explosives) during the Philippines campaign. Naval History and Heritage Command photograph #NH 44316

One of the ships damaged by a suicide boat was the transport USS *War Hawk* (AP-168), which was anchored in a berth adjacent to the cargo ship USS *Alcyone* (AKA-7). In answer to *War Hawk*'s call for assistance, *Alcyone* sent a fire and rescue party which remained aboard her for twelve hours. Earlier at 0410, a suicide boat, laden with explosives and going full-throttle, had crashed into her port side, blasting a 25-foot hole in number three hold, killing sixty-one men. As she lay without power, and as an engine room began to flood, repair crews worked below decks to restore electrical power and to patch the gash in the ship's side. Christened the "sitting duck" by her crew, *War*

*Hawk* remained off Lingayen until 11 January, when she began a slow trek under her own power to Leyte Gulf.[15]

Following the explosive boat attack, there were other threats to deal with. *War Hawk*'s gunners were kept busy repelling Japanese aircraft as they attacked her. Throughout the invasion fleet, concurrent threats both by air and sea that night were common, as described in a USS *Alshain* war diary entry:

> During one attack by enemy boats, a combatant ship illuminated the area. A plane dived very low over this ship, and a few moments later three explosions, apparently light bombs, were heard, one forward and two aft of the ship, in quick succession. Again, on the 12th, an enemy fighter dropped light bombs, one aft and one forward of the ship. Most plane attacks, however were of the suicide type and aimed at large combatant ships.[16]

During an air raid shortly after dawn, *Alcyone*'s gun crews scored a definite hit on a "Dinah" (Army reconnaissance plane), which crashed several hundred yards astern of the cargo ship. Two bombs had been dropped close aboard *Alcyone*, one each on her port and starboard quarters, wounding two crewmen with shrapnel.[17]

Withdrawal of assault shipping from Lingayen Gulf involved sailing one fast and one slow convoy for Leyte Gulf each day, commencing on S-Day. No ships were lost during these passages, but several were attacked by suicide planes with the following results.[18]

| Date | Ship | Damage | Casualties |
|---|---|---|---|
| 10 Jan 45 | USS *DuPage* (APA-41) | minor | 171 (35 KIA, 136 WIA) |
| 10 Jan 45 | USS *War Hawk* (AP-168) | near miss | |
| 12 Jan 45 | USS *LST-700* | (own fire) | 6 |
| 13 Jan 45 | USS *Zeilin* (APA-3) | minor | 40 (7 KIA, 3 MIA, 30 WIA) |
| 13 Jan 45 | USS *LST-700* | serious | 4 |

KIA: Killed in Action MIA: Missing in Action WIA: Wounded in Action

# 20

# Loss of Cargo Ship *Serpens* with Nearly all Hands

*I felt and saw two flashes after which only the bow of the ship was visible. The rest had disintegrated and the bow sank soon afterwards.*

—Coast Guard Lt. Comdr. Perry L. Stinson, commanding officer, USS *Serpens* (AK-97), who tragically witnessed from ashore, the obliteration of his ship by an explosion of munitions she carried, which killed all but two of those aboard, who miraculously survived the massive blast.[1]

Photo 20-1

Cargo ship USS *Serpens* (AK-97) in harbor during World War II.
Naval History and Heritage Command photograph #NH 89186

Tragedy struck the U.S. Navy, and particularly U.S. Coast Guard families when, on 29 January 1945, the cargo ship USS *Serpens* (AK-97) exploded and sank off Lunga Point, Guadalcanal, with the resultant loss of nearly her entire crew. *Serpens* had been acquired by the Navy, and commissioned, on 28 May 1943, but was manned by a U.S. Coast Guard crew under a Coast Guard commander. She was one of sixty *Crater*-class AKs converted from standard Maritime Commission E-C2-S-C1 Liberty ships, to Navy standards that permitted them to operate in forward areas if necessary.[2]

## SHIP MANNING

During World War II, the U.S. Coast Guard came under the control of the U.S. Navy, and Guardsmen fully manned more than 350 ships, including 76 tank landing ships, 21 cargo and attack-cargo ships, 75 frigates, and 31 transports. Additionally, the Coast Guard operated more than 800 of its own cutters (a collective term for USCG vessels), nearly 300 ships for the U.S. Army, and thousands of amphibious assault craft.[3]

*Serpens*' crew referred to her at various times as a "sea-going moving van," the "South Pacific Slow Freight," or a "floating packing case, with booms." An officer remarked that serving on her was like being in the Merchant Marine, without hitting State-side ports. Unlike attack cargo ships (AKAs), AKs did not carry landing craft. They were not intended to steam into an enemy-held beachhead on D-Day, and disembark personnel and supplies under fire. That didn't mean that AKs didn't see action—many did—but ordinarily, their job was to keep the supply lines running, following initial amphibious assaults.[4]

The Navy's manpower organization (also responsible for the Coast Guard) tried to fill *Serpens*' crew complement (as it sought to do for other AKs) with as many cargo-wise, ex-merchant marine sailors as possible. *Serpens* had to settle for twenty percent experienced crewmen, with the remainder a cross-section of young Americans from every walk of life. However, sixty percent of her wardroom were former Merchant Marine officers. In one year, the seasoned men aboard the ship were able to make real sailors out of the ex-civilians, who were credited with shooting down an enemy plane.[5]

## ACTION, BOREDOM, AND "DEAR JOHN LETTERS"

On 10 February 1944, *Serpens* arrived in Empress Augusta Bay, Bougainville Island, in the Solomons. At that time, U.S troops ashore had secured only a three- by ten-mile peninsula, and Japanese planes were bombing the harbor systematically, but ineffectually, every night.

During the eight days that *Serpens* remained in the harbor unloading, there were more than thirty air raids. With a cargo of gasoline and tons of bombs, a hit would have blown her into a million shards of steel.[6]

Every time the enemy planes came over, the ships in the harbor opened fire, but the bombers were too high for effective anti-aircraft fire. However, on the fourth night, a daring attacker came in low on its run, and *Serpens'* forward three-inch gun cut loose. A round hit the plane squarely and, in an instant, a whirling mass of flames plummeted into the sea with a hissing roar. Army Air Corps Intelligence personnel ashore credited *Serpens* with the hit, and her crew proudly painted a small Japanese plane with the rising sun above it, on the three-inch gun, as well as both the sides of the bridge.[7]

Typical life aboard *Serpens* was mostly tedious days of slow cruising, broken only by an occasional air raid in some atoll harbor, and rugged work loading and unloading the ship. The heat, particularly in the southwest Pacific, where she operated most of the time, was intense. Nearly everyone on board had one or more forms of heat rash, known by a variety of exotic names: Oriental crud, South Seas scabies, Jungle rot, or the Asiatic itch. To add to these woes, many in the crew experienced an epidemic of "heart trouble," resulting in their referring to themselves as, "the most popular ex-darling club in the Pacific," or the "You-Went-Away-And-Stayed-Away-Too-Long-Association."[8]

Owing to their long absences from home, the men had been losing sweethearts at increasing regularity. In their letters home, their general explanation for this to their "heart's desires," and as a matter of their own rationalizations, was that they couldn't make any promises (or even offer guesses) as to when they might return. There were also some pointed remarks amongst themselves about amorous war-workers back home, but on the whole, the men took their jilts good-naturedly. Finally, at a point of dejection, all the unlucky lovers got together, each with a photograph of his ex-girlfriend. In unison, they ripped the pictures to shreds, then—after repeating the girl's last words to them, and adding their own, colorful unprintable postscripts—tossed the remnants into the deep blue Pacific.[9]

## MODIFICATION TO CARRY ADDITIONAL MUNITIONS, AND SHIP LOSS AT LUNGA POINT, GUADALCANAL

In December 1944, *Serpens* underwent a drydocking availability at Wellington, New Zealand, for hull cleaning and important repairs, alterations, and additions. The latter work included the conversion of No. 1 hold (both upper and lower), No. 4 hold (lower), and No. 5 hold (lower) into magazines for the storage of munitions.[10]

Shortly before midnight on 29 January 1945, while anchored off Lunga Point, Guadalcanal, tragedy befell *Serpens* and her Coast Guard crew. She exploded at 2323, while handling bombs obtained from the Liberty ship SS *James B. Francis*. (*Serpens* was being loaded by the 492nd Port Battalion, U.S. Army, based at Guadalcanal, with bombs, plane parts, and general cargo.) The ship sank immediately, with the bow remaining above water for approximately two hours. Out of her complement of 206 officers and men, 196 were killed. The ship's commanding officer, Lt. Comdr. Perry L. Stinson USCG, one other officer, and six crewmembers ashore at the time, survived. So too, remarkably, did two members of the crew aboard the *Serpens*.[11]

Also killed were 57 Army personnel including a Public Health Service doctor, a soldier ashore hit by shrapnel, and a crewman aboard the patrol craft USS *YP-514* nearby. The 129-foot YP (ex-San Diego tuna clipper *American Beauty*), under Lt. George P. Paine, USNR, also suffered 4 injured, and damage to the vessel.[12]

Photo 20-2

Landing craft clustered offshore, during landing operations at Lunga Point, Guadalcanal, in November 1942.
National Archives photograph #80-G-30521

At the time of the explosion, Lieutenant Commander Stinson was in the quarters of Captain Conklin, 319th Fighter Control Squadron, preparatory to returning to the ship. He felt a shock, which he thought might be an earthquake, then came a great glare in the sky, followed by a tremendous explosion. Stinson immediately set out for the waterfront, while trying en route, to determine from the position of the smoke, whether it could have been the *Serpens*.[13]

Upon learning that *Serpens* had exploded, and with his captain's gig wrecked by the blast, Stinson commandeered a boat and in company with Lt. Comdr. Audrey A. Scott, USCGR, searched the water covered by debris (along with many other boats) for survivors. When he returned to the dock, he located the other ship's company members that had been ashore, as well as two survivors of the obliteration of the ship.

- Lt. John R. Clark, USCG
- Chief Commissary Steward Stanley M. Jones, USCGR
- Steward 3rd Class Richard M. Figgs, USCGR
- Fireman 1st Class Morris M. Houseknecht, USCGR
- Seaman 1st Class William E. Hughes, USCGR
- Seaman 1st Class Kelsie K. Kemp, USCGR
- Seaman 1st Class Robert J. Smart, USCGR
- Seaman 2nd Class Fidel O. Carmana, USCGR
- Seaman 2nd Class George S. Kennedy, USCGR[14]

Both Seamen Kemp and Kennedy had been rescued from the bow of *Serpens* by Capt. Mark L. Hersey Jr., USN (commander, Naval Bases, South Solomons Sub-Area).[15]

It was sultry that night, at least 100 degrees, and Seaman Kelsie Kemp hadn't liked the idea of sleeping deep below in his assigned berthing space, so he placed his cot on the ship's bow near the boatswain's locker where he normally worked, issuing tools and supplies to his shipmates. Seaman George Kennedy also decided to sack out on the open deck. When it looked like it might rain, Kemp and Kennedy relocated inside the boatswain's locker.[16]

They were both asleep when the *Serpens* exploded. In an interview decades later, Kemp could not remember the blast that people on shore reported hearing as far as 65 miles away. He only remembered clinging to the twisted remains of the bobbing bow, with "lumber, bodies, pieces of fish, all that old oil and stuff" floating nearby. The lights of searching boats approaching, penetrated what Kemp remembered as a black hole, the area on the sea where he waited, listening to the groans of his friend. Although it had then seemed to him an eternity before help arrived, Kemp later estimated it was about an hour.[17]

## RESULTS OF INQUIRIES INTO *SERPENS*' LOSS

The loss of *Serpens* was initially attributed to enemy submarine action, and three Purple Heart Medals were awarded to the two survivors and posthumously to Harry Levin, the U.S. Public Health Service physician. A court of inquiry later determined that the cause of the explosion could not be established from surviving evidence. By 1949, the U.S. Navy officially closed the case deciding that the loss was not due to enemy action (perhaps as a result of examining post World War II, Japanese submarine deck logs and operating records) but an "accident intrinsic to the loading process."[18]

### The Gold Star

Written by the mother of S1c William Fossett Nivin, USCGR
who went down with his ship, the USS *Serpens*

A GOLD STAR hangs in my window
For passers-by to see—
To them it means so little,
Yet—that star is killing me.

"Missing in action," we regret to say,
That fatal message read,
Then followed, like the hour of doom,
The report, our son was dead.

My heart is full of pain, my son,
As I cry out for you:
Yet, it seems I hear you say,
As you were wont to do:

"Now please don't worry, Mother,
For I'm quite all right today,
And with any luck at all
I'll soon be home to stay."

You dreamed of home that awful night,
Your "leave" had just come through—
Not ever dreaming, my dear one,
There would be no dawn for you.

Where your body lies I do not know—
Or does it matter where?
Because you see, my darling—
My own heart's buried there.

They say the war is over,
There's peace—so dearly won—
Yet to me this war will *never* end
Until I meet my son.[19]

**21**

# Assault of Iwo Jima

*At great cost, you'd take a hill to find then the same enemy suddenly on your flank or rear. The Japanese were not on Iwo Jima. They were in it! I'd known combat in the Solomons with its sly ambushes and jungle firefights, but Iwo was another kind of war. On Iwo by the 8th day, only two officers of my second battalion (26th Marines, 5th Marine Division) were standing.... We had one prisoner—unconscious, his clothes blown off.*

—Col. Thomas M. Fields, USMC (Ret.)[1]

Photo 21-1

Shells explode ashore during the bombardment of Iwo Jima, on 17 February 1945. This view of Mount Suribachi's west side was probably taken during pre-invasion minesweeping activities, as the ship in the right foreground appears to be a YMS.
Naval History and Heritage Command photograph #NH 104142

The battles of Iwo Jima and Okinawa, fought by Allied forces between February and August 1945, were operations preliminary to securing island bases for a final B-29 bomber assault on Japan. The Iwo Jima operation was conducted first because it was expected to be easier than an assault on Okinawa. Because of the enemy's prolonged and bitter defense of Leyte and Luzon, the planned dates for both actions had slipped. Thus, the Fifth Fleet had to cover and support both invasions, while the Seventh Fleet and its amphibious forces were concurrently engaged in liberating the Philippines.[2]

Preliminary bombings of Iwo and the minor airbase at Chichi Jima were conducted by shore-based aircraft from the Marianas. Supporting operations for the invasion were begun, on 16 and 17 February, when Fast Carrier raids were made on the Tokyo area of the Japanese home islands. During these raids, and one ensuing on the 25th, 420 enemy planes were shot down, 228 were destroyed on the ground, and some effort was directed against aircraft engine plants and airplane factories. These raids eliminated enemy aircraft that otherwise would have been available for defending Iwo Jima and Okinawa. [3]

On 19 February, as naval gunfire pounded the island, more than 450 ships massed off Iwo Jima. Marines of the 4th and 5th Divisions hit the four assault beaches shortly after 0900, initially finding little enemy resistance. Coarse volcanic sand hampered their movement as they struggled to move up the beach from the surf zone. As the protective naval gunfire subsided to allow for troop advancement, the Japanese emerged from fortified underground positions to begin a heavy barrage against the invading force. The 4th Marines, continuing to push forward against heavy opposition, took the Quarry, a Japanese strong point, while the 5th Marine Division's 28th Marines isolated Mount Suribachi that same day.[4]

The 3rd Marine Division joined the fighting on the fifth day, charged with securing the center sector of the island. The fortified enemy defenses linked miles of interlocking caves, concrete blockhouses and pillboxes, which required frontal assaults to gain nearly every inch of ground. Maj. Gen. Harry Schmidt, commanding the Fifth Amphibious Corps—of which the 3rd, 4th, and 5th Marines were a part—declared Iwo Jima secured on 16 March. However, isolated ground fighting continued between then and the official completion of the operation, on 26 March 1945.[5]

## NO ENEMY-CAUSED DAMAGE TO CARGO SHIPS

Owing to the Fast Carrier Force raids on the Tokyo area, and associated large numbers of enemy planes shot down, or destroyed on the ground, there was little enemy air opposition at Iwo Jima. Lacking availability of reinforcements, there was a paucity of Japanese aircraft in the Bonin Islands-Iwo Jima area. Forty enemy aircraft were reported destroyed by U.S. forces; 8 planes destroyed in the air, 16 shot down by shipboard AA fire, 11 destroyed on the ground, and 5 by suicide crashes.[6]

The only opposition encountered by cargo ships assigned to Assault Force transport divisions, was enemy fire on their parties on landing beaches, and the occasional bomb dropped in transport areas. Because no cargo ships suffered damage, and there were no personnel casualties as a result of enemy action, there is little to report of their activities during the landings. However, the AKs/AKAs associated with the assault and subsequent occupation of Iwo Jima are identified in the table, with the periods for which they earned battle stars, and the command structure associated with the cargo ships provided.[7]

Admiral Spruance was in overall command of the Central Pacific Task Forces which comprised nearly all elements engaged in the operation. The Joint Expeditionary Force (Task Force 51), which actually landed the troops and effected the capture of the island, was under the direct command of Vice Admiral Turner.

Rear Admiral Hill commanded the Attack Force (Task Force 53). It included the transports, cargo vessels, and dock landing ships of Transport Groups Able and Baker, which carried the troops, equipment, and supplies to the objectives.

**Attack Force (TF 53): Rear Adm. Harry W. Hill**

**Transport Group Able (TG 53.1): Commodore John B. McGovern**

| **Transport Division 46 (TU 53.1.1): Commodore Henry C. Flanagan** | | |
|---|---|---|
| USS *Tolland* (AKA-64) | 19-28 Feb 45 | C2-S-AJ3 hull |
| USS *Whiteside* (AKA-90) | 19 Feb-7 Mar 45 | C2-S-B1 hull |
| **Transport Division 47 (TU 53.1.2): Capt. A. S. Wotherspoon** | | |
| USS *Whitley* (AKA-91) | 19-27 Feb 45 | C2-S-B1 hull |
| USS *Yancey* (AKA-93) | 19 Feb-2 Mar 45 | C2-S-B1 hull |
| **Transport Division 48 (TU 53.1.3): Capt. C. L. Andrews Jr.** | | |
| USS *Athene* (AKA-22) | 19-28 Feb 45 | S4-SE2-BE1 hull |
| USS *Stokes* (AKA-68) | 19 Feb-2 Mar 45 | C2-S-AJ3 hull[8] |

**Transport Group Baker (TG 53.2): Commodore Henry C. Flanagan**

| **Transport Division 43 (TU 53.2.1): Commodore Henry C. Flanagan** | | |
|---|---|---|
| USS *Artemis* (AKA-21) | 19-27 Feb 45 | S4-SE2-BE1 hull |
| USS *Shoshone* (AKA-65) | 19 Feb-1 Mar 45 | C2-S-AJ3 hull |
| **Transport Division 44 (TU 53.2.2): Capt. J. H. Seyfried** | | |
| USS *Southampton* (AKA-66) | 19 Feb-1 Mar 45 | C2-S-AJ3 hull |
| USS *Starr* (AKA-67) | 19 Feb-5 Mar 45 | C2-S-AJ3 hull |

| **Transport Division 45 (TU 53.2.3): Capt. A. C. J. Sabalot** | | |
|---|---|---|
| USS *Leo* (AKA-60) | 19-28 Feb 45 | C2-S-B1 hull |
| USS *Muliphen* (AKA-61) | 19 Feb-4 Mar 45 | C2-S-B1 hull[9] |

The cargo ships assigned to Task Force 51 (along with the transport ships of their divisions, which are not listed below) were a part of the reserve forces. This term refers to combat forces afloat offshore (an "ace in the hole"), ready to land, should they be needed in the event enemy resistance proved greater than anticipated.

| **Joint Expeditionary Force Reserves (TG 51.1): Commodore Lewis**<br>**Transport Squadron 11 (TU 51.1.1): Commodore Donald W. Loomis** | | |
|---|---|---|
| USS *Almaack* (AKA-10) | 19 Feb-6 Mar 45 | C3-E hull |
| USS *Warrick* (AKA-89) | no battle star | C2-S-B1 hull |
| **Transport Division 32 (TU 51.1.3): Capt. W. S. Popham** | | |
| USS *Jupiter* (AK-43) | 19 Feb-14 Mar 45 | C2 hull |
| USS *Libra* (AKA-12) | 19 Feb 45 | C2-F hull |
| **Transport Division 33 (TU 51.1.4): Capt. S. M. Haight** | | |
| USS *Alhena* (AKA-9) | 19 Feb-5 Mar 45 | C2-S hull |
| USS *Hercules* (AK-41) | 19 Feb-14 Mar 45 | C3-E hull[10] |

The Logistics Group of Rear Adm. Donald B. Beary (commander, Service Squadron Six) was composed of many types of ships, including cargo vessels, responsible for furnishing support at sea to the Central Pacific Task Forces. The attack cargo ship *Virgo* was one of the ships supporting the Okinawa operation.

| **Logistics Support Group (TG 50.8): Rear Adm. Donald B. Beary**<br>**Division 1 (50.8.12): Comdr. Herbert E. Randall** | | |
|---|---|---|
| USS *Virgo* (AKA-20) | 17 Feb-2 Mar 45 | C2-S-B1 hull |

All of the AKs/AKAs listed in the preceding tables earned battle stars for the Assault and Occupation of Iwo Jima (with the exception of USS *Warrick*), as did the cargo ships identified below.

| **Iwo Jima operation: Assault and occupation of Iwo Jima** | | |
|---|---|---|
| USS *Lakewood Victory* (AK-236) | 28 Feb-8 Mar 45 | VC2-S-AP2 hull |
| USS *Electra* (AKA-4) | 9-16 Mar 45 | C2-T hull |
| USS *Thuban* (AKA-19) | 9-16 Mar 45 | C2-S-B1 hull |
| USS *Woodford* (AKA-86) | 24-30 Jun 45 | C2-S-AJ3 hull |
| USS *Wyandot* (AKA-92) | 26 Mar-29 Apr 45 | C2-S-AJ3 hull |

22

# Assault on Okinawa

*The effectiveness of radar pickets in protecting an amphibious operation against enemy air attacks was demonstrated. They provided air warning service, shot down many planes by AA and, by controlling their own CAP's [combat air patrols], contributed to the destruction of many more. Although the radar picket ships suffered heavy losses from these attacks, it is believed that the enemy committed a serious error in concentrating upon them instead of avoiding or ignoring them in favor of the transports.*

—Commander in Chief, U.S. Pacific Fleet and Pacific Ocean Areas, Operations in the Pacific Ocean Areas, 16 October 1945.

Operations in the Pacific, during 1944, were directed toward the establishment of air and supply bases for the final assault on the Japanese homeland. By November 1944, B29 bombers were operating from bases in the Marianas. Earlier sea battles in the Philippines had inflicted severe damage on the enemy fleet, so that U.S. Navy Fast Carrier Forces were in a position to carry out air raids on Japan almost at will. The capture and occupation of Okinawa in the Nansei Shoto (islands) was undertaken to secure a position for further attacks against the Empire. Okinawa offered numerous sites for airfields from which almost any type of aircraft could bomb Japan under fighter escort, and could attack the enemy's lifelines between the home islands and conquered territories to the south.[1]

The capture and occupation of Okinawa involved an amphibious operation larger than any other in the Pacific during the war; one to which were committed: 318 combatant vessels, and 1,139 auxiliaries (including amphibious ships and craft), as well as approximately 548,000 U.S. Soldiers, Sailors, and Marines. This offensive carried U.S. and UK flags, and those of other Commonwealth ships that were a part of the British Pacific Fleet, to within 350 miles of the enemy homeland.[2]

Map 22-1

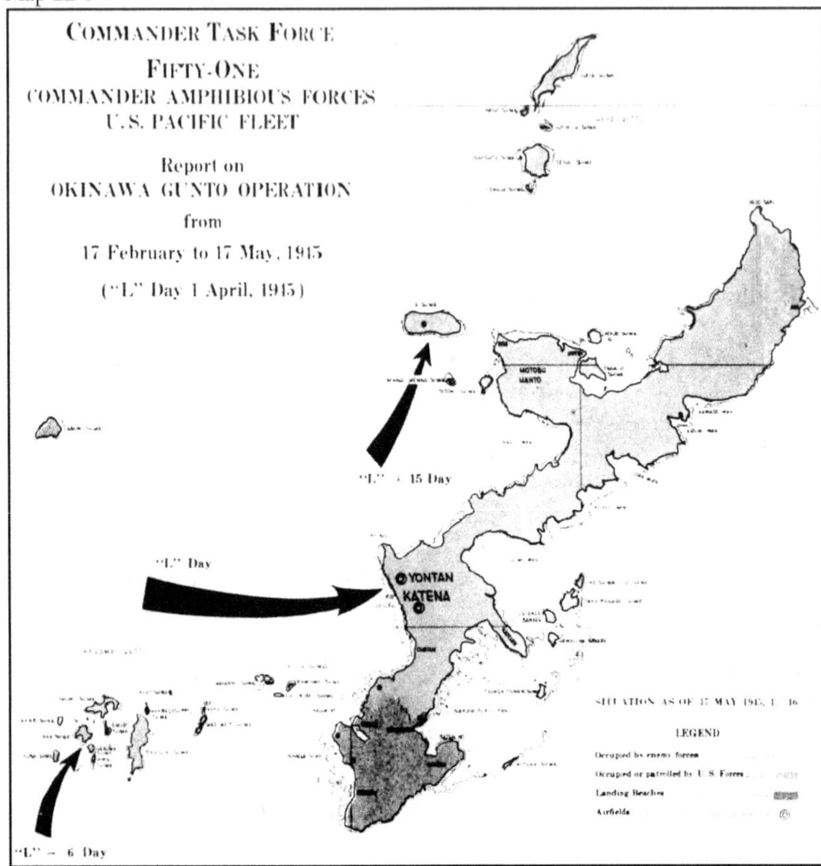


Amphibious landings at Kerama Retto preceded those at Okinawa to the northeast Commander Amphibious Forces, U.S. Pacific Fleet, General Action Report, Capture of Okinawa Gunto, Phases I and II, 17 February 1945 to 17 May 1945 – Submission of, 25 July 1945

The Battle of Okinawa (1 April-22 June 1945) was so massive that it is impossible to capture the scope of the U.S. Navy's valor, and that of the British carrier force involved in supporting combat operations in this short chapter. Accordingly, the next few pages merely sketch out the backdrop against which the combat took place, contributing fuller dimension for a few actions involving AKAs.

## CAPTURE AND OCCUPATION OF KERAMA RETTO

On 26 March, six days before the scheduled landings on Okinawa, Rear Admiral Kiland's Task Group 51.1 (Western Islands Attack Group) captured Kerama Retto, a group of small islands fifteen miles to the southwest of the main objective. The occupation of this island group provided the Joint Expeditionary Force a protected anchorage, and necessary base for logistics support, and from which seaplanes could carry out search and anti-submarine patrols. Task Group 51.1 was composed of one transport squadron with sufficient landing ships to transport the 7th Infantry Division (reinforced), and necessary screening ships and control craft to provide independent movement and amphibious assault by that single division.[3]

The five attack cargo ships assigned to the transport squadron are identified in the table, as well as the qualifying periods for battle stars they earned.

**Western Islands Attack Group (TG 51.1): Rear Adm. Ingolf N. Kiland (U.S. 7th Infantry Division: Maj. Gen. Andrew Davis Bruce)**

| **Transport Division 49 (TU 51.1.3): Commodore Thomas B. Brittain** | | |
|---|---|---|
| USS *Oberon* (AKA-14) | 26 Mar-26 Apr 45 | C2-F hull |
| USS *Torrance* (AKA-76) | 26 Mar-30 Apr 45 | C2-S-AJ3 hull |
| **Transport Division 50 (TU 51.1.4): Capt. Elmer Kiehl** | | |
| USS *Tate* (AKA-70) | 26 Mar-22 Apr 45 | C2-S-AJ3 hull |
| **Transport Division 51 (TU 51.1.5): Capt. J. H. Willis** | | |
| USS *Wyandot* (AKA-92) | 26 Mar-29 Apr 45 | C2-S-AJ3 hull |
| **Transport Division 10 (TU 51.1.2): Capt. H. D. McIntosh** | | |
| USS *Suffolk* (AKA-69) | 26 Mar-30 Apr 45 | C2-S-AJ3 hull[4] |

Photo 22-1

Painting *Kerama Retto* by C. W. Smith, 1945, depicting a destroyer—perhaps the USS *William F. Porter* (DD-579)—firing on what may be "suicide boats" in the roadstead. Naval History and Heritage Command photograph #NH 92591-KN

No opposition was encountered by the Western Islands Attack Group during movement from Leyte (assembly and loading point) to the Kerama Retto islands. Nor was there any serious opposition or interference by the enemy encountered in the actual landings on 26 March, and casualties to troops were correspondingly low.[5]

Apparently, the assault on Kerama Retto came as a surprise to the Japanese, as evidenced by the small preparation for defense of the islands, which were held by relatively weak forces. An unintended consequence, but added benefit of their capture, was the discovery that the islands were a base for suicide boats, used to attack any Allied naval forces which might assault Okinawa. More than 350 of these craft were destroyed in Kerama Retto before effective use of the potentially dangerous weapons could be made—likely reducing naval losses and damage to ships and craft of the main assault forces.[6]

Because of the rapid location and destruction of suicide boats, enemy reaction to Allied offense at Kerama Retto came mainly in the form of suicide plane attacks. In Task Group 51.1, these caused serious damage to one destroyer and two transports, and minor damage to another two transports. Additionally, the attack cargo ship USS *Wyandot* (AKA-92), received significant underwater hull damage from a bomb delivered in a conventional aircraft attack.[7]

## *WYANDOT* DAMAGED BY ENEMY BOMBER ATTACK

> *It was a masthead-height bombing attack by a single KATE, skillfully executed from the standpoint of the safe escape of the attacking plane but not skillful from the standpoint of damage inflicted.*
>
> —Commander Transport Division 41 describing in a report, a single plane attack on the USS *Wyandot*, carried out by a Japanese Nakajima B5N Navy type 97 carrier attack bomber.[8]

On the night of 29 March, *Wyandot* was anchored at Kerama Retto in company with various ships of Task Group 51.1. Periodically (since before midnight on the 28th), smoke was ordered made, following receipt of reports of enemy aircraft activity, then ceased once "all clear" was passed. Between 0050 and 0808, several reports of enemy activity were received, one involving an aircraft attack on *Wyandot*:

- 0050: Minesweeper USS *Design* (AM-219) reported a twin-engine plane came over low and fast, and dropped two objects near a special reflector buoy laid several days earlier
- 0123: Destroyer USS *Twiggs* (DD-591) reported a torpedo dropped across her bow
- 0213: Transport USS *Crosley* (APD-87) reported a "Betty" crashed off her starboard beam
- 0304: Minesweeper USS *Hazard* (AM-240) reported a low flying aircraft dropped object in the water
- 0405: Attack cargo ship USS *Wyandot* (AKA-92) reported a near miss by a dropped bomb or mine on her starboard bow
- 0450: Tank landing ship USS *LST-130* reported men swimming from southeast corner of Yakabi Shima to small boats
- 0633: Report of transports USS *Bunch* (APD-79) and *Crosley* (APD-87) having intercepted two small suicide boats; one of the boats exploded, the second, after dropping a depth charge, was still afloat
- 0808: Transport USS *Scribner* (APD-122) reported three enemy suicide craft destroyed; no survivors[9]

Investigation aboard *Wyandot* revealed that she'd been damaged by the second of two aircraft-dropped bombs, which hit the water ten to twenty yards off her starboard side. The bomb detonated opposite bulkhead 41, and hold No. 1, and the forward, starboard deep tank of hold No. 2 were penetrated. (The first bomb fell about fifty yards off the starboard side of hatches four and five, and did no damage other than to splash picric acid on the hatches.) *Wyandot* took an eight-degree list to starboard, and settled by the bow, but pumping, shifting of liquid load, and counterflooding of the forward deep tank of hold three, corrected this condition.[10]

On the evening of 2 April, Rear Admiral Kiland transferred command of Task Group 51.1 to Commodore Thomas Brittain, and Kiland took up new duties as commander, Task Group 51.15 (Senior Officer Present Afloat Kerama Retto). During the first half of April, local repair work was carried out to make *Wyandot* seaworthy for continued participation in the Okinawa operation. While this effort was in progress, the cargo ship served as unofficial flagship for Comdr. Carl H. Holm, USN, then in charge of salvage in the Kerama Retto area.[11]

## SHIP LOSES / DAMAGE TO KAMIKAZE ATTACKS

The 1 April landings on Okinawa by the Northern and Southern Attack Forces, occurred in concert with a feint (a deceptive amphibious demonstration), discussed later, in an effort to draw defenders away from the main site. The actual landings brought, as expected, a massive air response on the part of the enemy. Counter-response by the Joint Expeditionary Force and Fast Carrier Force through 30 April, cost the Japanese 1,124 planes. Of this total, 461 were shot down by ships' gunfire, 530 by combat air patrol (CAP), and 133 as a result of suicide crashes.[12]

In this same period, the Allies suffered 15 ships sunk, and another 137 damaged by enemy air attacks (see Appendix F.) The majority of the ship damage was the result of Kamikaze attacks. However, owing to the excellent performance of destroyers on radar picket stations and CAP engaging enemy air raids, few air attacks in force reached the transport area off Okinawa.[13]

Map 22-2

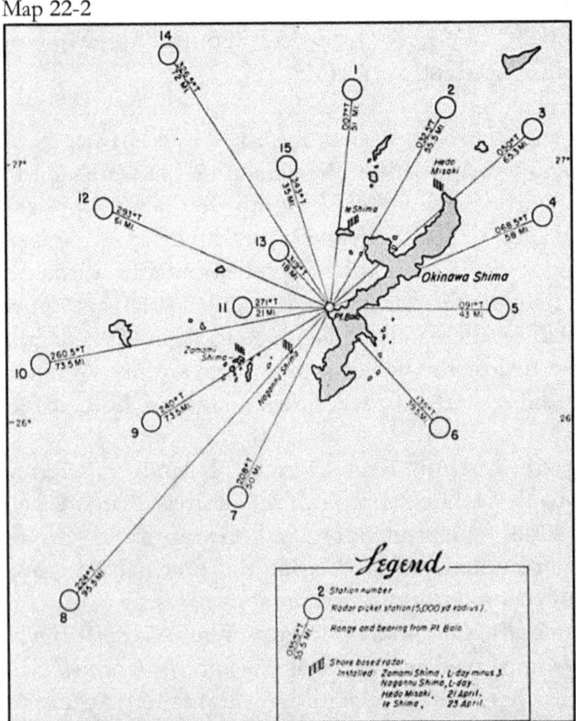


Okinawa Operation Radar Picket Stations
Commander in Chief, U.S. Pacific Fleet and Pacific Ocean Areas, Operations in Pacific Ocean Areas – April 1945, 16 October 1945

In the early stages of the campaign at Okinawa, destroyers and minelayers (converted destroyers) were assigned to picket stations to detect and destroy, through the use of fighter aircraft under their control, or by their own gunfire, approaching waves of enemy aircraft. As the scope and frequency of air attacks on these exposed "sentries" increased, it became necessary to augment the picket stations with LCSs and LSM(R)s. These support landing craft, and medium landing ships with rocket launchers, added firepower and were capable of providing immediate assistance to stricken picket ships.[14]

Of the original nineteen destroyers and minelayers pressed into "picket" duties, one was sunk by a mine, and four by Kamikaze aircraft. Eight other ships were seriously damaged, and three received minor damage as a result of attacks by suicide aircraft.[15]

After bolstering the number of ships assigned, a radar picket line, consisting of two or three destroyer divisions, was typically stationed 30 to 40 miles on the flank or flanks of the anticipated direction of enemy approach. The destroyers along this picket line normally controlled two to four divisions of fighter aircraft. The purpose of the picket line was to facilitate interceptions at as great a distance as possible from the transport area. A minimum of nine radar picket stations was necessary for complete effective coverage (see Map 22-2 on preceding page, which also identifies Okinawa shore-based air warning stations).[16]

The heaviest enemy air attacks came on 6, 11, 12, 16, 22, 27, and 28 April. Attacks the morning and night of 2 April resulted in minor damage to USS *Tyrrell* (AKA-80), and extensive structural damage to USS *Achernar* (AKA-53). Both attack cargo ships had arrived in the transport area off the west coast of Okinawa the previous day. The Southern Attack Force, of which they were a part, was tasked with putting its assault troops ashore along the southern part of a five- to six-mile strip of beach off Hagashi.[17]

This area was chosen because it provided ready access to airfields at Yontan and Kadena. However, owing to the geography of the site, the planned landings were expected to be strongly contested. Cliff faces at the rear of the beaches were honeycombed with caves and tunnels from which the areas below could be enfiladed by gunfire (allow engagement of the entire depth of soldiers advancing upward from the shoreline). Behind the beaches was a thick, six- to ten-foot-high seawall of masonry and concrete, and inland from it, the area surrounding the airfields was dominated by tunneled and fortified hills.[18]

However, what had been anticipated to be hard, proved relatively easy. The landings on Easter Sunday (1 April) by the Northern and Southern Attack Forces went off like clockwork. They were made on schedule against light opposition. The airfields were captured that same day and, on 3 April, U.S. troops reached the east coast of Okinawa.[19]

**Southern Attack Force (TF 55): Rear Adm. Harry W. Hill (U.S. Army 7th Infantry Division: Maj. Gen. Archibald V. Arnold, and 96th Infantry Division: Maj. Gen. James L. Bradley)**

| | | |
|---|---|---|
| **TransDiv 13** | | |
| USS *Aurelia* (AKA-23) | 1-14 Apr 45 | S4-SE2-BE1 hull |
| USS *Corvus* (AKA-26) | 1-10 Apr 45 | S4-SE2-BE1 hull |
| **TransDiv 14** | | |
| USS *Achernar* (AKA-53) | 1-19 Apr 45 | C2-S-B1 hull |
| **TransDiv 37** | | |
| USS *Algorab* (AKA-8) | 1-9 Apr 45 | C2 hull |
| USS *Tyrell* (AKA-80) | 1-10 Apr 45 | C2-S-AJ3 hull |
| **TransDiv 38** | | |
| USS *Alshain* (AKA-55) | 1-5 Apr 45 | C2-S-B1 hull |
| **TransDiv 39** | | |
| USS *Algol* (AKA-54) | 1-10 Apr 45 | C2-S-B1 hull |
| USS *Arneb* (AKA-56) | 1-10 Apr 45 | C2-S-B1 hull |
| **TransDiv 40** | | |
| USS *Capricornus* (AKA-57) | 1-9 Apr 45 | C2-S-B1 hull |
| USS *Chara* (AKA-58) | 1-6 Apr 45 | C2-S-B1 hull |
| **TransDiv 41** | | |
| USS *Diphda* (AKA-59) | 1-10 Apr 45 | C2-S-B1 hull |
| USS *Uvalde* (AKA-88) | 1-9 Apr 45 | C2-S-B1 hull |
| **TransDiv 42** | | |
| USS *Starr* (AKA-67) | 1-10 Apr 45 | C2-S-AJ3 hull[20] |

## KAMIKAZES CRASH *TYRELL* AND *ACHERNAR*

Shortly after dawn on 2 April, a twin-engine suicide plane attempted to first strafe, then crash into USS *Tyrell* (AKA-80). Fortune smiled on the attack cargo ship that day, as she was apparently the target of an inexperienced pilot. No machine gun hits were made on her, and while trying to crash the ship's bridge, the aircraft sheared off *Tyrell*'s radio antennae, then impacted her main mast, lower starboard yard arm support, and starboard five-ton boom at No 5. hatch. The fuselage of the demolished aircraft fell into the sea off the ship's starboard quarter, leaving parts of the wing and tail strewn across the after deck. In memory of the failed attack, *Tyrell*'s crew painted a miniature Japanese flag on the cargo boom, in the same manner the USS *Mercy* (AK-42) crew had done earlier in the war, when one of her booms had similarly dispatched an attacking aircraft (see Chapter 16).[21]

That night (0043 on 3 April), a determined, and more proficient Japanese pilot sacrificed his life upon hitting the port side of USS

*Achernar* (AKA-53) at No. 3 hatch, killing five members of ship's company, wounding another twenty-seven, and also four passengers aboard her. At the time of the attack, *Achernar* was steaming in company with Transport Division 14 during night retirement from Okinawa. This practice of interrupted cargo discharge was intended to afford ships greater safety during the hours of darkness before their return to the transport area at, or shortly after daybreak.[22]

The suicide aircraft that crashed *Achernar* was identified as a "Sonia" (Mitsubishi Ki-51 Army Type 99 assault plane), thought to have come in at an altitude of thirty feet. The bomb it carried was estimated to be a 500 to 1,000-pound fragmentation type with a delayed action fuse. Harm done by its high order detonation would have been worse, had not an LCM medium landing craft (located on No. 3 hatch, starboard side) absorbed much of the explosion. As it was, bomb fragments of ½ to 4 inches in diameter caused damage to the port side of No. 3 hatch, deck winches, forward midship superstructure, and ship's hull between frames 65 and 80.[23]

Photo 22-2

Japanese Army Ki-51 "Sonia" light bomber over Bataan, Leyte, Philippines, 1942. Naval History and Heritage Command photograph #NH 73534

A resultant fire caused by the bomb detonation was extinguished at 0100, and *Achernar* was able to continue her mission. That morning, she returned to the transport area off Okinawa and in mid-afternoon, transferred her casualties to the hospital ship USS *Solace* (AH-5). The

Next day, 3 April, *Achernar* proceeded to the Kerama Retto anchorage, and transferred her dead to Graves Registration Service, Kerama Retto:

- Seaman 1st Class Marion D. Freeman, USNR
- Seaman 1st Class Harold E. Harding, USNR
- Seaman 1st Class James W. Maiorana, USNR
- Seaman 2nd Class Ogle D. Shipley, USNR
- Seaman 1st Class Arthur O. Tobasco, USNR[24]

## *STARR* IS TARGET OF SUICIDE BOAT ATTACK

Photo 22-3

Attack cargo ship USS *Starr* (AKA-67) at Bath, Maine, in 1945.
Naval History and Heritage Command photograph #NH 77111

During her nine days at Okinawa, USS *Starr* (AKA-67), also a member of the Southern Attack Force, had come through a great number of air raids unscathed. This changed at 0420 on 9 April, when a tremendous explosion rocked her from stem to stern. At first, its source was believed to be a torpedo. However, it was soon ascertained, even on that moonless night, that the cause was a suicide boat.[25]

Fortunately for all aboard the cargo ship, damage was minimal and personnel injuries were light, apparently as a result of the two Japanese aboard the boat having a last-minute change of heart, regarding the need to sacrifice their lives for their emperor. They abandoned the boat before impact and, unsteered, it crashed into some landing craft moored alongside *Starr*, instead of the ship itself. Because the explosion was premature, and a sufficient distance from her side, seawater absorbed the shock. The sailors on watch in their landing crafts were knocked unconscious by the explosion, and sustained minor injuries. Two enemy personnel were sighted and shot (presumably by sailors aboard the cargo ship) as they attempted to swim to safety.[26]

As described by an entry in *Starr*'s ship's history, a thorough investigation of the ship and adjacent area at first light, revealed some interesting aspects of the attack:

> The strangest part of the incident was that most of the boat carrying the depth charge (with which the *Starr* was to have been sunk) landed in pieces in the small craft alongside. The engine was recovered from an LCM as were the rudder, screw and various non-metal parts including most of the bow. Another strange element was the discovery, on the *Starr*'s flying bridge many feet above the water, of a box wrapped in cloth with Jap[anese] characters printed on it which contained a dried fish and two large net sacks of biscuits. The Japs apparently intended to take only lodging wherever they went and supply their own board. [Clearly, their decision not to make the ultimate sacrifice was preplanned].[27]

This was *Starr*'s only contact with the enemy at Okinawa. The next day she left there in convoy, bound for Guam.[28]

## BRITISH PACIFIC FLEET CONTRIBUTIONS

> *Defense against suiciders has not been perfected and they remain a major threat. The best defense is destruction before takeoff or by the CAP [combat air patrol] before they reach their targets. Those which evade the CAP must be made to penetrate a large volume of [shipboard] fire from all guns which bear. [Ship] evasive action and speed should be used, but their effect is limited by the rapidity of the action. The volume of fire cannot be discontinued when the attacker bursts into flame, but must continue until he disintegrates or crashes into the sea.*
>
> *The operations of the British Carrier Force (TF 57) was carried out with commendable initiative and reflected credit on the British personnel concerned and their leadership.*
>
> —Adm. Raymond A. Spruance, USN, commander, U.S. Fifth Fleet.[29]

As previously highlighted, destroyers and destroyer-type ships on picket stations at Okinawa bore the brunt of Kamikaze attacks, while directing combat action patrol fighter aircraft to enemy air targets, or shooting attackers down themselves. The great sacrifices of these ships and their crews, significantly reduced the numbers of planes otherwise available to attack transports and cargo ships. The sailors aboard all U.S. ships and craft at Okinawa, including those of the AKAs, also owed a great debt of gratitude to the British Pacific Fleet (BPF).

The British Pacific Fleet under the command of Vice Adm. Sir Bernard Rawlings, KCB RN, was the most powerful strike force ever assembled by the Royal Navy. It included 6 fleet carriers and their squadrons, 4 light carriers, 2 maintenance carriers, 9 escort carriers, 4 battleships and dozens of cruisers, destroyers and lesser combat ships, as well as a huge train of supply and maintenance ships, oilers and assorted auxiliaries.[30]

The combatant ship component of the BPF had been assigned to the U.S. Fifth Fleet, on 23 March, and designated Task Force 57. Its mission was to neutralize enemy air installations in Sakishima Gunto. Between 26 March and 25 May, Fleet Air Arm pilots aboard the carriers carried out twenty-two days of strikes against Sakishima Gunto and two days against Formosa (Taiwan) airfields.[31]

As noted by Admiral Spruance, commander Fifth Fleet, in his report on the Ryukyus operation:

> Complete neutralization of SAKISHIMA GUNTO airfields by this means was not to be expected, but the efforts of the British Carrier Force in that direction unquestionably greatly reduced the magnitude of enemy attacks upon our forces in the OKINAWA area by aircraft from or staging through SAKISHIMA GUNTO, thereby reducing our losses and damage and contributing significantly to our success.[32]

During these operations, suicide plane attacks were made on the carriers HMS *Indefatigable*, *Formidable*, *Indomitable*, and *Victorious*. Their armored flight decks reduced damage caused by these hits, and the carriers remained operational.[33]

## DOZENS OF AKS/AKAS AWARDED BATTLE STARS

A total of 2,586 battle stars were awarded to the 2,253 U.S. Navy ships and 374 supporting units for the assault and occupation of Okinawa. The Battle of Okinawa, which extended into late June 1945, was one of the most ferocious of the war. American casualties reached a staggering 49,151, with 12,520 killed or missing. The cost to the Navy was 34 ships and craft sunk, and another 368 damaged, with over 4,900 sailors killed or missing in action, and over 4,800 wounded.[34]

Among the ships earning battle stars were eighty-eight AKs/AKAs/AKSs. The eighteen AKAs associated with the Western Islands Attack Group, and the Southern Attack Group, have been identified in preceding pages. The remaining seventy ships are listed in the tables which follow, beginning with six attack cargo ships that participated in an amphibious demonstration carried out on 1 April, concurrent with landings by the Northern and Southern Attack Forces.

## AMPHIBIOUS FEINT CONCURRENT WITH LANDINGS

On 1 April, and again the following day, a Demonstration Group under the command of Rear Adm. Jerauld Wright (Task Group 51.2) made mock landings on the southeastern beaches of Okinawa, in coordination with the actual assault landings on the Haguchi beaches by the Northern and Southern Attack Forces. Wright's task group consisting of one Transport Group (51.2.1) and one Tractor Flotilla (51.2.6), carrying the 2nd Marine Division plus the Army Reserve. (The title Tractor Flotilla referred to the tank landing ships, and associated landing craft assigned.) The demonstration was a feint intending to deceive Japanese defense forces into believing that landings were being made on southeastern beaches, as well as the western beaches of Okinawa.[34]

The demonstration landing, on 1 October, involved dispatching 7 boat waves of 24 LCVPs (Higgins boats) each, from the line of departure at 5-minute intervals. Each boat carried about one squad of Marines. When the fourth wave crossed the line of departure at 0830 (H-Hour), all boat waves reversed course and returned to their parent ships. No enemy reaction to the demonstration was observed, and the group was not taken under enemy fire. The next day, 2 April, the demonstration was repeated in its entirety. Task Group 51.2 remained in the Okinawa area until 9 April, when ordered to depart for Saipan. Upon its arrival there, on 14 April, the task group was dissolved.

**Demonstration Group (Task Group 51.2): Rear Adm. Jerauld Wright (2nd Marine Division and Army troops: Maj. Gen. Thomas E. Watson)**

**Transport Group Charlie (TU 51.2.1): Commodore Henry C. Flanagan**

| **TransDiv 43 (51.2.11): Commodore Henry C. Flanigan** | | |
|---|---|---|
| USS *Shoshone* (AKA-65) | 1-11 Apr 45 | C2-S-AJ3 hull |
| USS *Theenim* (AKA-63) | 1-15 Apr 45 | C2-S-B1 hull |
| **TransDiv 44 (51.2.12): Capt. J. H. Seyfried** | | |
| USS *Leo* (AKA-60) | 1-14 Apr 45 | C2-S-B1 hull |
| USS *Southampton* (AKA-66) | 1-11 Apr 45 | C2-S-AJ3 hull |

**TransDiv 45 (51.2.13): Capt. A. C. J. Sabalot**

| | | |
|---|---|---|
| USS *Lacerta* (AKA-29) | 1-9 Apr 45 | S4-SE2-BE1 hull |
| USS *Muliphen* (AKA-61) | 1-11 Apr 45 | C2-S-B1 hull[36] |

Overall explanations of operations by the AKs/AKAs/AKSs of the Northern Attack Force, Expeditionary Force Floating Reserve, and follow-on echelons of ships are not included in this book owing to space limitations. However, observations regarding the Navy's utilization of 88 cargo vessels for the assault and occupation of Okinawa are in order as this is the book's focus. Of these vessels:

- 42 assault cargo ships (AKA) were allocated among the Western Islands Attack Group, Northern Attack Force, Southern Attack Force, Demonstration Group, and the Expeditionary Floating Reserve
- 46 cargo vessels provided follow-on support; including 14 assault cargo ships (AKA), 29 cargo ships (AK), and 3 general stores issue ships (AKS)

The 46 cargo vessels which provided support subsequent to the landings, included 21 Liberty ships (EC2-S-C1 type hulls), and 7 Victory ships (VC2-S-AP2 hulls). The Victories, successors of Liberty ships, entered service late in the war. Both ship types at Okinawa were from among those which the Navy had acquired, modified for naval service, commissioned, and put to sea.

There has been little previous reference in this book, other than in the preface, to the Navy's AKSs (general stores issue ships). They were relatively few in number and only those involved in the Okinawa operation—USS *Antares* (AKS-3), *Castor* (AKS-1), and *Kochab* (AKS-6)—earned battle stars in World War II. All three earned one each for Okinawa, *Antares* and *Castor* had earlier earned one each for the attack on Pearl Harbor, and *Castor* one for the later Gilbert Islands operation.

### Northern Attack Force (TF 53): Rear Adm. Lawrence Reifsnider (1st Marine Division: Maj. Gen. Pedro del Valle, and 6th Marine Division: Maj. Gen. Lemuel C. Shepherd Jr.)

**Transport Group Able (TG 53.1): Commodore Herbert B. Knowles**

**TransDiv 34 (53.1.1): Commodore Herbert B. Knowles**

| | | |
|---|---|---|
| USS *Hydrus* (AKA-28) | 1-15 Apr 45 | S4-SE2-BE1 hull |
| USS *Sheliak* (AKA-62) | 1-19 Apr 45 | C2-S-B1 hull |

**TransDiv 35 (53.1.2): Capt. Samuel P. Jenkins**

| | | |
|---|---|---|
| USS *Caswell* (AKA-72) | 1-9 Apr 45 | C2-S-AJ3 hull |
| USS *Devosa* (AKA-27) | 1-10 Apr 45 | S4-SE2-BE1 hull |

| **TransDiv 36 (53.1.3): Capt. G. W. Johnson** | | |
|---|---|---|
| USS *Aquarius* (AKA-16) | 1-19 Apr 45 | C2-S-B1 hull |
| USS *Circe* (AKA-25) | 1-6 Apr 45 | S4-SE2-BE1 hull |
| **Transport Group Baker (TG 53.2): Commodore J. G. Moyer** | | |
| **TransDiv 52 (TU 53.2.1): Commodore J. G. Moyer** | | |
| USS *Andromeda* (AKA-15) | 1 Apr-14 Jun 45 | C2-S-B1 hull |
| USS *Cepheus* (AKA-18) | 1-15 Apr 45 | C2-S-B1 hull |
| **TransDiv 53 (TU 53.2.2): Capt. W. N. Thornton** | | |
| USS *Arcturus* (AKA-1) | 1-14 Apr 45 | C2 hull |
| USS *Centaurus* (AKA-17) | 1 Apr-14 Jun 45 | C2-S-B1 hull |
| **TransDiv 54 (TU 53.2.3): Capt. Joseph R. Lannom** | | |
| USS *Betelgeuse* (AKA-11) | 1-9 Apr 45 | C2 hull |
| USS *Procyon* (AKA-2) | 1-7 Apr 45 | C2F hull[37] |

### Expeditionary Force Floating Reserve (TF 51.3): Commodore John B. McGovern (27th Infantry Division: Maj. Gen. George W. Griner Jr.)

| **TransDiv 46 (TU 51.3.12): Commodore John B. McGovern** | | |
|---|---|---|
| USS *Tolland* (AKA-64) | 9-15 Apr 45 | C2-S-AJ3 hull |
| USS *Whiteside* (AKA-90) | 9-15 Apr 45 | C2-S-B1 hull |
| **TransDiv 47 (TU 51.3.13): Capt. A. S. Wotherspoon** | | |
| USS *New Hanover* (AKA-73) | 9-19 Apr 45 | C2-S-AJ3 hull |
| **TransDiv 48 (TU 51.3.14): Capt. C. L. Andrews Jr.** | | |
| USS *Yancey* (AKA-93) | 9-15 Apr 45 | C2-S-B1 hull |
| **TransDiv 27 (TU 51.3.15): Capt. August J. Detzer Jr.** | | |
| USS *Athene* (AKA-22) | 15 Apr-11 Jun 45 | S4-SE2-BE1 hull |
| USS *Stokes* (AKA-68) | 9-19 Apr 45 | C2-S-AJ3 hull[38] |

### Other U.S. Navy Cargo Vessels Which Earned Battle Stars for the assault and occupation of Okinawa

| | | |
|---|---|---|
| USS *Adhara* (AK-71) | 8-27 May 45 | EC2-S-C1 hull |
| USS *Alkaid* (AK-114) | 21 May-13 Jun 45 | EC2-S-C1 hull |
| USS *Alkes* (AK-110) | 28 May-30 Jun 45 | EC2-S-C1 hull |
| USS *Appanoose* (AK-226) | 3 May-30 Jun 45 | EC2-S-C1 hull |
| USS *Arided* (AK-73) | 18-30 Jun 45 | EC2-S-C1 hull |
| USS *Auriga* (AK-98) | 10-19 Apr 45 | C1-B hull |
| USS *Azimech* (AK-124) | 18 Apr-19 May 45 | EC2-S-C1 hull |
| USS *Bedford Victory* (AK-231) | 25 Apr-7 Jun 45 | VC2-S-AP2 hull |
| USS *Bucyrus Victory* (AK-234) | 24 Mar-30 Jun 45 | VC2-S-AP2 hull |
| USS *Carina* (AK-74) | 26 Apr-3 May 45 | EC2-S-C1 hull |
| USS *Celeno* (AK-76) | 18 Apr-27 Jun 45 | EC2-S-C1 hull |
| USS *Cetus* (AK-77) | 26 Apr-5 May 45 | EC2-S-C1 hull |
| USS *Degrasse* (AK-223) | 26 Apr-5 May 45 | EC2-S-C1 hull |
| USS *Draco* (AK-79) | 26-30 Jun 45 | EC2-S-C1 hull |
| USS *Hyperion* (AK-107) | 8-27 May 45 | EC2-S-C1 hull |
| USS *Jupiter* (AK-43) | 24-30 Jun 45 | C2 hull |
| USS *Kenmore* (AK-221) | 26 Apr-5 May 45 | EC2-S-C1 hull |
| USS *Lakewood Victory* (AK-236) | 24 Mar-30 Jun 45 | VC2-S-AP2 hull |

| | | |
|---|---|---|
| USS *Las Vegas Victory* (AK-229) | 24 Mar-30 Jun 45 | VC2-S-AP2 hull |
| USS *Lynx* (AK-100) | 26 Apr-5 May 45 | EC2-S-C1 hull |
| USS *Manderson Victory* (AK-230) | 24 Mar-30 Jun 45 | VC2-S-AP2 |
| USS *Matar* (AK-119) | 29 May-25 Jun 45 | EC2-S-C1 hull |
| USS *Mayfield Victory* (AK-232) | 18 May-30 Jun 45 | VC2-S-AP2 hull |
| USS *Media* (AK-83) | 10-19 Jun 45 | N3-M-A1 hull |
| USS *Menkar* (AK-123) | 10-19 May 45 | EC2-S-C1 hull |
| USS *Mintaka* (AK-94) | 21-31 May 45 | EC2-S-C1 hull |
| USS *Rotanin* (AK-108) | 21-31 May 45 | EC2-S-C1 hull |
| USS *Sterope* (AK-96) | 10-31 May 45 | EC2-S-C1 hull |
| USS *Zaurak* (AK-117) | 13-19 Jun 45 | EC2-S-C1 hull |
| USS *Allegan* (AKA-225) | 3 May-30 Jun 45 | EC2-S-C1 hull |
| USS *Fomalhaut* (AKA-5) | 24 Mar-30 Jun 45 | C1-A hull |
| USS *Lenoir* (AKA-74) | 17-22 Apr 45 | C2-S-AJ3 hull |
| USS *Lumen* (AKA-30) | 17-22 Apr 45 | S4-SE2-BE1 hull |
| USS *Merrick* (AKA-97) | 18 Jun 45 | C2-S-B1 hull, |
| USS *Pamina* (AKA-34) | 16-19 May 45 | S4-SE2-BE1 hull |
| USS *Prentis* (AKA-102) | 21-30 Jun 45 | C2-S-B1 hull |
| USS *Rankin* (AKA-103) | 11-28 Jun 45 | C2-S-AJ3 hull |
| USS *Trego* (AKA-78) | 3-11 Jun 45 | C2-S-AJ3 hull |
| USS *Trousdale* (AKA-79) | 11-22 Apr 45 | C2-S-AJ3 hull |
| USS *Union* (AKA-106) | 1-9 Apr 45 | C2-S-AJ3 hull |
| USS *Valencia* (AKA-81) | 17-22 Apr 45 | C2-SAJ3 hull |
| USS *Venango* (AKA-82) | 12-22 Apr 45 | C2-S-AJ3 hull |
| USS *Virgo* (AKA-20) | 1-16 May 45 | C2-S-B1 hull |
| USS *Antares* (AKS-3) | 9 May-18 Jun 45 | VC2-S-AP3 hull |
| USS *Castor* (AKS-1) | 1 Apr-30 Jun 45 | C2 hull |
| USS *Kochab* (AKS-6) | 8-30 Jun 45 | EC2-S-C1 hull |

**23**

# Fleet Train (British Pacific Fleet)

*In the long and proud history of the Royal Navy, the largest formation ever to see combat fought under the operational command not of Drake, Nelson, Jellicoe, or Cunningham, but rather of Americans Raymond Spruance and William Halsey.*

—Nicholas E. Sarantakes.[1]

*At the end of the war there were 142 ships in the British Pacific Fleet and 94 ships in the Fleet Train. There were some 500 first-line aircraft, 100 on ancillary services, and 1000 in reserve. The peak strength in personnel was about 125,000 officers and men. These figures were rapidly increasing, and by the end of 1945, had hostilities continued, there would have been 400 ships of all types, 900 first-line aircraft, and more than 200,000 officers and men.*

—Sydney D. Waters, *The Royal New Zealand Navy*, 1956.[2]

Photo 23-1

HMS *Indomitable*, flagship of the First Aircraft Carrier Squadron arriving at Sydney, NSW, on 10 February 1945, with units of the British Pacific Fleet. Australian War Memorial photograph 073533

The British Pacific Fleet (BPF) formally came into being on the morning of 22 November 1944, when Adm. Sir Bruce Fraser hauled down his flag at Trincomalee, Ceylon (today Sri Lanka), as commander in chief Eastern Fleet, and hoisted it in the gunboat HMS *Tarantula* as commander in chief British Pacific Fleet. On 2 December he shifted his flag to the battleship HMS *Howe*, which sailed eight days later for Australia with an escort of four destroyers.[3]

This action followed a Quebec conference in September 1944, at which American President Franklin D. Roosevelt and British Prime Minister Winston Churchill discussed the war in the Pacific. When Churchill offered the aid of a British fleet, Roosevelt leapt at the opportunity against the advice of Adm. Ernest J. King, commander in chief, U.S. Fleet. The British minutes recorded that Roosevelt said, "the British fleet was no sooner offered than accepted." Thus, the BPF was born and fought in the Pacific war from March to August 1945.[4]

Photo 23-2

Allied Conference, Quebec, Canada. Seated (l to r): Canadian Prime Minister Mackenzie King, President Franklin Roosevelt and Winston Churchill, British Prime Minister. Standing (l to r): General Henry H. Arnold, Chief of US Air Forces; Air Chief Marshall, Sir Charles Portal of Great Britain; General Sir Alan Brooke, Chief of the Imperial General Staff; Admiral Ernest J. King, Chief of US Naval Forces; Field Marshall Sir John Dill, Chief of Joint Staff Mission and Washington, DC; General George C. Marshall, Chief of Staff, US Army; Sir Dudley Pound, Admiral of the Fleet and First Sea Lord; Admiral William D. Leahy, Chief of Staff to the Commander in Chief of the Navy.
Naval History and Heritage Command photograph #80-G-178042

Admiral King, who did not wish to commit himself to specific future employment of the British fleet, strenuously objected to such an arrangement—stating that he was not "prepared to accept a British Fleet which he could not employ or support." In the end, the British received assurances that their "balanced and self-supporting" fleet would "participate in the main operations against Japan in the Pacific."[5]

## BRITISH TASK FORCE STRIKES AGAINST JAPAN

In January 1945, the British Pacific Fleet, while en route to Sydney and still under independent Royal Navy operating command (as British Task Force 63), struck its first blows at the Japanese from the Indian Ocean. Rear Adm. Sir Philip Vian's carrier force consisting of HMS *Indomitable*, *Indefatigable*, *Illustrious*, and *Victorious*; the cruisers *Suffolk*, *Ceylon*, *Argonaut*, and *Black Prince*, and eight destroyers, carried out three successful attacks on oil refineries in Sumatra (today Indonesia). These strikes on Japanese oil refineries south of Palembang, on 24 and 29 January 1945, eliminated the bulk of their supply of aviation fuel. This was the largest operation conducted by the Royal Navy Fleet Air Arm in World War II.[6]

Photo 23-3

Netherlands East Indies, 29 January 1945. Flames and smoke rising from the Sungei Gerong oil refinery in southern Sumatra, after a bombing attack by aircraft of the Fleet Air Arm from HMS *Illustrious* of the British Eastern Fleet.
Australian War Memorial photograph P02491.256

Rear Admiral Vian's carrier force, which then included battle ship HMS *King George V*, four cruisers, and nine destroyers, arrived at Fremantle, on 4 February 1945, and proceeded to Sydney to join the other ships of the British Pacific Fleet.[7]

## FLEET TRAIN BASES AT MANUS, ADMIRALTIES

Photo 23-4

Lombrum Naval Base on Manus Island in the Admiralty Islands, September 1949. Australian War Memorial photograph 041524

On 16 February 1945, in preparation for establishing an Intermediate Base (between Australian and BPF operating forces) at Manus Island in the Admiralties, Capt. Henry Francis Waight, OBE RN, flew in to these islands from Sydney, landing at Momote Airfield on adjacent Los Negros Island. Waight's title was Senior British Naval Officer, Manus. A British Naval Camp was prepared by resident U.S. military personnel on a high ridge overlooking Lorengau Harbor, on the NE Coast of Manus Island, three miles from the main U.S. base. British accommodation was in Quonset huts in the camp, but messing was shared with the Americans. The camp was ready for occupation by the time personnel and stores arrived by sea, on 26 February, aboard the former liner SS *City of Paris*. She was unloaded alongside a large jetty, and all stores and equipment were moved inland with American assistance.[8]

Capt. Markham Henry Evelegh, RN, was appointed Naval Officer in Charge, Intermediate Base, on 28 February. The Fleet Train, under Rear Adm. Douglas Fisher in HMS *Lothian*, arrived at Seeadler Harbour on 2 March. Five days later on the 7th, the British Pacific Fleet arrived. Under the command of Vice Adm. Rawlings, it consisted of two battleships, including HMS *King George V* (flagship of Rawlings); the fleet carriers HMS *Indomitable* (flag of Vice Admiral Vian), *Indefatigable*, *Illustrious*, and *Victorious*; cruisers HMS *Swiftsure* (flag of Rear Admiral Brind), *Euryalis*, *Black Prince*, *Argonaut*, and *Gambia*; and a number of destroyers, under the command of Rear Admiral Edelston.[9]

Photo 23-5

Rear Adm. Douglas Fisher, RN, aboard HMS *Duke of York*, 1942.
Imperial War Museums photograph A12142

The Fleet Train of the British Pacific Fleet had been formed, on 10 October 1944, to provide an entirely self-sufficient support force, after Admiral King expressed concerns that the U.S. Navy had insufficient logistical support to supply a British Pacific Fleet. The Fleet Train, which initially consisted of 79 ships, grew to 94 ships, by September 1945. Under the command of Rear Adm. Douglas Fisher, RN, and designated Task Force 117 (TF 117) by the U.S. Navy, its primary role was to keep the British Fleet operating for weeks on end, thousands of miles from its bases.[10]

In other theaters of war, the Royal Navy utilized its extensive system of land bases to provide its ships with logistical support. In the forthcoming operations (supporting the invasion of Okinawa, and air strikes against mainland Japan), the BPF would be 17,000 miles from the United Kingdom, and a distant 5,000 miles from the main support base in Australia. Requiring the Fleet Train to be closer to the fighting forces, Task Force 117 moved their operating base forward from Sydney to Manus, in the Admiralty Islands, which lay only 2,500 miles from Japan.[11]

A host of different type vessels comprised the Fleet Train. These included aircraft repair ships, armament stores carriers, armament stores-issuing ships, destroyer depot ships, hospital ships, naval stores carriers, distilling ships, netlayers, oilers, repair ships, replenishment carriers, tugs, and victualling stores-issuing ships. A list of the stores carriers and stores-issuing ships, and their associated pennant numbers, is provided in the table. These ships were generally the equivalent of U.S. Navy cargo ships. The Fort ships and Hickory ships were built in Canada and United States, respectively, and under charter to the British, as were other ships not British-owned. They were manned by civilian Merchant Marine crews.[12]

**British Fleet Train Stores Ships (August 1945)**

| | |
|---|---|
| **Stores-Issuing Ships (Victualling)** | **Air Stores-Issuing Ships** |
| British SS *Fort Alabama* (B577) | British SS *Fort Corville* (B531) |
| British SS *Fort Constantine* (B578) | British SS *Fort Langley* (B532) |
| British SS Fort *Dunvegan* (B579) | **Mine-Issuing Ship** |
| British SS *Fort Edmonton* (B580) | British SS *Prome* (B432) |
| British SS *City of Dieppe* (B558) | **Armament Stores Carriers** |
| **Stores-Issuing Ships (Naval)** | Danish MS *Gudrun Maersk* (B538) |
| Norwegian MS *Bosphorus* (B557) | British MV *Kistna* (B542) |
| British SS *Glenartney* (B584) | British MV *Kola* (B543) |
| British SS *Fort Providence* (B582) | **Armament Stores-Issuing Ships** |
| British SS *Fort Wrangell* (B583) | Australian MV *Corinda* (B536) |
| British MV *Hickory Burn* | British SS *Darvel* (B537) |
| British MV *Hickory Dale* | Norwegian MS *Hermelin* (B539) |
| British MV *Hickory Glen* | British SS *Heron* (B540) |
| British MV *Hickory Stream* | British SS *Kheti* (B541) |
| Dutch SS *Jaarstroom* (B562) | British MV *Pacheco* (B544) |
| SS *Marudu* (B563) | Belgian SS *Prince de Liege* (B545) |
| Norwegian MS *San Andres* (B564) | Belgian SS *Prinses Maria-Pia* (B546) |
| SS *Slesvig* (B565) | Danish MS *Robert Maersk* (B547) |
| | Danish MS *Thyra S.* (B548) |

## SUSTAINMENT OF BPF CARRIER OPERATIONS

During the spring and summer of 1945, the British Pacific Fleet supported the invasion of Okinawa, codenamed Operation ICEBERG, between 26 March-25 May, and deployed off the Japanese coast in the July/August operations from 17 July-10 August. With four fleet carriers on station at any one time, carrying between 230-250 aircraft, First Aircraft Carrier Squadron was the spearhead of British effort in a theatre dominated by air power. Its operations had no precedent in the Royal Navy's history, and were made possible by an air logistic chain which by July 1945 extended some 17,000 miles from the United Kingdom to south of Honshu Island.[13]

The first requirement of the Fleet Train was transportation of aircraft, either packed or assembled, and their associated spare parts out to Sydney from Britain, bases in the Indian Ocean, or factories in America. It was 12,000 miles from Britain to Sydney. Following arrival in Australia, there were six Mobile Operational Naval Air Bases and a Transportable Aircraft Repair Yard to take possession of the material, and store reserve aircraft. A Forward Area Naval Air Station, and an afloat Fleet Air Maintenance Group were at Manus in the Admiralty Islands. Royal Navy escort carriers were used to ferry aircraft from Australia to the forward areas and, aid as backups, in replenishment roles with the fleet, particularly when fleet carriers were being refueled.[14]

Photo 23-6

Seafire II aircraft, with wings folded, on the deck of the aircraft carrier HMS *Slinger*. She is at Brisbane, preparing to depart for Manus Island. Australian War Memorial photograph P00965.005

The first ships of the "Air Train" were the maintenance carrier HMS *Unicorn*, and the air stores-issuing ship SS *Fort Colville*, which arrived with other units of the Fleet Train at Sydney, in February. The air units proceeded to Manus in early March. HMS *Deer Sound*, an aircraft component repair ship, reached Manus, on 9 April, when ICEBERG was in progress. A second maintenance carrier, HMS *Pioneer*, and air stores-issuing ship, SS *Fort Langley*, arrived in time to support the July/August operations, which improved the situation considerably.[15]

## EXPERIENCES OF A RFA "TANKER SAILOR"

As mentioned in the book's preface, sailors aboard British Pacific Fleet ships arriving in Australia greatly enjoyed the liberty available ashore and abundant hospitality afforded them. Gerry Baxter, a former crewman aboard the Royal Fleet Auxiliary (RFA) oiler *Serbol*, aptly expressed these sentiments years later:

> In January 1945 I signed on RFA *Serbol* at Greenock. RFA *Serbol* arrived at Darwin on the 3rd May, the port was a mess, with overturned ships everywhere, it had been hit badly but the Aussie's greeted us with everything they could and we paid for nothing. We got hold of meat, vegetables and fruit, and even some beer![16]

*Serbol* was an old ship, having been put in service in January 1918. The small, 2,000-ton *Belgol*-class tanker had been built with a tall mast near her upright funnel, but it was removed in World War II prior to her arrival in Australia. She left the Clyde in England, on 16 February 1945, and made her way to Darwin via Gibraltar, Port Said, the Suez Canal, Aden, and Columbo, arriving at the "top end of Down Under," a little over six weeks later.[17]

After two weeks at the rustic port in accordance with Baxter's narrative, *Serbol* left Darwin for Brisbane, which lay about half-way down Australia's east coast:

> The ship sailed from Darwin on the 17th May for Brisbane, where we arrived on the 18th for repairs to the engine, which was a triple expansion one … and a 110-volt generator, every time we fired the big gun all the electrics went. One of my memories is sailing through the Great Barrier Reef and getting paid for it![18]

The RFA oiler remained at Brisbane, until 6 July, when she stood out of the river port, bound for Manus. During their stay, Baxter and the other merchant marines had once again, been warmly welcomed:

> When we arrived in Brisbane, well what can I say, the Australian people invited us out to dinner, to shows and we were not allowed to pay a cent for anything, even the drinks were 'on the house.'[19]

*Seabol* arrived at Manus, on 12 July, a place not beloved by Baxter and scores of other BPF sailors, particularly in comparison to Australia:

> What a hole Manus was in 1945 ... all we saw was a couple of storage tanks that were topped up by the large tankers, and from which we took our oil, at the time we simply came in and refueled, and then sailed again, we never went ashore. Manus is located at 2 degrees south, 147 degrees east in the Bismarck Sea, how the Admiralty could send a ship so badly equipped for the conditions there I will never understand.[20]

While operating out of Manus, *Serbol* supported the British Pacific Fleet along with the Royal Fleet Auxiliary (RFA) ships *Bacchus*, *Bishopdale*, *Brown Ranger*, *Cedardale*, *Dingledale*, *Easedale*, *Green Ranger*, *Rapidol*, *Wave Emperor*, *Wave Governor*, *Wave King*, *Wave Monarch*, and HMS (later RFA) *Olna*. On 2 September, the day the Japanese signed the formal surrender documents aboard the USS *Missouri*, ending the war, *Serbol* left Manus for the final time, arriving at Guam on 6 September.[21]

## SS *DARVEL* BARELY MAKES IT OUT OF SINGAPORE

> *Towards the end of the Malayan campaign, Captain Lutkin carried out two voyages from Singapore to Palembang, evacuating troops. When the fall of Singapore was imminent, Captain Lutkin was ordered to Batavia and while navigating the Banka Straits the ship was severely damaged by air attacks. Captain Lutkin displayed resourcefulness and courage in effecting temporary repairs and the* Darvel *was safely brought to Australia.*
>
> —Notification in *The London Gazette* of the appointment of Captain William Lutkin, Master, SS *Darvel*, Singapore Straits Steamship Company Ltd., to the Most Excellent Order of the British Empire, in the Merchant Navy, for his brave conduct.[22]

Photo 23-7

British merchant vessel SS *Darvel*, which evacuated Australian troops from Singapore to Batavia, in February 1942, before she escaped to Australia, escorted part of the way by HMAS *Yarra*. *Darvel* was later converted to an armament stores-issuing ship and operated in Australian waters.
Australian War Memorial photograph 303184

Many of the stores carriers and stores-issuing ships of the BPF had very active war service. Unfortunately, entries in the logs of merchant ships were generally limited to dates of departures and arrivals at ports, and disciplinary infractions by crewmembers. Notations about the latter activities were referred to as "being logged." There is a scarcity of information about their activities, particularly in comparison to Navy ships. However, examples of the mettle of merchant sailors in World War II may sometimes be revealed in accounts by individuals, when such exist and are found.

One such account involves the escape from Singapore in 1942 of the steamship SS *Darvel*, before the fall of the British Colony. After sustaining much damage from a Japanese air attack, *Darvel* was saved by heroic actions by her master, William Lutkin, and her ad hoc crew, assisted by members of the Royal New Zealand Air Force No. 1 Aerodrome Construction Squadron. This squadron, the first of its kind in the Air Forces of the British Empire, had been formed in New Zealand, in July 1941, and sent to Malaya to help in the building of airstrips, as part of the defense system of Singapore.[23]

On the afternoon of 6 February, the men of the construction squadron struck camp and were taken in trucks to the port. Two parties were formed of their members, one to be transported by the SS *City of Canterbury* and the other the SS *Darvel*. Amid the litter of bomb wreckage and in the glare of burning buildings, the New Zealanders loaded their personal kitbags, medical supplies, and their rifles and ammunition aboard the waiting ships.[24]

Both ships left their berths to join a small convoy, but the 1,929-ton *Darvel* was ordered back to port by the Naval authorities. This action was taken partly because she had insufficient crew, and partly because she was too slow (her best speed was only eleven knots) to keep pace with the other three, much faster ships that were to proceed together with naval escort. After lying at anchor all night, *Darvel* returned to the wharf, on the morning of the 7th. The squadron members were landed and were taken to an RAF transit camp near Seletar. Sited between the Japanese batteries on the southern tip of Johore and the British ones on Singapore, the roar of artillery rounds passing overhead filled the air—a most disturbing environment.[25]

The following afternoon, 8 February, the assigned passengers again went down to the port and embarked aboard the *Darvel.* After some hours, during which there were several enemy air raids, she eventually put to sea at dusk, but just as she had cleared the harbor she was again recalled and brought back to her berth. The bad weather brewing outside, and an associated decrease in visibility made it too risky for her to transit the protective minefields beyond the entrance. That night, the small, slow ship—seemingly unable to get out of Singapore—lay alongside the wharf and the men slept fitfully on her decks.[26]

## AT SEA, FINALLY

The next morning, 9 February, the New Zealanders were taken once more to the transit camp. While the men slept the previous night, the Japanese had landed on the western part of Singapore Island and, by morning, made considerable progress eastward. Toward midday, their artillery started shelling the camp and all personnel took refuge in shelter trenches.[27]

That afternoon, during a lull in the shelling, the RNZAF personnel scrambled into their trucks and once more made for the port. After quickly boarding *Darvel,* joining other troops already embarked, the ship immediately headed for the open sea. She cleared the port just in time to avoid a heavy dive-bombing and strafing attack on shipping in the docks. Her last view of Singapore was one of blazing wharf sheds, towering columns of smoke from burning oil tanks, and the sky full of Japanese planes and bursting anti-aircraft shells.[28]

*Darvel* proceeded through the night, and at daybreak anchored off the southern tip of a small island, in an effort to avoid observation by enemy aircraft. As she was still short-crewed, members of the squadron pitched-in and helped work her. Some took shifts in the engine room and stokehold (fireroom), some mounted and manned light anti-aircraft guns, and still others took over the messing (cooking) for all the troops on board.[29]

Bound for Java, *Darvel* next had to pass through the narrow Banka Strait between Sumatra and Banka Island, which Japanese bombers patrolled constantly during daylight. She got under way at dusk, with the hope of transiting the danger area in the night. It was a good plan but, just before entering the strait, she was delayed for two hours assisting the SS *Kintak*, which had run aground during the day. As a result, *Darvel* was still in the strait when the next day dawned. Seeking to avoid detection, she anchored in the shelter of a group of small islands, near where a small ship, bombed some days before, lay abandoned.[30]

## JAPANESE AIRCRAFT FIND / ATTACK THE *DARVEL*

Everything was peaceful until half past eleven, when a formation of enemy bombers appeared. They were too high for the ship's anti-aircraft guns to engage them, so the gun crews withheld their fire and took cover. The planes altered course slightly and, once overhead, bombs began to fall. They rained down all around the ship, and their explosions tossed her about like a cork and drenched her with spray. There were no direct hits, but damage and splinters from near misses left the ship in shambles and, during a brief interval of silence, the hiss of steam escaping from burst pipes could be heard. Five minutes later the bombers returned. This time, fortunately they concentrated their attack on the abandoned steamer a few hundred yards away, sank her and, having used up all their bombs, retired.[31]

*Darvel*—her hull riddled with holes by bomb splinters, and leaking badly, and with steering gear and all her lifeboats damaged—was in perilous condition. Fires had broken out in several areas, and many of the troops on board were killed or wounded. The New Zealand unit suffered one killed, seventeen wounded, and several slightly injured.[32]

The fires were quickly brought under control, and working parties from the Construction Squadron went below to fill in the scores of small holes with wooden plugs. Others set to work, repairing the lifeboats and rigging, and clearing up debris on deck. Medical orderlies cared for the wounded, there being no doctor on board.[33]

Rather than wait for another attack, *Darvel* continued her passage through the strait in daylight. At the southern entrance, she lay to in order to make repairs to the damaged steering gear. Finally, at 2030 in the evening when welcome darkness covered the ship, course was set for Batavia. By morning, 12 February, *Darvel* was listing badly to port.

All passengers with their baggage crowded to the starboard side, and squadron members went below and plugged more holes. After about two hours' effort, the leakage was brought under control. Finally, *Darvel* arrived off Batavia, Java Island, around noontime and berthed at 1400.[34]

The skeleton crew of mariners that had put to sea aboard *Darvel*, to save their ship and escape the Japanese, were insufficient to man her guns for self-defense, or deal with the battle damage from the air attack. Fortunately, there were skilled craftsmen in the form of RNZAF No. 1 Aerodrome Construction Squadron personnel aboard. Senior officers transported aboard *Darvel* credited the Kiwis with saving the ship while her small crew carried out vital functions necessary to safely navigate and operate her during passage to Java.[35]

Quoted material about Captain Lutkin's action may be found at the head of this account about the SS *Darvel*.

Importantly, the joint actions of the merchant mariners and military personnel made *Darvel* available for a new role as an armament stores-issuing ship, which she carried out for the duration of the war.

## SHIP MASTERS, AND A CHIEF ENGINEER LAUDED

As shown in the following table, a number of British Merchant Marine mariners were honored for their bravery in the evacuations of Singapore. William Lutkin and other employees of his company are included.

**Richard Edward Morton**
**Master, SS *Vyner Brooke***

**Charles Edward Cleaver**
**Master, MV *Larut***

**Alec Butler Durrant**
**Master, SS *Kinta***

**MBE Medal**

**William Lutkin**
**Master, SS *Darvel***

**Kenneth Drummond Thomas**
**Chief Engineer,**
**SS *Empire Hamble***

William Lutkin, master of the SS *Darvel* was appointed by King George VI of England to the Most Excellent Order of the British Empire, in the Merchant Navy, and received the MBE Medal. Lutkin was an employee of the Singapore Straits Steamship Company Ltd., as were others similarly honored with the same appointment. Other examples of such bravery are noted below; one was Richard Edward Morton of the Sarawak Steamship Company Ltd.[36]

Morton was the master of SS *Vyner Brooke*, which prior to the fall of Singapore was engaged in evacuating troops to Sumatra. The

Scottish-built steamship had previously functioned both as the royal yacht of Sarawak in Borneo, and as a merchant ship frequently used between Singapore and Kuching, Sarawak. She was named after Sir Charles Vyner Brooke, the 3rd Rajah of Sarawak. Upon outbreak of war with Japan, the Royal Navy requisitioned and armed her.

The SS *Vyner Brooke* did not survive the war. On 14 February 1942, while evacuating nurses and wounded servicemen from Singapore, she was continuously attacked by enemy aircraft while passing through the Banka Straits, between Sumatra and Banka Island. "Captain Morton displayed coolness and courage in defending his ship which was eventually sunk. He was picked up by the Japanese and held as a prisoner until the end of the war."[37]

The motor vessel MV *Larut* was bombed by enemy air attack, on 23 January 1942, and set on fire. She was beached at Sebang, northern Java, and burned. Her crew was saved, and her master, Captain Cleaver, found his way to Batavia, where he took over the abandoned ship *Perak*. The SS *Kinta* was badly damaged at Singapore during an enemy air attack, but her master, Captain Durrant, brought her through further attacks to Batavia. There, he joined Cleaver on the *Perak* and, together, they effected such repairs as they could, and brought her to Colombo in the face of enemy action. "Both masters displayed courage and determination in getting the ship to port."[38]

The cargo ship SS *Empire Hamble*, while anchored at Singapore, sustained serious damage during enemy air attacks. Kenneth Thomas, her chief engineer officer, "though faced with a partially dismantled engine room, and severe bomb damage, set a splendid example and worked continuously to render the vessel fit for sea. It was due to his efforts that the ship was enabled to reach Australia."[39]

## BRITISH RETURN TO HONG KONG AND SINGAPORE

On 15 August 1945, Emperor Hirohito announced the surrender of Imperial Japan. Two weeks later, the combined U.S./British fleets made their triumphal entry into Tokyo Bay. There, on the morning of 2 September 1945, the formal surrender of Japan to the Allies was made on board USS *Missouri*. The United States flag purposefully flown by the battleship that day carried thirty-one stars, in lieu of the forty-eight of the present time; it was the flag flown by Commodore Perry, USN, when he entered Japanese waters ninety-two years before. (Perry had headed an expedition, a squadron of ships, that forced Japan in 1853–54 to enter into trade and diplomatic relations with the West after more than two centuries of isolation.)[40]

The British Pacific Fleet was given the honor via the new light aircraft carrier HMS *Glory* to receive the surrender of Rabaul on 6 September 1945. This was the powerful enemy base that the Allies had elected to bypass on their long road to the Japanese homeland.

The Fleet Train remained at Manus until early September 1945, when many of its ships sailed for Hong Kong and some to Singapore to support the reoccupation forces tasked with liberating the former colonies, and support, as well, the British Pacific Fleet operating on the China Coast and in Japan.[41]

On 14 September, Rear Adm. Douglas Fisher, RN (Rear Admiral, Fleet Train), was appointed Flag Officer, Western Area, British Pacific Fleet, a new command set up to coordinate the movements of the BPF on the China Coast. Capt. Henry Francis Waight, RN, was appointed as Captain Superintendent (Designate), Hong Kong, and he departed Manus aboard the submarine depot ship HMS *Maidstone* with an advance party to begin surveying the dockyard and area.[42]

The British Pacific Fleet, which formally came into being on 22 November 1944, remained in existence until 15 September 1948. On 15 September 1946, the fleet's final commander in chief, Vice Adm. Sir Denis Boyd, RN, based in Sydney, moved his headquarters to Hong Kong, which became known as Far East Station. Two years later, Boyd's title was changed to commander in chief, Far East Station, at which point the British Pacific Fleet officially ceased to exist.[43]

Photo 23-8

British Pacific Fleet Air Arm farewell salute to Sydney, on 12 November 1945, with up to 80 planes flying in formation over the Sydney Harbour bridge. Australian War Memorial photograph P05489.001

Photo 23-9

Cover of a magazine produced by the British Pacific Fleet, saying goodbye to Australia when the war ended.
Australian War Memorial photograph 107013

# 24

# Australian Borneo Campaign

Map 24-1

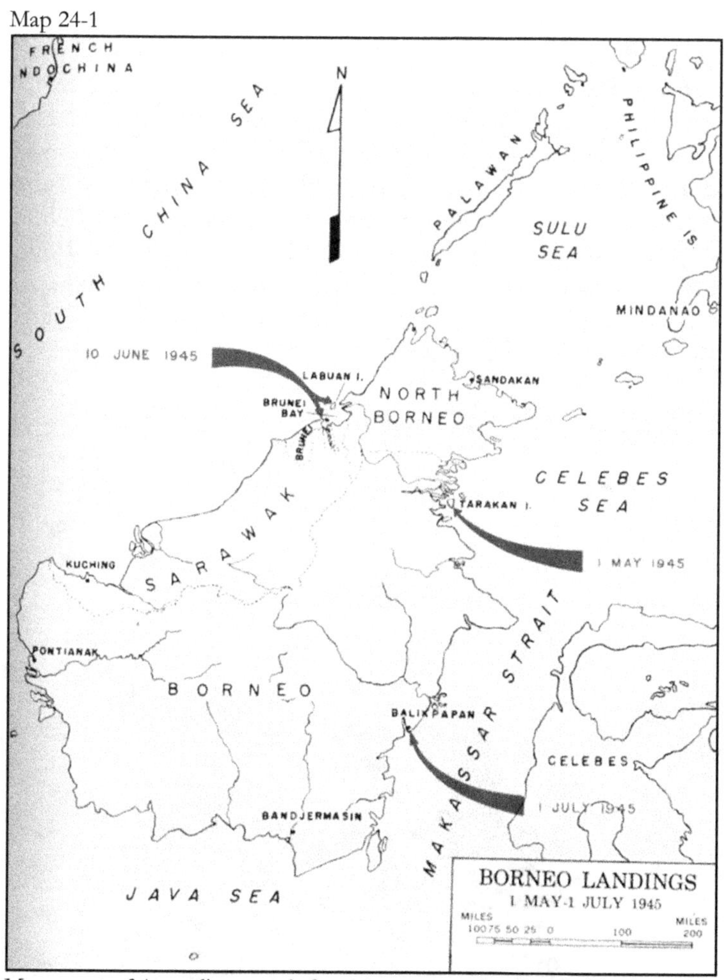


Movement of Australian assault forces en route to landings at Tarakan Island, at Labuan Island in Brunei Bay, and at Balikpapan during the 1945 Borneo Campaign
www.lib.utexas.edu/maps/historical/engineers_v1_1947/borneo_landings_1945.jpg

The Allied Borneo Campaign of 1945 (1 May-15 August) involved the final amphibious assaults of the last major Allied campaign in the Southwest Pacific area in World War II. Its objectives were to deny Japan the continued fruits of its conquests in the Netherlands East Indies, and use of the approaches to those areas. To achieve these aims, an Australian-led force captured Tarakan Island to provide an airfield for support of an assault on Balikpapan, and seized Brunei Bay for an advance fleet base that could protect resources in the area. The final phase was to occupy Balikpapan for its naval air and logistic facilities as well as its petroleum installations.[1]

The totality of RAN infantry landing ships—HMAS *Manoora*, *Kanimbla*, and *Westralia* (all former passenger ships) took part in the operations at Borneo, as did the attack cargo ship USS *Titania* (AKA-13) and the cargo ship *Poinsett* (AK-205). *Titania* was an old "war horse," having already garnered five battle stars, and an astounding five Navy Unit Commendations, earlier in the war. *Poinsett*—a 338-foot, diesel-powered ship capable of 11 knots—had just been commissioned on 7 February 1945. Her presence with the invasion forces at Balikpapan appears to have been an afterthought, arising from an unanticipated need for the ammunition she carried.

## INVASION OF TARIKAN ISLAND

Photo 24-1

Second wave of 2/48 Infantry Battalion troops leaving HMAS *Manoora*, during the D-Day landings at Tarakan Island, 1 May 1945.
Australian War Memorial photograph 090812

On 11 April 1945, USS *Titania*—under the command of recalled retired Navy commander Malcolm Whitfield Callahan (U.S. Naval Academy Class of 1914)—stood out of Subic Bay with the infantry landing ship HMAS *Manoora*, bound for Morotai, Halmahera Islands, via a stop en route at Leyte Gulf, Philippines. The commanding officer of the *Manoora*, Acting Capt. Allan Cousin, RANR(S), was the officer in tactical command during the transit, and his ship the formation guide.[2]

The ships arrived at Morotai the afternoon of 16 April and, the next day, *Titania* began loading supplies and equipment of the 26th Australian Infantry Brigade (Reinforced)—known as the "The Rats of Tobruk," following their heroic exploits in North Africa. *Manoora* first loaded ammunition, stores, and equipment of the 2/48th Battalion of the 26th Brigade 9th Australian Division, then the soldiers of the battalion. Units of the Oboe-One Attack Group (Task Group 78.1), of which *Titania* and *Manoora*, were a part, lay at anchor nearby. Rear Adm. Forrest B. Royal, USN, the task group commander, was embarked aboard the amphibious command ship USS *Rocky Mount* (AGC-3).[3]

Photo 24-2

Acting Comdr. Eric W. Livingston RANR(S) (left), commanding officer of HMAS *Westralia*, on the bridge of his ship. Livingston was Gazetted (awarded) a DSC, on 9 October 1945, for "Gallantry, fortitude and skill, S. W. P. [Southwest Pacific Area]." Australian War Memorial photograph 018441

On the afternoon of 21 April, the ships of Task Group 78.1 stood out of the anchorage, en route to Tarakan Island off northeast Borneo.

The Australian landing ships *Manoora* and *Westralia*, and USS *Titania* each had in tow astern, a 119-foot tank landing craft (LCT). They and the dock landing ship USS *Rushmore* (LSD-14) comprised the Transport Unit (Task Unit 78.1.1) under Captain Cousin, RANR(S). At 1802, the ships formed a cruising disposition with *Rocky Mount* as the guide. The transport force arrived off Tarakan in the late afternoon on 30 April and the LCTs were slipped. The following morning, 1 May, the group stood into the Muara Batagau (estuary) in a violent rain squall, and each vessel proceeded independently, due to low visibility, to their assigned anchorage in the transport area.[4]

*Titania* anchored in the transport area at 0603 on 1 May, and surface and air bombardment of the landing area commenced at 0700. At 0756, the first wave of amphibious craft proceeded toward landing beaches. The LVTs (Amtraks) forming the first boat waves were followed by waves of LCMs, LCTs, and LCVPs. The only opposition to the landings were Japanese small arms and mortar fire. *Titania* began unloading cargo from her anchorage at 0940, and continued until going alongside Langkas dock, on 5 May.[5]

Although the assault forces of the 9th Australian Division met little resistance the first day, American minesweepers found plenty of it in the form of mines the second day. The sea area was heavily mined by both enemy and Allied mines (laid earlier in the war) and, as a result, their participation was an essential element in the landings. On that day, in mid-afternoon, a battery of camouflaged coastal guns opened fire on four YMS minesweepers sweeping north of Tarakan, sinking *YMS-481*, and damaging YMSs *334* and *364*. Counterbattery fire from the high-speed transport USS *Cofer* (APD-62) and the large support landing craft *LCS(L)-8* and *LCS(L)-28* silenced the battery.[6]

After mooring alongside the dock, *Titania* resumed discharging cargo. Completing this work on the morning of 9 May, she disembarked all passengers, and embarked 7 officers and 49 enlisted Australian Army personnel for transportation to Morotai. She left Tarakan that afternoon with Task Unit 78.1.97.[7]

Ashore, there was fierce fighting as the Australians pushed inland to initially take the airfield, then all of Tarakan Island. More than 200 troops were killed before the last Japanese positions fell, on 20 June. The dead included one of the most famous Australian soldiers of the war, Lt. Thomas Currie "Diver" Derrick, 2/48th Battalion, awarded the Distinguished Conduct Medal for his bravery at Tel el Eisa in North Africa and the Victoria Cross at Sattleberg, New Guinea. Also killed was Corp. John Mackey, 2/3rd Pioneer Battalion, the sole recipient of the Victoria Cross for actions on Tarakan.[8]

## BRUNEI BAY OPERATION

*Rarely was such a prize obtained at such a low cost.*

—Gen. Douglas MacArthur in the book *Reminiscences*

Photo 24-3

A USN attack transport (left) and HMAS *Westralia* off Labuan Island, Borneo, 1945. Australian War Memorial photograph 134253

The second operation, codenamed OBOE 6, was a landing on Labuan Island in Brunei Bay on northwest Borneo. The Australian 9th Division had orders to secure the Brunei Bay area so it could be used as an advance naval base. The secondary objective was to capture the area's oil fields and rubber plantations and production plants. On 10 June, the Australians augmented by American troops pushed ashore at the main objective while also making smaller landings on nearby Muara Island and Brunei Peninsula.[9]

Naval bombardment and air attacks helped clear the way as the Allied troops advanced on each front. On Labuan Island, a force of several hundred Japanese made a determined stand in an area known as "the pocket," consisting of a number or ridges and spurs covered with heavy timber and dense undergrowth almost entirely surrounded by swamp. After this enemy position was subjected to concentrated fire from ship and shore, the infantry was sent in to finish the job. While the battle for Labuan was soon as good as over, sporadic fighting continued until war's end, by which time more than 100 Australians had been killed.[10]

The transport unit under Capt. Alan P. Cousin, RANR(S), consisted of five vessels—HMAS *Manoora*, *Westralia*, and *Kanimbla*; USS *Titania*; and the dock landing ship USS *Carter Hall*.

As at Tarakan, the biggest threat to Allied sea forces were Japanese mines. Through 16 June, a total of 440 moored, contact mines were swept in the Brunei Bay and Miri areas. Their neutralization came at the cost of the minesweeper USS *Salute* (AM-294) which was sunk, and much damage to sweep gear of other units of Lt. Comdr. Thomas R. Fonick's Mine Division 34, both from mines and underwater obstructions.[12]

## BALIKPAPAN LANDING

> *Ammunition expenditures were far in excess of those anticipated. The Covering and Support Group alone fired 908 eight-inch, 12,189 six-inch, 18,820 five-inch, and 600 British 4.7-inch projectiles during the period including 1 July. The amount of naval gunfire employed exceeded any previous bombardments in the SWPA [Southwest Pacific Area], even the major landings in the Philippines. In general, bombardments against enemy shore installations were effective in neutralization, but were not always completely destructive, in view of the long range and dispersal of enemy targets.*
>
> —Commander in Chief, U.S. Pacific Fleet and Pacific Ocean Areas, Operations in Pacific Ocean Areas – June 1945, 21 November 1945

Photo 24-4

Cargo ship USS *Poinsett* (AK-205) under way in 1945, location unknown. Naval History and Heritage Command photograph #NH 86706

Troops of the 7th Australian Division landed unopposed at the large oil region of Balikpapan, on 1 July 1944, in the final amphibious operation of the Borneo Campaign. Against scattered resistance and in quick succession, these troops captured Sepinggan Aerodrome on the 2nd, the town of Balikpapan on the 3rd, and Manggar Airstrip on the 6th, followed by further objectives. By month's end, the Australians established a perimeter encompassing the Bodja oilfields, 28 miles from Balikpapan.[13]

Naval operations preliminary to the landings, on 1 July, began in the Balikpapan area, on 15 June, and were largely completed by the end of that month. Shallow water and extensive enemy minefields kept cruisers 12,000-14,000 yards away from the landing beaches, presenting a difficult gunnery problem. It was not until 24 June that enough minesweeping had been done to make it reasonably safe for destroyers to enter the shallow approaches to the harbor, and deliver counter-battery fire at effective ranges against defensive installations.[14]

Sweeping operations cost four members of Fonick's minesweeping unit sunk—*YMS-50*, *YMS-365*, *YMS-39*, and *YMS-84*—and others damaged by mines and enemy gunfire. Including *YMS-481* lost at Tarakan, and *Salute* (AM-294) at Brunei, Borneo, the U.S. Navy lost six sweepers. As Arnold Lott in *Most Dangerous Sea* so adroitly noted, "*YMS-84* was the last surface ship of the U.S. Navy sunk in the far corner of the Pacific, where the realm of the Golden Dragon sweeps down under the Southern Cross."[15]

## TRANSPORT UNIT

At Balikpapan, the same five RN and USN ships comprised the Transport Unit (Task Unit 78.2.2) under Captain Cousin, RANR(S), as had participated earlier in the Brunei landing. These were the *Manoora*, *Westralia*, *Kanimbla*, *Titania*, and *Carter Hall.* A second cargo ship ordered to Balikpapan, the USS *Poinsett* (AK-205), was part of the Service Unit (Task Unit 78.2.14).[16]

*Poinsett* had been commissioned, on 7 February 1945, at Houston, Texas, with Lt. Comdr. Robert M. Baughman, USNR, in command. Loaded with 120 tons of frozen provision cargo from New Orleans, and the balance of her full load consisting of various types of ammunition picked up at Theodore, Alabama, she left there on 21 March, bound for the far Pacific for her assigned duty with the 7th Fleet Service Force.[17]

The *Poinsett* reached Zamboanga, Philippines, on 10 May, under escort by the sub-chaser USS *PC-492* (after detaching from her accompanying convoy) and issued ammunition there until 10 June. On that date, she received orders, directing her to proceed to Tawi Tawi in the Sulu Archipelago, to issue ammunition to the supporting naval forces for the invasion of Borneo. *Poinsett* arrived at Tawi Tawi the following day, and remained there until 28 June, when she was ordered to join the main invasion force scheduled to invade Balikpapan, on 1 July.[18]

She joined up in the Celebes Sea, and continued with the force through the Strait of Makassar to Balikpapan, arriving there at 0700 on 1 July. Per orders, *Poinsett* anchored about five miles off Balikpapan, and immediately began issuing ammunition to cruisers, destroyers, and smaller supporting naval units. On 10 July, when the bulk of her ammo had been discharged, she received orders to proceed to Morotai. The cargo ship arrived there, on the 13th, escorted by *PC-1134*.[19]

Orders received, on 21 July, sent her to Leyte to discharge her remaining munitions to Floating Ammunition Storage Unit #1. Offload of ammo continued until 6 August, then new orders sent her to Manus, Admiralty Islands, to load fleet-issue clothing and small stores for onward distribution. It was there, on 15 August, while laying alongside the oiler USS *Trinity* (AO-13), that she received official word that Japan had accepted the terms of the Potsdam Conference, and made a public announcement of its surrender that day. The Pacific war formally ended, on 2 September 1945, when Japanese representatives signed surrender documents aboard the USS *Missouri* (BB-63) in Tokyo Bay.[20]

## DISTINGUISHED SERVICE ORDER AND CROSSES

Following the war, King George VI awarded Acting Captain Cousin, the Distinguished Service Order and Commanders Bunyan and Livingston the Distinguished Service Cross. All three were Mercantile Marine officers who had served as RANR(S)—Royal Australian Naval Reserve (Sea-going)—officers during the war.

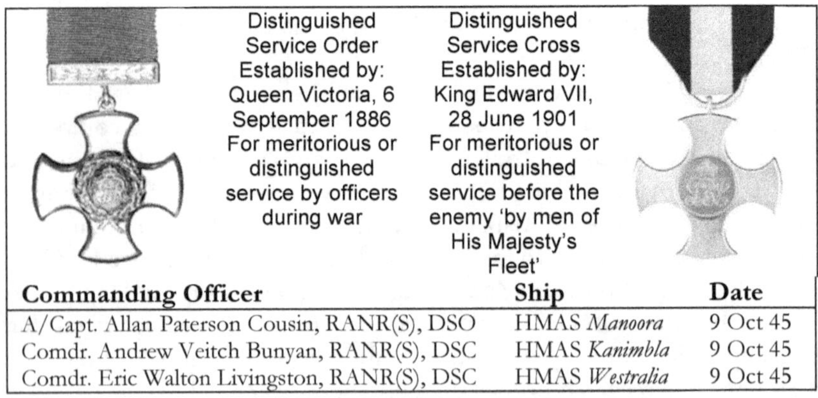


| Commanding Officer | Ship | Date |
|---|---|---|
| A/Capt. Allan Paterson Cousin, RANR(S), DSO | HMAS *Manoora* | 9 Oct 45 |
| Comdr. Andrew Veitch Bunyan, RANR(S), DSC | HMAS *Kanimbla* | 9 Oct 45 |
| Comdr. Eric Walton Livingston, RANR(S), DSC | HMAS *Westralia* | 9 Oct 45 |

# The RAN's Contribution to Amphibious Warfare in the Pacific

**By Commodore Hector Donohue AM RAN (Rtd)**

In March 1942, the Australian Government, recognising the importance of an amphibious capability in any effort to drive the Japanese out of the South West Pacific, began exploring the requirements for combined operations training in Australia. Navy and Army planners, both Australian and US, began to examine both the training and organisation requirements. In addition, in late 1942 the RAN commenced converting HMA (His Majesty's Australian) Ships *Manoora*, *Kanimbla* and *Westralia* to infantry landing ships (LSI). The three passenger liners had been requisitioned in late 1939 and following conversion were deployed as armed merchant cruisers.

## HMAS ASSAULT

Photo Postscript-1

Training at HMAS Assault. (AWM)

Approval was granted by the Australian War Cabinet in August 1942 to establish HMAS Assault, a RAN Training Centre, in the Port Stephens area of New South Wales. Assault was initially commissioned onboard the armed merchant cruiser *Westralia* on 1 September 1942 and began providing instruction for landing craft crews, beach parties (naval commandos) and combined operations signals teams. Assault transferred ashore on 10 December 1942 with *Westralia* allocated as a temporary accommodation ship and HMAS *Ping Wo* as her tender.

An American Amphibious Training Group was also established nearby and the two facilities were combined as the Amphibious Training Centre (ATC) in February 1943 under Rear Admiral Daniel E. Barbey, USN, the commander of the South West Pacific Amphibious Force, who answered directly to General MacArthur. (The South West Pacific Force was renamed the US 7th Fleet on 15 March 1943, and Barbey's title became Commander, 7th Amphibious Force.) This brought all combined amphibious training in Australia under US command. Training at Assault, from then on, included US soldiers and Marines, as well as Australian army and navy personnel. Training was intense, covering every aspect of landing operations on hostile shores. Assault also provided operational and logistics support to RAN amphibious units.

Meanwhile, the Australian Army had established training facilities in Queensland, initially a Combined Training Centre at Toorbul Point near Bribie Island north of Brisbane. Cairns Trinity Beach was also used to train troops in all aspects of amphibious warfare before heading into war zones north of Australia.

By October 1943, there were 141 ships and landing craft based at Port Stephens. Thirty-six of these ships were controlled by Assault and 105 by the US Navy.

In early March 1944, with operations underway in New Guinea, training at Assault ceased. It had served its purpose well. In August the base was reduced to a 'care and maintenance' status, decommissioning on 7 April 1945. During its short three-year commission more than 22,000 personnel undertook training there. The training expertise was transferred to a 'mobile team' and moved to Milne Bay in mid-1944 and ultimately to Subic Bay in the Philippines the following year.

## INTRODUCTION OF LANDING SHIP INFANTRY

Rear Admiral Daniel E. Barbey, USN was appointed in command of the South West (later 7th Fleet) Amphibious Force at the end of 1942 but because of the pressing need for amphibious ships in other theatres none were immediately available.

HMAS *Manoora*, recommissioned as an infantry landing ship (LSI) in February 1943, was the first ship to join the force. The two other former Australian armed merchant cruisers were converted to LSI with *Westralia* and *Kanimbla* recommissioning in June 1943.

*Manoora* and *Kanimbla* were motor vessels of some 11,000 gross registered tons (GRT) and 480 feet long, whilst *Westralia* was slightly smaller at 8,000 GRT and 455 feet long. The three ships carried US landing craft – 20-22 LCVP (Higgins boats) plus 2-3 LCMs (mechanized landing craft) and 1,280 troops. As customary for other USN amphibious ships, landing craft were lowered to the waterline, and the troops embarked in these down rope ladders.

Photo Postscript-2

HMAS *Westralia* configured as an infantry landing ship. (AWM)

In March 1943 the US attack transport USS *Henry T Allen* joined the force; in June the destroyer tender USS *Rigel* joined, and for the remainder of the year was Barbey's flagship.

From the middle of 1943 until April 1944, these three LSIs, with the *Henry T Allen*, were the only transports available to the Seventh Fleet Amphibious Force. They continued to operate with the Force through the South West Pacific Area operations and accomplished all missions assigned them in a most creditable manner.

## 7TH FLEET AMPHIBIOUS FORCE OPERATIONS

Operations of the Seventh Fleet Amphibious Force in the South West Pacific Area can be divided into three general phases:

- September 1943 - September 1944: Amphibious landings on the Eastern and Northern Coast of New Guinea; in New Britain, Admiralty Islands; and at Morotai in Indonesia's Maluku Islands
- October 1944 - February 1945: Amphibious landings at Leyte, Mindoro, Lingayen and supporting operations in Luzon culminating in the Bataan-Corregidor landings on 15-16 February 1945
- February 1945 - July 1945: Amphibious landings together with extensive minesweeping operations in the Central, and Southern Philippines including Sulu Archipelago; and Borneo

# PHASE I: SEPTEMBER 1943 – SEPTEMBER 1944

Map Postscript-1

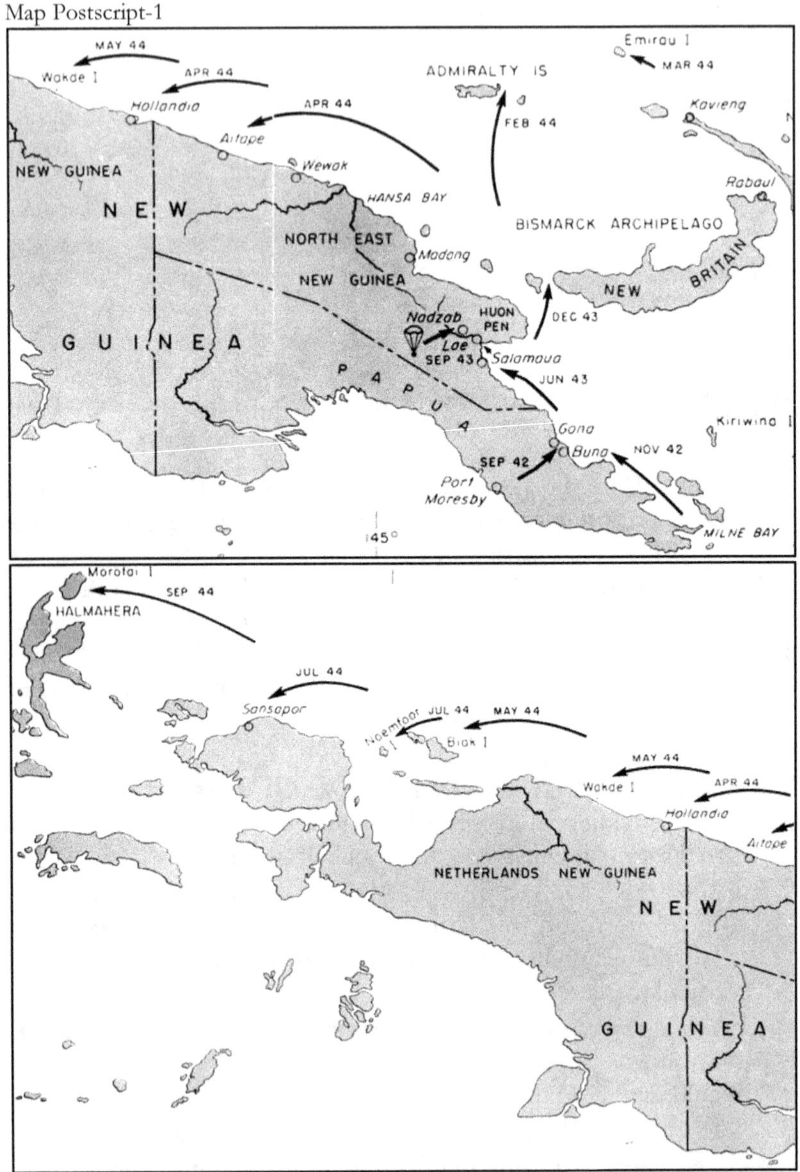


Movement of Allied forces up the Papua New Guinea coast, from Milne Bay (top map, lower right) to Morotai Island (bottom map, upper left)

The amphibious landing at Lae (which may be found on the top map, just to the left of Salamaua) by the 9th Australian Division on 4

September 1943 was the first of a series of assault landings along the New Guinea coast, in the Bismarck Archipelago and in the Halmaheras designed to provide staging areas, minor naval bases, and airfields to support the major assault on the Philippine Islands. From September 1943 to September 1944, the Seventh Amphibious Force made 14 major landings in these areas involving the movement of approximately 300,000 men and 350,000 tons of supplies and equipment.

Following conversion in March 1943, *Manoora* spent the first few months off Port Stephens and Cairns before transporting Australian troops to Milne Bay mid-year. (Following the defeat of the Japanese by Australian forces in September 1942, the first major battle of the war in the Pacific in which Allied troops decisively defeated Japanese land forces, Milne Bay had been developed into a major Allied Base.) *Manoora* transported troops in company with *Kanimbla* to Oro Bay east New Guinea in late August and again in early November. During this period, both ships carried out landing exercises in the Cairns area. *Manoora* returned to Sydney for refit whilst *Kanimbla* remained north.

*Westralia* conducted a series of training exercises at Port Stephens before transporting US Marines to Goodenough Island near Milne Bay in September. She then conducted further landing exercises and transport of troops to Goodenough Island before being assigned to transport US troops for the landings at Arawe, New Britain on 15 December 1943.

*Westralia* landed US troops at Cape Cretin (New Guinea) in January 1944. En route from the area on 28 January she was attacked by Japanese aircraft and suffered casualties and damage which was, fortunately, insufficient to put her out of commission. On 5 February she was back at the US base at Cape Cretin south of Finschaven with reinforcements. March and April were spent operating in New Guinea waters including the landings at Hollandia. *Westralia* returned to Sydney on 31 May 1944 for refit.

*Kanimbla* proceeded to New Guinea in March for exercises with elements of the US 24th Infantry Division at Goodenough Island in preparation for the landings at Hollandia. Allied forces undertook an amphibious landing on 22 April 1944 at Aitape on northern coast of New Guinea. The amphibious landing was undertaken simultaneously with the landings at Humboldt and Tanahmerah Bays to secure Hollandia to isolate the Japanese 18th Army at Wewak. On 22 April, in company with *Manoora*, five other transports, 16 infantry landing craft and seven tank landing ships, both LSIs participated in the landings at Tanahmerah Bay, Hollandia without incident.

Photo Postscript-3

Landing craft approach the beaches at Tanahmerah Bay, 22 April 1944. (AWM)

In July 1944 *Westralia* again began transporting troops to the forward areas operating from New Guinea bases. August was spent mainly in the Solomon Islands on training exercises, followed by similar duties in the Aitape area in September.

*Manoora* and *Kanimbla* prepared for the Morotai landings in early September 1944. On 9 September both ships each embarked 1,272 men at Maffin Bay on the northern coast of New Guinea. On 10 September, with 36 other landing ships and supporting vessels of the White Beach Attack Group, they departed for Morotai. The landings took place on 15 September with little opposition and few casualties to the Allied forces. *Manoora* and *Kanimbla* departed for Humboldt Bay (located on the north coast of New Guinea to the east of Hollandia.) arriving 18 September. They were joined by the third Australian LSI, *Westralia*, where they overhauled equipment and embarked troops and supplies for the landings on Leyte.

## PHASE II: 2 OCTOBER 1944 – FEBRUARY 1945

The retaking of the Philippines began with an assault on the Leyte Gulf-Surigao Strait area. Planning was complicated by the huge distances involved, for while the Normandy landings on 6 June 1944 were conducted 50 nautical miles across the English Channel, Leyte Gulf was more than 500 nautical miles from the main staging areas in Morotai and Palau. Much of the logistic support had to be sourced from the US west coast, more than 5,000 nautical miles from the front. Because the assault would also take place beyond the range of land-based aircraft, all air support would need to come from US Navy aircraft carriers. The advance from Morotai to Leyte in one bound was a calculated risk, as the Allied forces would be ringed by Japanese airfields and land-based aircraft with greater staying power (time on station) than the aircraft from US Navy aircraft carriers.

Commanded by Vice Admiral Thomas C. Kinkaid, USN, the US Seventh Fleet and assigned elements of the US Third Fleet together formed Task Force 77 and the Central Philippines Attack Force, and comprised 157 combat ships (including 6 battleships, 11 cruisers and 18 escort carriers), 420 amphibious ships and 84 patrol, minesweeping and hydrographic vessels. Another 17 aircraft carriers, 6 battleships, 16 cruisers and 56 destroyers of the Third Fleet, under Admiral William F. Halsey, USN, were tasked with covering the invasion.

The RAN's contribution to Kinkaid's force, under the command of Commodore John A Collins (later Vice Admiral), consisted of the heavy cruisers HMA Ships *Australia* and *Shropshire*; the destroyers *Arunta* and *Warramunga*; the infantry landing ships *Manoora*, *Kanimbla* and *Westralia*; the frigate *Gascoyne*; and the motor launch *HDML 1074*. The RAN was also represented in Task Group 77.7 (the Leyte Gulf Service Force of the Seventh Fleet) by the oiler *Bishopdale*, the provision ship *Merkur* and the ammunition ships *Poyang* and *Yunnan*.

A full-scale rehearsal was carried out at Tanahmerah Bay on 10 October. On 13 October the three Australian LSIs departed for Leyte as part of a large assault convoy escorted by a covering force of US and Australian cruisers and destroyers. The Australian landing ships were part of the Panaon Attack Group which detached from the main group at 0200 on the morning of 20 October arriving off Panaon at 0845. No Japanese resistance was encountered. Within 45 minutes the three Australian ships had disembarked over 2,800 troops of the US 21st Regimental Combat Team on the undefended island. The LSIs then sailed for Humboldt Bay arriving on 25 October 1944.

During November, the three LSIs were engaged in transporting troops from Humboldt Bay to Leyte. On 30 November they, and 15 other ships designated Transport Group 'A', commenced embarking troops and stores for the Lingayen landings. Transport Group 'A' then proceeded to Lae where, in company with Landing Group 'B', practice landings were carried out. The ships then sailed for Manus Island. On

31 December they departed Manus to execute Assault Mike I on Luzon Island in Lingayen Gulf.

Photo Postscript-4

Crewmembers of HMAS *Kanimbla*, watch the ship's landing craft as they proceed towards the beach in the first wave of the landing in Brunei Bay, June 1945. (AWM)

As part of Task force 79, the Lingayen Attack Force, *Manoora*, *Kanimbla* and *Westralia* passed through Surigao Strait and proceeded up the western side of the Philippine Archipelago to Lingayen Gulf, arriving on 8 January 1945. As the convoy made its final approach to the Gulf, it was subjected to air attack. Following a suicide attack which damaged the escort carrier USS *Kitkun Bay*, *Westralia* was attacked by a Kamikaze (Zero fighter) from astern. The final steep dive indicated it was aimed at the bridge, but the gunners onboard maintained a strong barrage and shot the aircraft down some 10 feet from the stern. There were no casualties, but the steering gear was disabled for a short time. The troops were landed on 9 January, supported by a heavy bombardment. The LSIs discharged their cargoes rapidly and left the area that evening to avoid further air attack, returning via Leyte to Morotai, arriving on 21 January 1945.

The LSIs were now wearing a new paint scheme. Following the landings at Lingayen Gulf a decision was made to apply camouflage paint schemes to all three Australian LSIs. The ships continued conducting transport duties in the area.

Photo Postscript-5

HMAS *Manoora* at Morotai Island, January 1945, with new paint scheme. (AWM)

*Kanimbla* returned to Sydney for refit on 7 March 1945 and on completion of the refit, in late April she embarked Australian troops in Brisbane and discharged them in Morotai mid-May.

## PHASE III: 3 FEBRUARY-JULY 1945 – BORNEO

The Borneo campaign of 1945 was one of the most complex operations involving Australian land, air and sea forces in the war. It was also the last Australian campaign to be planned and undertaken. Initial plans called for six OBOE operations (code name for a series of amphibious assaults). However, as the Allied offensives progressed closer to Japan, OBOE THREE, FOUR and FIVE were cancelled. The remaining three amphibious landings were: OBOE ONE, the invasion of Tarakan Island; OBOE SIX, the invasion of north Borneo at Labuan and Brunei; and OBOE TWO, the invasion of Balikpapan.

*Manoora* and *Westralia* were engaged in the Australian landings on Tarakan (OBOE ONE). Embarking Australian units for the first time, the LSIs sailed from Morotai on 27 April 1945, each ship towing a tank landing craft (LCT). The transport force arrived off Tarakan on 30 April and the LCTs were slipped. The troops were successfully landed on 1 May and the ships finished discharging their cargoes the next day. They then departed for Morotai where further stores were loaded and brought forward to the landing beaches. By mid-May *Kanimbla* had re-joined the group.

The next operation was the invasion of Brunei, OBOE SIX. The three LSIs departed Morotai on 4 June, in company with a large group of US Navy ships, mostly landing ships and landing craft. They arrived off Brunei on 10 June and commenced landing troops on Green Beach just before 0900 with little or no opposition. After unloading cargo that day they set sail for Morotai on 11 June, arriving on 14 June 1945.

Map Postscript-2

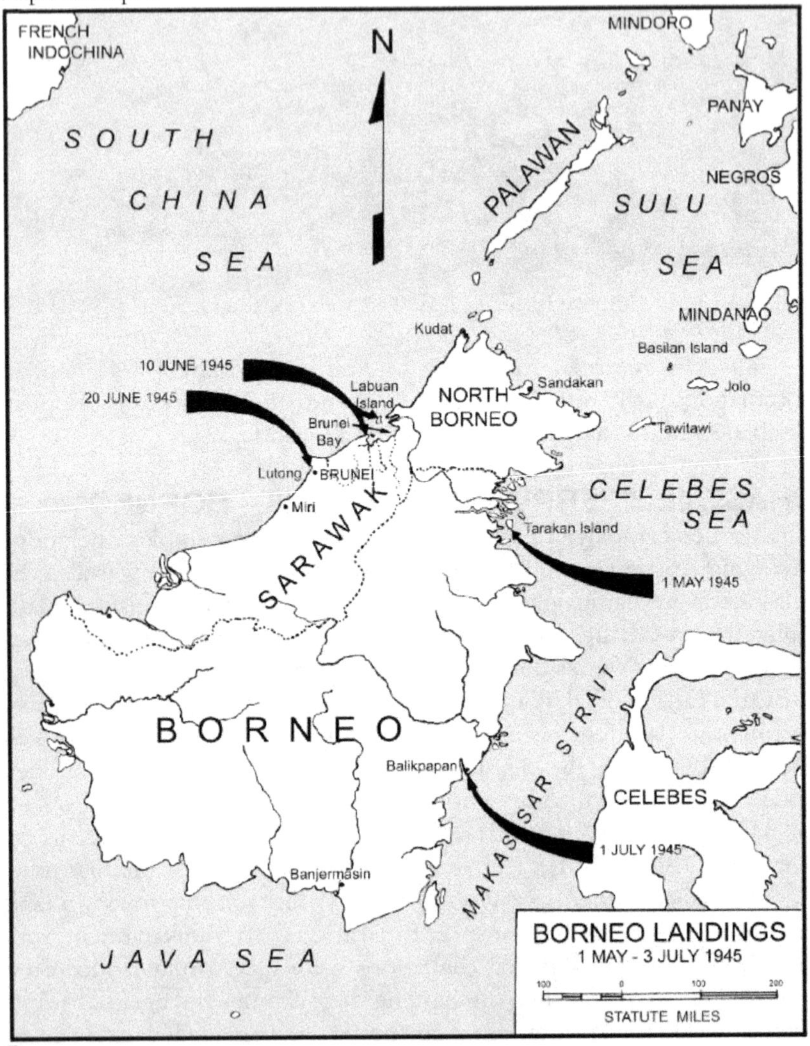

Borneo

The final amphibious operation in which the LSIs took part was the Balikpapan landing, OBOE TWO. It was the largest ever amphibious assault by Australian forces with more than 33,000 personnel landed. After embarking troops and cargo at Morotai, *Manoora*, *Kanimbla* and *Westralia* sailed on 26 June 1945, arriving off Balikpapan on 1 July 1945. That day was spent disembarking troops and unloading the cargo after which the ships sailed for Morotai at 1930. Arriving on 4 July, they embarked reinforcements and departed the same day, returning to Balikpapan on 7 July. It was the LSIs last operation together and the remainder of the war was spent on transport duties around New Guinea, the Philippines and Borneo.

Photo Postscript-6

HMAS *Manoora*, 1944, showing the LCVPs (Higgins boats) carried by davits along her ship's side and in the deck between the bridge and the funnel. (AWM)

Photo Postscript-7

Troops scrambling down nets slung over the side of HMAS *Kanimbla* into landing craft (Higgins boats), for the OBOE SIX Operation Landing, 10 June 1945. (AWM)

## THE LSI COMMANDING OFFICERS

During the period they were active as LSIs, each ship had two commanding officers, all being members of the RAN Reserve (Seagoing). It is interesting to briefly review their careers as it offers a snapshot of a typical wartime service for qualified seagoing officers. Of the six officers, three had joined the second entry of the newly formed Royal Australian Naval College in 1914, served briefly in the North Sea in 1918 and ultimately transferred to the reserve, before returning to full time service at the outbreak of World War II. Given the adverse impact on the RAN during the Depression, early retirement of RAN officers was common. Of the remaining three, one was ex-Royal Navy Reserve and merchant marine officer, whilst the other two were ex-merchant marine officers who had enlisted in the RANR(S) in the early 1930s. Four of the six received awards for their time in command of an LSI.

### Cecil Claude Baldwin

Cecil Claude Baldwin was born in Kempsey, NSW on 30 September 1900 and joined the RAN College in 1914 as a member of the second college intake, graduating as a midshipman in January 1918. He joined the heavy cruiser HMAS *Australia* in March 1918 for training. He was promoted to Lieutenant in March 1922 and following further training in UK served in the RAN cruisers *Melbourne*, *Sydney*, *Adelaide* and *Australia* during the 1920s and mid-1930s. He was appointed Personal aide-de-camp to the Duke of Gloucester during his Australian visit October/December 1934 and in April 1935 was awarded an MVO (Member Royal Victorian Order) medal.

Baldwin was promoted commander in September 1936 and transferred to the Reserve the next month. At the outbreak of war, Baldwin joined Flinders Naval Depot, the primary RAN Training Establishment (HMAS Cerberus) and assumed command of *Manoora* and as Senior Naval Officer Australian Landing Ships, after her conversion as an LSI on 29 June 1943, as an acting captain. He was involved in training and transporting of troops until her refit in December 1944. He was then appointed as Naval Officer in Charge, Darwin until his death by illness on 10 April 1945.

### Norman Hamon Shaw

Norman Hamon Shaw was born in Perth, WA on 9 July 1900 and joined the RAN College in 1914 graduating as a midshipman in January 1918. Following training at sea he was promoted lieutenant in September 1921 and joined the cruiser *Melbourne*. He returned to the UK in November 1924 to qualify as a submariner. In mid-1926 he commanded the RN submarine *H27* before standing by (assigned as part of the crew of) the first RAN submarine HMAS *Oxley* in early 1927. She commissioned in June 1928 and Shaw was posted in command in February 1929. Given the financial constraints at the time the RAN determined not to continue to introduce a submarine capability and returned them to RN service in April 1931. Shaw spent three years exchange service in the RN before returning to Australia in 1934 and shortly after transferred to the RANR(S).

On the outbreak of war, he mobilised and was posted as executive officer of the newly commissioned armed merchant cruiser HMAS *Manoora* and was promoted acting commander in September 1940. Following a short period ashore in Brisbane from mid-1942, he was appointed in command of HMAS *Kanimbla* on commissioning as an LSI in June 1943. He remained in command until June 1944, participating in the amphibious assault by US troops at Hollandia in April. He commanded the naval depot HMAS Kuttabul in Sydney and was confirmed as a commander in January 1946. In July 1949 he was appointed to command the Australian Naval Dockyard Police with the rank of Superintendent – a role he fulfilled until his retirement on 5 August 1958. He was awarded an OBE (Order of the British Empire) in the 1951 New Year's Honours List for his long and distinguished service to the RAN and concurrently the Naval Dockyard Police.

## Alfred Victor Knight

Photo Postscript-8

Commander Alfred Victor Knight on the bridge of HMAS *Westralia* holding the ship's mascot, 1943. (AWM)

Alfred Victor Knight was born in Dover, England on 20 February 1897 and joined the merchant marine as a cadet in 1912. Following the outbreak of World War I, he joined the Royal Navy Reserve. As a sub lieutenant, he was awarded the DSC (Distinguished Service Cross) for his conspicuous bravery under heavy enemy fire whilst involved in an assault on Ostende Harbour in April 1918. He also received a MID (Mention in Despatches) as a lieutenant after the war for his service in minesweeping. Post war he returned to the merchant marine and after moving to Australia, he joined the RANR(S) as a lieutenant in 1923. Promotion to lieutenant commander followed in 1931 and commander in 1937.

He mobilised in March 1940, joining the Naval Staff in Melbourne before serving as Commanding Officer of the newly commissioned *Bathurst*-class corvette, HMAS *Lithgow* in June 1941. During this time *Lithgow* swept German laid mines in Bass Strait, assisted in the sinking of a Japanese submarine off Darwin, escorted the first contingent of Allied troops from Townsville to Milne Bay, and took part in the campaign to recapture Buna in northern New Guinea. For his services in *Lithgow*, Knight was awarded the OBE.

Knight was appointed Commanding Officer of HMAS *Westralia* in February 1943, remaining there until December 1943, taking part in the Allied landings at Arawe, New Britain; Humbolt Bay, New Guinea; and Panoan in the Philippines. For exceptional service in command Knight was awarded the US Legion of Merit. The citation described him as a 'forceful leader,' and by his 'splendid cooperation in the conduct of a vital training programme, aggressive determination and untiring energies' he had 'contributed materially to combined large-scale operations and the successful prosecution of the war' in the South West Pacific Area. He returned to shore duty and was promoted captain on 31 December 1946. He retired from the RANR(S) on 19 February 1952.

## Allan Paterson Cousin

Allan Paterson Cousin was born in Clifton, Queensland on 29 March 1900 and joined the RAN College in 1914 graduating as a midshipman in January 1918. Following training at sea he was promoted lieutenant in October 1922 but resigned his commission on 23 April 1923 and joined the Union Steam Ship Co. of New Zealand Ltd. He transferred to the RANR(S) as a lieutenant in April 1925. He was promoted to lieutenant commander in 1930 and commander on 30 June 1936.

He mobilised in March 1941 and in December took command of the newly commissioned *Bathurst*-class corvette, HMAS *Katoomba*. She joined the 24th Minesweeping Flotilla at Darwin and participated in the action in which the Japanese submarine *I-124* was sunk on 20 January 1942. *Katoomba* then began a period of escort duty to New Guinea, shepherding convoys between Townsville and Port Moresby, Milne Bay and Oro Bay.

On 27 January 1944 Cousin was appointed to command *Manoora*, and as Senior Naval Officer, Australian Landing Ships. Between April 1944 and July 1945 *Manoora* landed troops in Netherlands New Guinea at Tanahmerah Bay, Wakde and Morotai; in the Philippines at Leyte and Lingayen Gulf, Luzon; and in Borneo and Brunei at Tarakan, Labuan and Balikpapan. Cousin was awarded the DSO (Distinguished Service Order) in October 1945 for his 'gallantry, fortitude and skill' during the amphibious assaults. He was promoted acting captain in February 1945 and confirmed in the rank on 30 June 1946.

Cousin transferred to command *Kanimbla* in January 1948. She conducted two trips to Japan repatriating prisoners of war and transporting personnel of the British Commonwealth Occupation Force in the first half of 1948. In July, she took personnel and stores to the UK to commission HMAS *Sydney*, returning with British personnel who had enlisted in the RAN as well as 432 displaced persons. After one more voyage to Japan she was decommissioned in Sydney on 25 March 1949. Cousin was demobilised in May 1949.

## Andrew Veitch Bunyan

Andrew Veitch Bunyan, born in Leith, Scotland on 17 February 1902, qualified as a master mariner and joined the RANR(S) as a lieutenant in September 1930. He regularly completed Reserve training during the 1930s and mobilised in February 1940. He was promoted to acting lieutenant commander in December 1939 and confirmed in November 1940. In March 1940 he joined the sloop HMAS *Swan* as executive officer and on 6 December assumed command of the newly commissioned corvette HMAS *Bathurst*. After a short period sweeping German mines on Australia's east coast, *Bathurst* deployed to Colombo mid-1941, joining the RN Eastern Fleet for escort and patrol duties in the Indian Ocean, Persian Gulf and Arabian Sea.

Bunyan was relieved in October and after leave joined *Manoora* as executive officer in April 1943. In September 1943, he moved on to command of *Swan* and was promoted to commander in December. *Swan* was assigned to New Guinea waters for escort and patrol duties, and as a fire support ship for military operations ashore. Bunyan was appointed in command of *Kanimbla* in August 1944. She participated in the landings at Morotai, Leyte and Lingayen Gulf and at Tarakan,

Labuan and Balikpapan. In October 1945 he was awarded the DSC 'for gallantry, fortitude and skill whilst serving in HMAS *Kanimbla* in numerous amphibious assaults in the South West Pacific Area, including operations in New Guinea, the Philippines and Borneo.' In September 1945 he was posted ashore and demobilised from the RANR(S).

## Eric Walton Livingston

Eric Walton Livingston was born in Balmain, Sydney on 27 January 1903 and qualified as a master mariner before joining the RANR(S) as a lieutenant in August 1931. He regularly completed Reserve training during the 1930s and mobilised in February 1940. He was appointed in command of HMAS *Bingen*, an auxiliary anti-submarine vessel and was promoted lieutenant commander in March. In September 1940 he was posted in command of the newly commissioned anti-submarine auxiliary vessel, HMAS *Wyrallah*, which was re-named *Wilcannia* in February 1942, to avoid confusion with the new *Bathurst*-class corvette, HMAS *Whyalla*.

*Wyrallah* commenced service as an anti-submarine/patrol vessel on the west coast where in November 1941 she participated in the search for HMAS *Sydney* survivors. As *Wilcannia*, she was moved to New Guinea and South Pacific waters to carry out patrols, pilot rescues and transport of stores, returning to the Darwin area in early 1943.

Livingston took command of *Westralia* in December 1943 and continued reinforcing the build-up of troops in Leyte before participating in the Lingayen Gulf landings and ultimately the landings in Borneo. He was in command during the Kamikaze attack in January 1945 when well-aimed fire from *Westralia*'s guns caused the aircraft to disintegrate and crash ten feet astern. He was promoted to acting commander in February 1945. In October 1945 he was awarded the DSC 'for gallantry, fortitude and skill whilst serving in HMAS *Westralia* in numerous amphibious assaults in the South West Pacific Area, including operations in the Philippines and Borneo.' He remained in command of *Westralia* until November 1945, when he was demobilised but remained in the Reserve. He was confirmed as a commander RANR(S) in December 1945 and retired from the Reserve in 1953.

# Appendix A: Maritime Commission Cargo Ship Hull Types

| Hull Type | Length (feet) | Displ. (tons) FL or DW | Max Draft (feet) | Speed (knots) | Prop |
|---|---|---|---|---|---|
| C1-A | 412'3" | 11,653 FL | 23'6" | 14.5 | diesel |
| C1-B | 417'9" | 9,104 FL | 27'6" | 14 | steam |
| C2 Cargo | 459'1" | 10,850 FL | 25'9" | 15.5 | diesel |
| | 459'2" | 13,900 FL | 25'9" | 15.5 | steam |
| C2-F | 459'2" | 13,898 FL | 25'10" | 15.5 | steam |
| C2-S | 479'8" | 11,154 FL | 20'9" | 16 | steam |
| C2-S-AJ3 | 459'2" | 13,050 FL | 24'6" | 15.5 | steam |
| C2-S-B1 | 459'2" | 13,893 FL | 25'9" | 15.5 | steam |
| C2-T | 459'1" | 10,432 FL | 25'10" | 15.5 | diesel |
| C3 Cargo | 492' | 12,349 FL | 21'1" | 16.5 | diesel |
| | 492' | 14,907 FL | 28'6" | 16.5 | steam |
| C3-E | 473'1" | 14,480 FL | 27'2" | 16.5 | steam |
| EC2-S-C1 | 441'7" | 14,250 FL | 27'7" | 11 | steam |
| N3-M-A1 | 269'10" | 5,200 FL | 20'8" | 10.25 | diesel |
| S4-SE2-BE1 | 426' | 6,800 FL | 15'6" | 16.5 | steam |
| VC-2-S-AP2 | 455'3" | 15,580 FL | 28'6" | 17 | steam[1] |

FL: Full load
DW: Dead weight

Prop: Type propulsion

# Appendix B: U.S. Navy Cargo Vessels (271 AK, AKA, and AKS) in World War II

### Miscellaneous Cargo Ships (No Class) – 47 Cargo Ships (AK)

Fourteen converted to Attack Cargo Ships (AKA) in 1943
Ten of the forty-seven AKs were former Victory ships

| Cargo Ship | Comm | Hull Type | Reclass as AKA |
|---|---|---|---|
| *Alchiba* (AK-23) | 1941<br>1943 | C2 | <br>(AKA-6) |
| *Alcyone* (AK-24) | 1941<br>1943 | C2 | <br>(AKA-7) |
| *Algorab* (AK-25) | 1941<br>1943 | C2 | <br>(AKA-8) |
| *Alhena* (AK-26) | 1941<br>1943 | C2-S | <br>(AKA-9) |
| *Almaack* (AK-27) | 1942<br>1943 | C3-E | <br>(AKA-10) |
| *Aquila* (AK-47) | 1941 | ex-MV *Tunis*/ MV *Aquila* | |
| *Arcturus* (AK-18) | 1940<br>1943 | C2 | <br>(AKA-1) |
| *Aries* (AK-51) | 1942 | ex-*Manomet* (AG-37) | |
| *Aroostook* (AK-44) | 1941 | converted from minelayer CM-3 | |
| *Asterion* (AK-63) | 1942 | ex-SS *Evelyn* | |
| *Atik* (AK-101) | 1942 | ex-SS *Carolyn* | |
| *Auriga* (AK-98) | 1943 | C1-B hull | |
| *Bedford Victory* (AK-231) | 1944 | VC-2-S-AP2 (Victory) | |
| *Bellatrix* (AK-20) | 1942<br>1943 | C2-T | <br>(AKA-3) |
| *Betelgeuse* (AK-28) | 1941<br>1934 | C2 | <br>(AKA-11) |
| *Bolder Victory* (AK-227) | 1944 | VC-2-S-AP2 (Victory) | |
| *Bucyrus Victory* (AK-234) | 1944 | VC-2-S-AP2 (Victory) | |
| *Capella* (AK-13) | 1921 | Ex-SS *Comerant* | |
| *Delta* (AK-29) | 1941 | C3 | |
| *Electra* (AK-21) | 1942<br>1943 | C2-T | <br>(AKA-4) |
| *Enceladus* (AK-80) | 1943 | N3-M-A1 | |
| *Fomalhaut* (AK-22) | 1942<br>1943 | C1-A | <br>(AKA-5) |
| *Gemini* (AK-52) | 1942 | ex-SS *Coperas* | |

| | | | |
|---|---|---|---|
| *Hamul* (AK-30) | 1941 | C3 | |
| *Hercules* (AK-41) | 1941 | C3-E | |
| *Hydra* (AK-82) | 1943 | N3-M-A1 | |
| *Jupiter* (AK-43) | 1942 | C2 | |
| *Lakewood Victory* (AK-236) | 1944 | VC-2-S-AP2 (Victory) | |
| *Las Vegas Victory* (AK-229) | 1944 | VC-2-S-AP2 (Victory) | |
| *Libra* (AK-53) | 1942 | C2-F | |
| | 1943 | | (AKA-12) |
| *Manderson Victory* (AK-230) | 1944 | VC-2-S-AP2 (Victory) | |
| *Markab* (AK-31) | 1941 | C3 | |
| *Mayfield Victory* (AK-232) | 1944 | VC-2-S-AP2 (Victory) | |
| *Mercury* (AK-42) | 1942 | C2 | |
| *Newcastle Victory* (AK-233) | 1944 | VC-2-S-AP2 (Victory) | |
| *Oberon* (AK-56) | 1942 | C2-F | |
| | 1943 | | (AKA-14) |
| *Pegasus* (AK-48) | 1941 | ex-SS *Rita Maersk* | |
| *Pleiades* (AK-46) | 1941 | ex-MV *Mangalia* | |
| *Procyon* (AK-19) | 1941 | C2 | |
| | 1943 | | (AKA-2) |
| *Provo Victory* (AK-228) | 1944 | VC-2-S-AP2 (Victory) | |
| *Red Oak Victory* (AK-235) | 1944 | VC-2-S-AP2 (Victory) | |
| *Regulus* (AK-14) | 1940 | ex-SS *Glenora* | |
| *Saturn* (AK-49) | 1942 | ex-SS *Arauca* | |
| *Sirius* (AK-15) | 1922 | ex-SS *Saluda* | |
| *Spica* (AK-16) | 1940 | ex-SS *Shannock* | |
| *Titania* (AK-55) | 1942 | C2-F | |
| | 1943 | | (AKA-13) |
| *Vega* (AK-17) | 1921 | ex-SS *Lebanon*[1] | |

## *Crater*-class Cargo Ships (AK) – 66

Maritime Commission EC2-S-C1 type (Liberty)
441 feet, 11,565 tons (full load), 11.5 knots, 206 crew complement
One 5-inch/38, one 3-inch/50, two 40mm, six 20mm guns,
or one 5-inch/38, four 40mm, twelve 20mm guns

| Cargo Ship | Comm | Cargo Ship | Comm |
|---|---|---|---|
| *Adhara* (AK-71) | 1942 | *Grumium* (AK-112) | 1943 |
| *Albireo* (AK-90) | 1943 | *Hyperion* (AK-107) | 1943 |
| *Alderamin* (AK-116) | 1943 | *Kenmore* (AK-221) | 1944 |
| *Alioth* (AK-109) | 1943 | *Leonis* (AK-128) | 1943 |
| *Alkaid* (AK-114) | 1943 | *Lesuth* (AK-125) | 1943 |
| *Alkes* (AK-110) | 1943 | *Livingston* (AK-222) | 1944 |
| *Allegan* (AK-225) | 1944 | *Lynx* (AK-100) | 1943 |
| *Alnitah* (AK-127) | 1943 | *Lyra* (AK-101) | 1943 |
| *Aludra* (AK-72) | 1942 | *Matar* (AK-119) | 1944 |
| *Appanoose* (AK-226) | 1944 | *Megrez* (AK-126) | 1943 |
| *Ara* (AK-136) | 1944 | *Melucta* (AK-131) | 1944 |
| *Arided* (AK-73) | 1942 | *Menkar* (AK-123) | 1944 |
| *Arkab* (AK-130) | 1944 | *Mintaka* (AK-94) | 1943 |
| *Ascella* (AK-137) | 1944 | *Murzim* (AK-95) | 1943 |

| | | | |
|---|---|---|---|
| *Azimech* (AK-124) | 1944 | *Naos* (AK-105) | 1943 |
| *Baham* (AK-122) | 1944 | *Pavo* (AK-139) | 1944 |
| *Bootes* (AK-99) | 1943 | *Phobos* (AK-129) | 1944 |
| *Caelum* (AK-99) | 1943 | *Prince Georges* (AK-224) | 1944 |
| *Carina* (AK-74) | 1942 | *Propos* (AK-132) | 1944 |
| *Cassiopeia* (AK-75) | 1942 | *Rotanin* (AK-108) | 1943 |
| *Celeno* (AK-76) | 1943 | *Rutilicus* (AK-113) | 1943 |
| *Cetus* (AK-77) | 1943 | *Sabik* (AK-121) | 1943 |
| *Cheleb* (AK-138) | 1944 | *Sculptor* (AK-103) | 1943 |
| *Cor Caroli* (AK-91) | 1943 | *Seginus* (AK-133) | 1944 |
| *Crater* (AK-70) | 1942 | *Serpens* (AK-97) | 1943 |
| *Crux* (AK-115) | 1944 | *Shaula* (AK-118) | 1943 |
| *De Grasse* (AK-223) | 1944 | *Situla* (AK-140) | 1944 |
| *Deimos* (AK-78) | 1943 | *Sterope* (AK-140) | 1944 |
| *Draco* (AK-79) | 1943 | *Syrma* (AK-134) | 1944 |
| *Eridanus* (AK-92) | 1943 | *Triangulum* (AK-102) | 1943 |
| *Etamin* (AK-93) | 1943 | *Venus* (AK-135) | 1944 |
| *Ganymede* (AK-104) | 1943 | *Zaniah* (AK-120) | 1943 |
| *Giansar* (AK-104) | 1943 | *Zaniah* (AK-120)[2] | 1944 |

## *Alamosa*-class Cargo Ships (AK) – 48

Maritime Commission C1-M-AV1 type
339 feet, 7,435 tons (full load), 12 knots, 79 crew complement
One 3-inch/50, six 20mm guns

| Cargo Ship | Comm | Cargo Ship | Comm |
|---|---|---|---|
| *Alamosa* (AK-156) | 1945 | *Gadsden* (AK-182) | 1945 |
| *Alcona* (AK-157) | 1945 | *Glacier* (AK-183) | 1945 |
| *Amador* (AK-158) | 1945 | *Grainger* (AK-184) | 1945 |
| *Antrim* (AK-159) | 1945 | *Habersham* (AK-186) | 1945 |
| *Autauga* (AK-160) | 1945 | *Hennepin* (AK-187) | 1945 |
| *Beaverhead* (AK-161) | 1945 | *Herkimer* (AK-188) | 1945 |
| *Beltrami* (AK-162) | 1945 | *Hidalgo* (AK-189) | 1945 |
| *Blount* (AK-163) | 1945 | *Kenosha* (AK-190) | 1945 |
| *Brevard* (AK-164) | 1945 | *Lebanon* (AK-191) | 1945 |
| *Bullock* (AK-165) | 1945 | *Lehigh* (AK-192) | 1945 |
| *Cabell* (AK-166) | 1945 | *Marengo* (AK-194) | 1945 |
| *Caledonia* (AK-167) | 1945 | *Midland* (AK-195) | 1945 |
| *Charlevoix* (AK-168) | 1945 | *Muscatine* (AK-197) | 1945 |
| *Chatham* (AK-169) | 1945 | *Muksingum* (AK-198) | 1945 |
| *Chicot* (AK-170) | 1945 | *Pembina* (AK-200) | 1945 |
| *Claiborne* (AK-171) | 1945 | *Poinsett* (AK-205) | 1945 |
| *Clarion* (AK-172) | 1945 | *Pontotoc* (AK-206) | 1945 |
| *Codington* (AK-173) | 1945 | *Richland* (AK-207) | 1945 |
| *Colquitt* (AK-174) | 1945 | *Rockdale* (AK-208) | 1945 |
| *Craighead* (AK-175) | 1945 | *Schuyler* (AK-209) | 1945 |
| *Fairhead* (AK-178) | 1945 | *Screven* (AK-210) | 1945 |
| *Faribault* (AK-179) | 1945 | *Sebastoam* (AK-211) | 1945 |
| *Fentress* (AK-180) | 1945 | *Sussex* (AK-213) | 1945 |
| *Flagler* (AK-181) | 1945 | *Tarrant* (AK-214)[3] | 1945 |

### *Andromeda*-class Attack Cargo Ships (AKA) – 30

Maritime Commission C2-S-B1 type
459 feet, 13,905 tons (full load), 16.5 knots, 380 crew complement
One 5-inch/38, four twin-40mm, twelve 20mm guns
One LCP(L), eight LCM(3), fifteen to sixteen LCVP

| Cargo Ship | Comm | Cargo Ship | Comm |
|---|---|---|---|
| *Achernar* (AKA-53) | 1944 | *Montague* (AKA-98) | 1945 |
| *Algol* (AKA-54) | 1944 | *Muliphen* (AKA-61) | 1944 |
| *Alshain* (AKA-55) | 1944 | *Oglethorpe* (AKA-100) | 1944 |
| *Andromeda* (AKA-15) | 1944 | *Rolette* (AKA-99) | 1945 |
| *Aquarius* (AKA-16) | 1943 | *Sheliak* (AKA-62) | 1944 |
| *Arneb* (AKA-56) | 1943 | *Theenim* (AKA-63) | 1944 |
| *Capricornus* (AKA-57) | 1943 | *Thuban* (AKA-19) | 1943 |
| *Centaurus* (AKA-17) | 1944 | *Uvalde* (AKA-88) | 1944 |
| *Cepheus* (AKA-18) | 1944 | *Virgo* (AKA-20) | 1943 |
| *Chara* (AKA-58) | 1944 | *Warrick* (AKA-89) | 1944 |
| *Diphda* (AKA-59) | 1944 | *Whiteside* (AKA-90) | 1944 |
| *Leo* (AKA-60) | 1944 | *Whitley* (AKA-91) | 1944 |
| *Marquette* (AKA-95) | 1945 | *Winston* (AKA-94) | 1945 |
| *Matthews* (AKA-96) | 1945 | *Wyandot* (AKA-92) | 1944 |
| *Merrick* (AKA-97) | 1945 | *Yancey* (AKA-93)[4] | 1944 |

### *Artemis*-class Attack Cargo Ships – 32

Maritime Commission S4-SE2-BE1 type
426 feet, 4,087 tons, 16.9 knots, 303 crew complement
One 5-inch, eight 40mm guns

| Cargo Ship | Comm | Cargo Ship | Comm |
|---|---|---|---|
| *Artemis* (AKA-21) | 1944 | *Roxane* (AKA-37) | 1945 |
| *Athene* (AK-22) | 1944 | *Sappho* (AKA-38) | 1945 |
| *Aurelia* (AKA-23) | 1944 | *Sarita* (AKA-39) | 1945 |
| *Birgit* (AKA-24) | 1944 | *Scania* (AKA-40) | 1945 |
| *Circe* (AKA-25) | 1944 | *Selinur* (AKA-41) | 1945 |
| *Corvus* (AKA-26) | 1944 | *Sidonia* (AKA-42) | 1945 |
| *Devosa* (AKA-27) | 1944 | *Sirona* (AKA-43) | 1945 |
| *Hydrus* (AKA-28) | 1944 | *Sylvania* (AKA-44) | 1945 |
| *Lacerta* (AKA-29) | 1944 | *Tabora* (AKA-45) | 1945 |
| *Lumen* (AKA-30) | 1944 | *Troilus* (AKA-46) | 1945 |
| *Medea* (AKA-31) | 1945 | *Turandot* (AKA-47) | 1945 |
| *Mellena* (AKA-32) | 1945 | *Valeria* (AKA-48) | 1945 |
| *Ostara* (AKA-33) | 1945 | *Vanadis* (AKA-49) | 1945 |
| *Pamina* (AKA-34) | 1945 | *Veritas* (AKA-50) | 1945 |
| *Polana* (AKA-35) | 1945 | *Xenia* (AKA-51) | 1945 |
| *Renate* (AKA-36) | 1945 | *Zenobia* (AKA-52)[5] | 1945 |

## *Tolland*-class Attack Cargo Ships (AKA) – 32

Maritime Commission C2-S-AJ3 type
459 feet, 13,905 (full load), 165 knots, 395 crew complement
One 5-inch/38, four twin-40mm, twelve 20mm guns
One LCP(L), eight LCM(3), fifteen to sixteen LCVP

| Cargo Ship | Comm | Cargo Ship | Comm |
|---|---|---|---|
| *Alamance* (AKA-75) | 1944 | *Todd* (AKA-71) | 1944 |
| *Caswell* (AKA-72) | 1944 | *Tolland* (AKA-64) | 1944 |
| *Duplin* (AKA-87) | 1945 | *Torrance* (AKA-76) | 1944 |
| *Lenoir* (AKA-74) | 1944 | *Towner* (AKA-77) | 1944 |
| *New Hanover* (AKA-73) | 1944 | *Trego* (AKA-78) | 1944 |
| *Ottawa* (AKA-101) | 1945 | *Trousdale* (AKA-79) | 1944 |
| *Prentiss* (AKA-102) | 1945 | *Tyrell* (AKA-80) | 1944 |
| *Rankin* (AKA-103) | 1945 | *Union* (AKA-106) | 1945 |
| *Seminole* (AKA-104) | 1945 | *Valencia* (AKA-81) | 1945 |
| *Shoshone* (AKA-65) | 1944 | *Vanango* (AKA-82) | 1945 |
| *Skagit* (AKA-105) | 1945 | *Vermilion* (AKA-107) | 1945 |
| *Southampton* (AKA-66) | 1944 | *Vinton* (AKA-83) | 1944 |
| *Starr* (AKA-67) | 1944 | *Washburn* (AKA-108) | 1945 |
| *Stokes* (AKA-68) | 1944 | *Waukesha* (AKA-84) | 1944 |
| *Suffolk* (AKA-69) | 1944 | *Wheatland* (AKA-85) | 1944 |
| *Tate* (AKA-70) | 1944 | *Woodford* (AKA-86)[6] | 1944 |

## General Stores-Issue Ships (AKS) – 16

Variety of hull types

| Stores-Issue Ship | Comm | Stores-Issue Ship | Comm |
|---|---|---|---|
| *Acubens* (AKS-5) | 1944 | *Kochab* (AKS-6) | 1944 |
| *Antares* (AKS-3) | 1940 | *Liguria* (AKS-15) | 1944 |
| *Castor* (AKS-1) | 1941 | *Luna* (AKS-7) | 1944 |
| *Cybele* (AKS-10) | 1944 | *Mercury* (AKS-20) | 1945 |
| *Gratia* (AKS-11) | 1944 | *Pollux* (AKS-2) | 1941 |
| *Hecuba* (AKS-12) | 1945 | *Pollux* (AKS-4) | 1944 |
| *Hesperia* (AKS-13) | 1945 | *Talita* (AKS-8) | 1944 |
| *Iolanda* (AKS-14) | 1944 | *Volans* (AKS-9)[7] | 1944 |

# Appendix C: Canada's "Fort" and "Park" Merchant Vessels

In 1939, at the outset of the war, Canada's ocean-going merchant fleet numbered only 38 vessels. By war's end, in 1945, Canadian shipyards had produced more than 400 merchant vessels, an impressive achievement given that the country's shipbuilding industry also yielded thousands of naval vessels, including escort ships, minesweepers, tugs, and landing craft.[1]

The eight "Fort" ships identified below were a part of the British Fleet Train (logistics force) that supported the British Pacific Fleet during the Battle of Okinawa, and other actions. These British stores issuing-ships provided the Royal Navy fleet with dry cargo (food, naval stores, and aviation stores). All these vessels, built in British Columbia yards were of the "Victory" type with oil fired water tube boilers. All eight shared the same dimensions: 7,130 gross tons, length 424.5 feet, beam 57.0 feet. The biggest difference between them and the earlier North Sands ships was they burned fuel oil while the North Sands burned coal. All these vessels like their other merchant vessel sisters were registered as "British" vessels on completion.

| Stores-Issuing Ship | Builders Yard |
|---|---|
| British SS *Fort Alabama* | Burrard Dry Dock Co. Ltd, North Vancouver, BC |
| British SS *Fort Constantine* | Burrard Dry Dock Co. Ltd, North Vancouver, BC |
| British SS *Fort Dunvegan* | Burrard Dry Dock Co. Ltd, North Vancouver, BC |
| British SS *Fort Edmonton* | Burrard Dry Dock Co. Ltd, North Vancouver, BC |
| British SS *Fort Providence* | Burrard Dry Dock Co. Ltd, North Vancouver, BC |
| British SS *Fort Wrangell* | Burrard Dry Dock Co. Ltd, North Vancouver, BC |
| British SS *Fort Colville* | North Van. Ship Repairs Ltd, North Vancouver, BC |
| British SS *Fort Langley* | Victoria Machinery Depot Co. Ltd, Victoria, BC[2] |

## EARLY DEVELOPMENT OF OCEAN / LIBERTY SHIPS

Britain's huge merchant fleet losses in the early part of the war with Germany resulted in the creation of the British Merchant Shipbuilding Mission which left Britain in September 1940 to tour the U.S. and Canada. Following its members' inspection of a number of shipbuilding and engineering works, orders were placed and contracts signed for sixty *Ocean*-class vessels to be built at U.S. yards. Thirty of the ships were subsequently constructed at Todd California Shipbuilding Corporation in Richmond, California, and the other thirty at Todd-Bath Iron Shipbuilding, South Portland, Maine.[3]

These U.S.-built ships were of steel construction with welded hulls, based on the British Sunderland Tramp, which originated in 1879 and was last built half a decade before becoming the basis for design of the *Ocean*-class freighter. They were all nominally 7,174 GRT with a length of 416 feet and a beam of 57 feet. Triple-expansion steam engines supplied with steam from three single-ended Scotch-type coal-fired boilers produced a design speed of 11 knots. The ships had two-word names, prefixed by "Ocean," such as *Ocean Wayfarer*, but at the time of construction were sometimes referred to as "British Victory ships."[4]

The identification of British, Canadian, and U.S. cargo ships built during the war, of essentially the same type, can be confusing. The design for the Ocean ships, occasionally mentioned as "British Victory" or victory ships, were later modified to become the famous U.S. Liberty ship. As a point of interest, the American yards that built the Oceans immediately began production of Liberty hulls following their completion of the British ships. The Liberty ships were slightly over 441 feet long, 57 feet wide, and could make 11 knots. The American Victory ships (designed and built as replacements for the slower Liberty ships) were a little over 455 feet long, 62 feet wide, and had a cruising speed of 15-17 knots. The first Victory ship, named the SS *United Victory*, was delivered on 28 February 1944.[5]

## CANADIAN-BUILT MERCHANT VESSELS

Canada produced its own "Victory ships" (a collective title for the Fort and Park ships), not to be confused with the later American-version Victory. Desperate for replacement merchant vessels, and with its own yards at full wartime capacity, Britain contracted with Canada as well as the United States to construct ships. By 1941, Canada's shipbuilders had begun to turn out merchant ships and, at the height of the war, Canadian shipyards were delivering one new ship every three days. Fort and Park ships were built at the following yards:

| West Coast Shipyards |
|---|
| Burrard Dry Dock Co. Ltd., North Vancouver, British Columbia |
| Burrard Dry Dock. Co. Ltd., Vancouver Harbour, South Shore |
| North Vancouver Ship Repairs Ltd., North Vancouver, British Columbia |
| Prince Rupert Drydock and Shipyard, Prince Rupert, British Columbia |
| Victoria Machinery Depot Co. Ltd., Victoria, British Columbia |
| West Coast Shipbuilders Ltd., Vancouver, British Columbia |
| Yarrows Ltd., Esquimalt, Victoria, British Columbia |

| East Coast Shipyards |
|---|
| Canadian Vickers Ltd., Montreal, Quebec |
| Davie SB & Rep. Co. Ltd., Lauzon, Quebec |
| Foundation Maritime Ltd., Picton, Nova Scotia |
| Marine Industries Ltée, Tracy, Quebec |
| Morton Engineering & Drydock Co. Ltd., Quebec |
| Saint John Dry Dock & Shipbuilding Co., East Saint John, New Brunswick |
| United Shipyards Ltd., Montreal, Quebec |
| **Great Lakes Shipyards** |
| Canadian SB & Eng. Co. Ltd., Kingston, Ontario |
| Collingwood Shipyards Ltd., Collingwood, Ontario |
| Midland Shipyards Ltd., Midland, Ontario |
| Port Arthur Shipbuilding, Thunder Bay, Ontario[6] |

## THE FORT AND PARK SHIPS

Photo Appendix-1

British stores-issuing ship SS *Fort Langley*.
Imperial War Museum © IWM (FL 16222)

Canadian-built cargo ships destined for service with the British Ministry of War Transport (MoWT) were given "Fort" names, while ships retained for Canada had "Park" names. Painted "Admiralty Grey," these vessels were deployed the world over.[7]

The first Canadian-produced Fort ships were the "North Sands" type, so called because they were built based on British working drawings supplied by the J. L. Thompson & Sons North Sands shipyard in Sunderland, England. The Canadian ships, which differed little from similar British-built vessels, were designed with a dead-weight of 9,300 tons. Subsequently all ships of this size were referred to as 10,000 tonners as the war-time regulations allowed deeper loading.[8]

Canadian shipyards built 354 cargo ships during the war in three variants. All of them displaced 10,000 DWT (deadweight tonnage), the sum of the total weight of cargo, fuel, fresh water, ballast water, provisions, passengers, and crew, when loaded to the summer seawater "deep" condition. Burrard Dry Dockyard of Vancouver, the most prolific of the shipbuilders, produced 109 of the Fort and Park ships in their two yards, one in North Vancouver and the other on the south shore of Vancouver Harbour in the City of Vancouver. The first fifty it built were based on the North Sands design, which was suited to mass production by unsophisticated industries.[9]

Eventually, three variants of Fort and Park ships were built: "North Sands," "Victory," and "Canadian-classes." The biggest difference between these ships was the type of fuel they burned. The North Sands ships used coal, and the Victory ones burned fuel oil, whereas the Canadian-class could use either coal or oil. All the Fort and Park ships were patterned on the *Ocean*-class ships built in the United States for Britain. The Victory design made the ships more economical to operate by replacing the *North Sands'* three coal-fired Scotch marine boilers with two oil-fired water-tube boilers, saving fuel and requiring fewer firemen. Later Victory ships, the *Canadian*-class, were fitted with both oil storage tanks and coal bunkers, so that they could burn either fuel.[10]

The primary difference in construction between the Canadian and American-built cargo ships was that the Canadian freighters were of all riveted construction, while the American ones were welded.[11]

# Appendix D: Royal Australian Navy Auxiliary Ships

## Stores-Issuing Ships

| Ship | Built | Characteristics | RAN Service |
|---|---|---|---|
| HMAS *Adele* ex-*Franklin* | 1906 by Hawthorns & Co Ltd, Leith, Scotland | 350 tons, 145 feet<br>12 knots<br>2 x Vickers MG | 24 Oct 39- 7 May 43 |
| HMAS *Baralba* ex-*Nurnburg* ex-*Solskin* | 1921 at Stettin, Germany (now Poland) by Stettiner Oderwerke | 998 tons, 211 feet | 31 May 42- 11 Feb 43 |
| HMAS *Falie* | 1919 in Holland by W. Richter Ultdenbogaardt | 110 feet | 17 Jul 40- 2 Aug 46 |
| HMAS *Gerard* ex-*Antilope* ex-*Therese* ex-*Palanga* | 1925 in Kiel, Germany by the Krupp Company | 8.5 knots<br>1 x 12 pounder<br>1 x 20mm Oerlikon<br>2 x Vickers MG | 1 Jul 41- 8 Apr 46 |
| HMAS *Maroubra* | 1930 at Brisbane by Norman Wright | 60 tons<br>7.5 knots | 20 Mar 42- 10 May 43 |
| HMAS *Matafele* Battle Honour NEW GUINEA 1942-44 | 1938 in Hong Kong for Burns Philp Pty Ltd | 186 tons, 115 feet<br>12-pounder anti-aircraft gun | 1 Jan 43- 24 Jun 44 |
| RAFA *Merkur* ex-*Rio Bravo* | 1924 in Kiel by Krupp | | 12 Dec 41- |
| HMAS *Mombah* | 1923 at Sydney by Cockatoo Island Dockyard | 3,440 GRT | Mar 44- 1948 |
| HMAS *Patricia Cam* Battle Honour DARWIN 1942-44 | 1940 at Daleys Point (Brisbane Water, SW) by G. Beattie Shipyard | 301 tons, 120 feet<br>1 x Oerlikon<br>2 x Vickers MG<br>1 x Browning MG | 3 Mar 42- 22 Jan 43 |
| HMAS *Ping Wo* | 1922 | 3,105 tons | 22 May 42- 26 Jun 46 |
| HMAS *Poyang* | 1941 in Hong Kong for the China Navigation Company | 2,873 tons, 299 feet<br>13 knots<br>1 x 4-inch Bofors<br>2 x 12-pounders<br>2 x Oerlikons | 6 May 43- 6 Mar 46 |

| | | | |
|---|---|---|---|
| HMAS *Wang Pu* | 1920 at Hong Kong by Taikoo Dockyard and Engineering Co. | 3,204 GRT, 228 feet<br>10½ knots<br>1 x 40mm Bofors<br>3 x Oerlikons<br>2 x twin Browning and Colt MGs | 1 Oct 43-22 Apr 46 |
| *Yampi Lass* | 1912 at Fremantle by A E Brown | 45 GRT, 68 feet | 22 Jan 42-11 Apr 43 |
| HMAS *Yunnan* | 1934 at Greenock, Scotland by Scotts Shipbuilding and Engineering Company | 2,812 tons, 299 feet<br>11 knots<br>1 x 4-inch gun<br>1 x 40mm Bofors<br>2 x Oerlikons<br>4 x water-cooled Lewis guns[1] | 20 Sep 44-31 Jan 46 |

## Fleet Attendant Tanker/Fleet Oilers

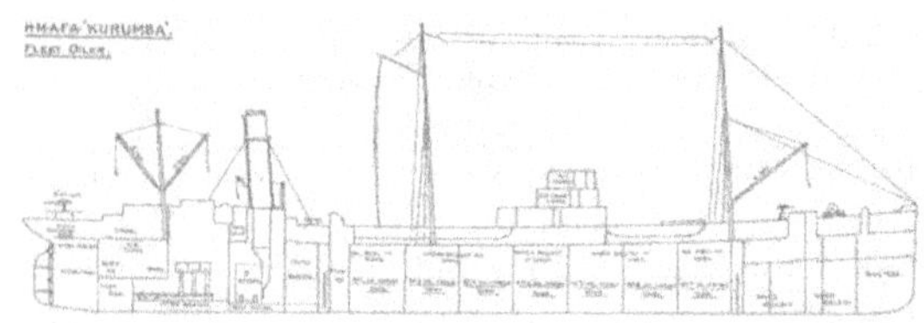

RAFA *Kurumba*: https://www.navy.gov.au/rafa-kurumba

| Ship | Built | Characteristics | RAN Service |
|---|---|---|---|
| RAFA *Aase Maersk*<br>Battle Honour OKINAWA 1945 | 1930 at Odense, Denmark by Odense Steel Shipyard | 6,184 tons, 407 feet<br>11 knots<br>1 x 4-inch gun<br>1 x 12 pounder<br>4 machine guns | Served until 1945 |
| RAFA *Bishopdale* | 1937 for RN | 17,350 tons<br>11.5 knots | 6 Apr 42-Aug 1945 |
| RAFA *Falkefjell* | 1931 Norway | 7,900 tons<br>11.5 knots | Dec 1941-April 1942 |
| RAFA *Kurumba* | 7 Dec 1916 at Wallsend-on-Tyne, England, by Swan Hunter and Wigham Richardson, Ltd. | 7,806 tons, 378 feet<br>10 knots<br>1 x 4-inch gun<br>2 x 6-pounders<br>1 x 4-inch QF gun<br>4 x machine guns | 4 Sep39-29 Jul 46 |

GRT: Gross register tonnage
RAFA: Royal Australian Fleet Auxiliary
MG: Machine gun
QF: Quick firing

# Appendix E: Cargo Ship (AK, AKA, AKS) Battle Stars

**Pearl Harbor**

| | | |
|---|---|---|
| *Vega* (AK-17) | 7 Dec 41 | Comdr. K. L. Forster |
| *Antares* (AKS-3) | 7 Dec 41 | Capt. Lawrence Charles Grannis |
| *Castor* (AKS-1) | 7 Dec 41 | Comdr. Harold James Wright |

**Guadalcanal-Tulagi Landings**

| | | |
|---|---|---|
| *Alchiba* (AK-23) | 7-9 Aug 42 | Comdr. James S. Freeman |
| *Alhena* (AK-26) | 7-9 Aug 42 | Comdr. Charles B. Hunt |
| *Bellatrix* (AK-20) | 7-9 Aug 42 | Comdr. William F. Dietrich |
| *Betelgeuse* (AK-28) | 7-9 Aug 42 | Capt. Harry. D. Power |
| *Fomalhaut* (AK-22) | 7-9 Aug 42 | Comdr. John D. Alvis |
| *Libra* (AK-53) | 7-9 Aug 42 | Comdr. William B. Fletcher Jr. |

**Capture and Defense of Guadalcanal**

| | | |
|---|---|---|
| *Carina* (AK-74) | 1 Feb 43 | Lt. Comdr. James Ian MacPherson |
| *Alchiba* (AK-23) | 15 Oct-28 Nov 42 | Comdr. James S. Freeman |
| *Alhena* (AK-26) | 29 Sep 42 | Comdr. Charles B. Hunt |
| *Bellatrix* (AK-20) | 8 Sep-25 Oct 42 | Comdr. William F. Dietrich |
| *Betelgeuse* (AK-28) | 11 Nov 42 | Capt. Harry D. Power |
| *Fomalhaut* (AK-22) | 23 Aug-27 Dec 42 | Comdr. Henry C. Flanagan |
| *Libra* (AK-53) | 11-12 Nov 42 | Comdr. William B. Fletcher Jr. |

**Guadalcanal – Third Strike**

| | | |
|---|---|---|
| *Betelgeuse* (AK-28) | 12-15 Nov 42 | Capt. Harry D. Power |
| *Libra* (AK-53) | 12-15 Nov 42 | Comdr. William B. Fletcher Jr. |

**Consolidation of Southern Solomons**

| | | |
|---|---|---|
| *Adhara* (AK-71) | 20 Mar-7 Apr 43 | Lt. Comdr. William W. Ball |
| *Aludra* (AK-72) | 16 Jun 43 | Lt. Comdr. Dale E. Collins |
| *Arided* (AK-73) | 11 Mar-23 May 43 | Lt. Comdr. John Boyer Blaine |
| *Cassiopeia* (AK-75) | 11 Mar 43 | Lt. Comdr. Werner E. Carlson |
| *Celeno* (AK-76) | 16 Jun 43 | Lt. Comdr. Niles Edward Lanphere |
| *Carina* (AK-74) | 3-5 Mar 43 | Lt. Comdr. James Ian MacPherson |
| *Deimos* (AK-78) | 16 Jun 43 | Lt. Comdr. Walter Louis Sorenson |
| *Fomalhaut* (AKA-5) | 20 May 43 | Comdr. Henry C. Flanagan |
| *Libra* (AKA-12) | 7 Apr 43 | Comdr. William B. Fletcher Jr. |
| *Titania* (AKA-13) | 13 May 43 | Comdr. Herbert Everett Berger |

**Aleutians Operation: Attu occupation**

| | | |
|---|---|---|
| *Spica* (AK-16) | 20 May-4 Jun 43 | Comdr. Joseph Wayland Long |

**New Georgia Group: New Georgia-Rendova-Vangunu occupation**

| | | |
|---|---|---|
| *Algorab* (AKA-8) | 30 Jun 43 | Comdr. Joseph R. Lannom |
| *Libra* (AKA-12) | 30 Jun 43 | Comdr. William B. Fletcher Jr. |

**Treasury-Bougainville: Occupation of Cape Torokina**

| | | |
|---|---|---|
| *Alchiba* (AKA-6) | 1-13 Nov 43 | Comdr. Howard R. Shaw |
| *Alhena* (AKA-9) | 1-13 Nov 43 | Comdr. Howard William Bradbury |
| *Libra* (AKA-12) | 1-9 Nov 43 | Comdr. Floyd F. Ferris |
| *Titania* (AKA-13) | 1-9 Nov 43 | Comdr. Herbert Everett Berger |

**Consolidation of Northern Solomons**

| | | |
|---|---|---|
| *Eridanus* (AK-92) | 1-7 Feb 44 | Lt. Comdr. F. W. Johnson |
| *Hyperion* (AK-107) | 14 Nov 43-5 Apr 44 | Lt. Comdr. Charles C. Newman |
| *Serpens* (AK-97) | 7 Dec 43-6 Nov 44 | Lt. Comdr. Magnus J. Johnson |

**Gilbert Islands Operation**

| | | |
|---|---|---|
| *Auriga* (AK-98) | 13 Nov-8 Dec 43 | Lt. Comdr. John Charles Hart |
| *Hercules* (AK-41) | 17 Nov-8 Dec 43 | Comdr. William Howard Turnquist |
| *Jupiter* (AK-43) | 21 Nov-8 Dec 43 | Lt. Comdr. Duncan Scott Baker |
| *Alcyone* (AKA-7) | 13 Nov-2 Dec 43 | Comdr. James Bertrand McVey |
| *Bellatrix* (AKA-3) | 20 Nov-7 Dec 43 | Comdr. Charles A. Joans |
| *Thuban* (AKA-19) | 20-29 Nov 43 | Comdr. James C. Campbell |
| *Virgo* (AKA-20) | 20-29 Nov 43 | Comdr. Clayton H. McLaughlin |
| *Castor* (AKS-1) | 17 Nov-8 Dec 43 | Comdr. Harold Burton Herty |

**Eastern New Guinea Operation: Finschhafen occupation**

| | | |
|---|---|---|
| *Ganymede* (AK-104) | 10-28 Jan 44 | Lt. Comdr. Glenn H. Melichar |

**Marshall Islands Operation: Occupation of Kwajalein/Majuro Atolls**

| | | |
|---|---|---|
| *Alkes* (AK-110) | 31 Jan-8 Feb 44 | Lt. Comdr. Charles L. Wickman |
| *Caelum* (AK-106) | 3-8 Feb 44 | Lt. Comdr. Edgar Johnson |
| *Mercury* (AK-42) | 31 Jan-8 Feb 44 | Lt. Comdr. George W. Graber |
| *Rutilicus* (AK-113) | 3-8 Feb 44 | Lt. Comdr. Hans Olaf Matthiesen |
| *Alcyone* (AKA-7) | 31 Jan-8 Feb 44 | Comdr. James Bertrand McVey |
| *Almaack* (AKA-10) | 31 Jan-8 Feb 44 | Lt. Comdr. John Y. Dannenberg |
| *Aquarius* (AKA-16) | 31 Jan-8 Feb 44 | Capt. Raymond Victor Marron |
| *Centaurus* (AKA-17) | 31 Jan-5 Feb 44 | Capt. George Evans McCabe |
| *Electra* (AKA-4) | 31 Jan-8 Feb 44 | Comdr. Charles Sprague Beightler |
| *Thuban* (AKA-19) | 31 Jan-8 Feb 44 | Comdr. James C. Campbell |
| *Virgo* (AKA-20) | 31 Jan-4 Feb 44 | Comdr. Clayton H. McLaughlin |

**Marshall Islands Operation: Occupation of Eniwetok Atoll**

| | | |
|---|---|---|
| *Mercury* (AK-42) no BS | 17-25 Feb 44 | Lt. Comdr. George W. Graber |
| *Electra* (AKA-4) no BS | 17-25 Feb 44 | Comdr. Charles Sprague Beightler |

**Bismarck Archipelago Operation: Admiralty Island landings**

| | | |
|---|---|---|
| *Celeno* (AK-76) | 24-28 Mar 44 | Lt. Comdr. John Paul Gately |
| *Aquarius* (AKA-16) | 11-12 Apr 44 | Capt. Raymond Victor Marron |
| *Centaurus* (AKA-17) | 5-9 Apr 44 | Capt. George Evans McCabe |

## Hollandia Operation: Aitape-Humboldt Bay-Tanahmerah Bay

| | | |
|---|---|---|
| *Bootes* (AK-99) | 21 Apr-1 May 44 | Lt. Comdr. Howard Perry Bacon |
| *Ganymede* (AK-104) | 21-27 Apr 44 | Lt. Comdr. Glenn H. Melichar |
| *Triangulum* (AK-102) | 21-27 Apr 44 | Comdr. Richard E. Loughborough |
| *Centaurus* (AKA-17) | 21 Apr-5 May 44 | Capt. George Evans McCabe |
| *Virgo* (AKA-20) | 21 Apr-5 May 44 | Comdr. Clayton H. McLaughlin |

## Marianas Operation: Capture and occupation of Saipan

| | | |
|---|---|---|
| *Auriga* (AK-98) | 16-28 Jun 44 | Lt. Comdr. John Charles Hart |
| *Cetus* (AK-77) | 27 Jul-9 Aug 44 | Lt. Charles B. Johnson |
| *Hercules* (AK-41) | 15-26 Jun 44 | Comdr. William Howard Turnquist |
| *Jupiter* (AK-43) | 15-26 Jun 44 | Lt. Comdr. Thurman A. Whitaker |
| *Leonis* (AK-128) | 20 Jun-3 Jul 44 | Lt. Comdr. Anthony J. Barkowsky |
| *Mercury* (AK-42) | 27 Jun-2 Jul 44 | Lt. Comdr. Nelson Denman Salmon |
| *Vega* (AK-17) | 25-31 Jul 44 | Lt. Comdr. William B. Dell |
| *Alcyone* (AKA-7) | 15-22 Jun 44 | Comdr. Hermann P. Knickerbocker |
| *Alhena* (AKA-9) | 15-22 Jun 44 | Comdr. Howard William Bradbury |
| *Almaack* (AKA-10) | 15-24 Jun 44 | Lt. Comdr. John Y. Dannenberg |
| *Aquarius* (AKA-16) | 15-25 Jun 44 | Capt. Raymond Victor Marron |
| *Bellatrix* (AKA-3) | 16-22 Jun 44 | Comdr. E. J. Anderson |
| *Electra* (AKA-4) | 15-26 Jun 44 | Comdr. Charles Sprague Beightler |
| *Fomalhaut* (AKA-5) | 16-24 Jun 44 | Comdr. Carter A. Printup |
| *Thuban* (AKA-19) | 15 Jun-28 Jul 44 | Comdr. James C. Campbell |
| *Titania* (AKA-13) | 21-26 Jul 45 | Comdr. Malcolm Whitfield Callahan |

## Tinian Capture and Occupation

| | | |
|---|---|---|
| *Thuban* (AKA-19) | 24-28 Jul 44 | Comdr. James C. Campbell |

## Marianas Operation: Capture and occupation of Guam

| | | |
|---|---|---|
| *Alkes* (AK-110) | 8-10 Aug 44 | Lt. Comdr. Charles L. Wickman |
| *Ara* (AK-136) | 27 Jul-15 Aug 44 | Lt. Comdr. William B. Hudgins |
| *Cor Caroli* (AK-91) | 27 Jul-15 Aug 44 | Comdr. Oscar C. B. Wev |
| *Draco* (AK-79) | 27 Jul-9 Aug 44 | Lt. Comdr. Earle Llewllyn Evey |
| *Rutilicus* (AK-113) | 10 Aug 44 | Lt. Comdr. Hans O. Matthiesen |
| *Sterope* (AK-96) | 21 Jul-9 Aug 44 | Lt. Comdr. John B. Krestensen |
| *Vega* (AK-17) no BS | 1-15 Aug 44 | Lt. Comdr. William B. Dell |
| *Alcyone* (AKA-7) no BS | 22-29 Jul 44 | Comdr. Hermann P. Knickerbocker |
| *Almaack* (AKA-10) no BS | 22-29 Jul 44 | Comdr. Clyde Oscar Hicks |
| *Alshain* (AKA-55) | 21 Jul-3 Aug 44 | Comdr. Roland Ernest Krause |
| *Aquarius* (AKA-16) no BS | 21-26 Jul 44 | Capt. Raymond Victor Marron |
| *Centaurus* (AKA-17) | 21-27 Jul 44 | Capt. George Evans McCabe |
| *Libra* (AKA-12) | 21-25 Jul 44 | Comdr. Floyd F. Ferris |
| *Titania* (AKA-13) no BS | 26 Jul 44 | Comdr. Herbert Everett Berger |
| *Virgo* (AKA-20) | 21-27 Jul 44 | Comdr. Clayton H. McLaughlin |

**Western Caroline Islands:**
**Capture and occupation of south Palau Islands**

| | | |
|---|---|---|
| *Alcyone* (AKA-7) | 6 Sep-14 Oct 44 | Comdr. Hermann P. Knickerbocker |
| *Hercules* (AK-41) | 6 Sep-14 Oct 44 | Comdr. William H. Turnquist |
| *Jupiter* (AK-43) | 6 Sep-14 Oct 44 | Comdr. John Morgan Bristol |
| *Leonis* (AK-128) | 6 Sep-14 Oct 44 | Lt. Comdr. Anthony J. Barkowsky |
| *Lesuth* (AK-125) | 6 Sep-14 Oct 44 | Lt. Comdr. Bertrand H. Bassett |
| *Matar* (AK-119) | 6 Sep-14 Oct 44 | Lt. Comdr. Erwin Ernest Smith |
| *Rotanin* (AK-108) | 6 Sep-14 Oct 44 | Lt. Comdr. John Bernard Blain |
| *Vega* (AK-17) | 6 Sep-14 Oct 44 | Comdr. William B. Dell |
| *Aquarius* (AKA-16) | 6 Sep-14 Oct 44 | Capt. Raymond Victor Marron |
| *Arneb* (AKA-56) | 6 Sep-14 Oct 44 | Comdr. Howard R. Shaw |
| *Centaurus* (AKA-17) | 6 Sep-14 Oct 44 | Capt. George Evans McCabe |
| *Electra* (AKA-4) | 6 Sep-14 Oct 44 | Comdr. Charles S. Beightler<br>Comdr. Dennis Slocum Holler |
| *Virgo* (AKA-20) | 9-24 Sep 44 | Lt. Comdr. Herbert E.Randall |

**Leyte Operation: Leyte landings**

| | | |
|---|---|---|
| *Auriga* (AK-98) | 20 Oct-18 Nov 44 | Lt. Comdr. John Charles Hart |
| *Hercules* (AK-41) | 13 Oct-29 Nov 44 | Comdr. William H. Turnquist |
| *Hyperion* (AK-107) | 29 Oct-12 Nov 44 | Lt. Comdr. Charles C. Newman |
| *Jupiter* (AK-43) | 16 Oct-29 Nov 44 | Comdr. John Morgan Bristol |
| *Mercury* (AK-42) | 20 Oct-18 Nov 44 | Lt. Comdr. Nelson Denman Salmon |
| *Murzim* (AK-95) | 23 Oct-29 Nov 44 | Lt. Comdr. DeWitt. S. Walton |
| *Triangulum* (AK-102) | 23-28 Nov 44 | Lt. Charles Kenneth Latus |
| *Alcyone* (AKA-7) | 20 Oct-18 Nov 44 | Comdr. Hermann P. Knickerbocker |
| *Almaack* (AKA-10) | 20 Oct-14 Nov 44 | Comdr. Clyde Oscar Hicks |
| *Alshain* (AKA-55) | 20 Oct-18 Nov 44 | Comdr. Roland Ernest Krause |
| *Aquarius* (AKA-16) | 13 Oct-20 Nov 44 | Capt. Raymond Victor Marron<br>Comdr. Ira Edwin Eskridge |
| *Arneb* (AKA-56) | 13 Oct-29 Nov 44 | Comdr. Howard R. Shaw<br>Lt. Comdr. Alan R. Barton |
| *Capricornus* (AKA-57) | 20 Oct-14 Nov 44 | Lt. Comdr. Benjamin F. McGuckin |
| *Chara* (AKA-58) | 20 Oct-18 Nov 44 | Comdr. John Pascoe Clark |
| *Electra* (AKA-4) | 13-30 Oct 44<br>19-29 Nov 44 | Comdr. Dennis Slocum Holler |
| *Thuban* (AKA-19) | 20 Oct-18 Nov 44 | Comdr. James C. Campbell |
| *Titania* (AKA-13) | 13-20 Nov 44 | Comdr. Malcolm Whitfield Callahan |

**Luzon Operation: Lingayen Gulf landing**

| | | |
|---|---|---|
| *Alcyone* (AKA-7) | 9 Jan 45 | Comdr. Hermann P. Knickerbocker |
| *Almaack* (AKA-10) | 9 Jan 45 | Comdr. Clyde Oscar Hicks |
| *Alshain* (AKA-55) | 9 Jan 45 | Comdr. Roland Ernest Krause |
| *Aquarius* (AKA-16) | 4-10 Jan 45 | Comdr. Ira Edwin Eskridge |
| *Chara* (AKA-58) | 9 Jan 45 | Comdr. John Pascoe Clark |
| *Libra* (AKA-12) | 11 Jan 45 | Comdr. George W. McCormick |
| *Titania* (AKA-13) | 9 Jan 44 | Comdr. Herbert Everett Berger |

| Manila Bay-Bicol Operations: Zambales-Subic Bay | | |
|---|---|---|
| *Auriga* (AK-98) | 29-31 Jan 45 | Lt. Comdr. Charles Henry Fonda |
| *Mercury* (AK-42) | 29-31 Jan 45 | Lt. Comdr. Nelson Denman Salmon |
| *Alcyone* (AKA-7) | 29-31 Jan 45 | Comdr. Hermann P. Knickerbocker |
| *Algol* (AKA-54) | 29-31 Jan 45 | Lt. Comdr. Axton Turner Jones |
| *Alshain* (AKA-55) | 29-31 Jan 45 | Comdr. Benjamin W. Strickland |
| *Aquarius* (AKA-16) | 29-31 Jan 45 | Comdr. Ira Edwin Eskridge |
| *Arneb* (AKA-56) | 29-31 Jan 45 | Lt. Comdr. Alan R. Barton |
| *Chara* (AKA-58) | 29-31 Jan 45 | Comdr. John Pascoe Clark |

| Iwo Jima Operation: Assault and occupation of Iwo Jima | | |
|---|---|---|
| *Hercules* (AK-41) | 19 Feb-14 Mar 45 | Comdr. William H. Turnquist |
| *Jupiter* (AK-43) | 19 Feb-14 Mar 45 | Comdr. John Morgan Bristol |
| *Lakewood Victory* (AK-236) | 28 Feb-8 Mar 45 | Lt. Comdr. Eric H. Petrelius |
| *Alhena* (AKA-9) | 19 Feb-5 Mar 45 | Comd. Raymond C. Ericson |
| *Almaack* (AKA-10) | 19 Feb-6 Mar 45 | Comdr. Clyde Oscar Hicks |
| *Artemis* (AKA-21) | 19-27 Feb 45 | Lt. Comdr. Thomas J. Rattray |
| *Athene* (AKA-22) | 19-28 Feb 45 | Comdr. Edward R. Nelson Jr. |
| *Electra* (AKA-4) | 9-16 Mar 45 | Comdr. Dennis S. Holler |
| *Leo* (AKA-60) | 19-28 Feb 45 | Comdr. Thomas E. Healy |
| *Libra* (AKA-12) | 19 Feb 45 | Comdr. George W. McCormick |
| *Muliphen* (AKA-61) | 19 Feb-4 Mar 45 | Lt. Comdr. Walter W. Williamson |
| *Shoshone* (AKA-65) | 19 Feb-1 Mar 45 | Lt. Comdr. Stanley E. Melville |
| *Southampton* (AKA-66) | 19 Feb-1 Mar 45 | Lt. Comdr. Lester V. Cooke |
| *Starr* (AKA-67) | 19 Feb-5 Mar 45 | Comdr. Frederick O. Goldsmith |
| *Stokes* (AKA-68) | 19 Feb-2 Mar 45 | Lt. Comdr. George W. Graber |
| *Thuban* (AKA-19) | 9-16 Mar 45 | Lt. Comdr. James H. Strudley |
| *Tolland* (AKA-64) | 19-28 Feb 45 | Comdr. Edward J. Kingsland |
| *Virgo* (AKA-20) | 17 Feb-2 Mar 45 | Lt. Comdr. Herbert E. Randall |
| *Whiteside* (AKA-90) | 19 Feb-7 Mar 45 | Capt. Charles P. Woodson |
| *Whitley* (AKA-91) | 19-27 Feb 45 | Comdr. Albert C. Thompson |
| *Woodford* (AKA-86) | 24-30 Jun 45 | Capt. Winston E. P. Folk |
| *Wyandot* (AKA-92) | 26 Mar-29 Apr 45 | Lt. Comdr. Robert B. Alderman |
| *Yancey* (AKA-93) | 19 Feb-2 Mar 45 | Lt. Comdr. Edward R. Rice |

| Okinawa Operation: Assault and occupation of Okinawa | | |
|---|---|---|
| *Adhara* (AK-71) | 8-27 May 45 | Lt. Comdr. Arvin Wilson Callaway |
| *Alkaid* (AK-114) | 21 May-13 Jun 45 | Lt. Comdr. Alfred O. Johansen |
| *Alkes* (AK-110) | 28 May-30 Jun 45 | Lt. Comdr. Charles L. Wickman |
| *Appanoose* (AK-226) | 3 May-30 Jun 45 | Lt. Comdr. Jack Joffre Hughes |
| *Arided* (AK-73) | 18-30 Jun 45 | Lt. Comdr. Jack J. Hughes |
| *Auriga* (AK-98) | 10-19 Apr 45 | Lt. Comdr. Charles H. Fonda |
| *Azimech* (AK-124) | 18 Apr-19 May 45 | Lt. Comdr. Earl Paul Gaither |
| *Bedford Victory* (AK-231) | 25 Apr-7 Jun 45 | Lt. Comdr. Dudley Albert Durrant |
| *Bucyrus Victory* (AK-234) | 24 Mar-30 Jun 45 | Lt. Comdr. Felix A. Geissert |
| *Carina* (AK-74) | 26 Apr-3 May 45 | Lt. Comdr. Ben Koerner |
| *Celeno* (AK-76) | 18 Apr-27 Jun 45 | Lt. Comdr. Arthur S. Haines Jr. |
| *Cetus* (AK-77) | 26 Apr-5 May 45 | Lt. Charles B. Johnson |

| | | |
|---|---|---|
| *De Grasse* (AK-223) | 26 Apr-5 May 45 | Lt. Comdr. Frank W. Schultz |
| *Draco* (AK-79) | 26-30 Jun 45 | Lt. Comdr. Robert M. Drysdale Jr. |
| *Hyperion* (AK-107) | 8-27 May 45 | Lt. Comdr. Charles C. Newman |
| *Jupiter* (AK-43) | 24-30 Jun 45 | Comdr. John Morgan Bristol |
| *Kenmore* (AK-221) | 26 Apr-5 May 45 | Lt. Comdr. Olin H. Pitts |
| *Lakewood Victory* (AK-236) | 24 Mar-30 Jun 45 | Lt. Comdr. Eric H. Petrelius |
| *Las Vegas Victory* (AK-229) | 24 Mar-30 Jun 45 | Comdr. William F. Lally |
| *Lynx* (AK-100) | 26 Apr-5 May 45 | Lt. Comdr. Julius Max Berrey |
| *Manderson Victory* (AK-230) | 24 Mar-30 Jun 45 | Lt. Comdr. John Larsen |
| *Matar* (AK-119) | 29 May-25 Jun 45 | Lt. Harold A. Weston |
| *Mayfield Victory* (AK-232) no BS | 18 May-30 Jun 45 | Lt. Comdr. Niels H. Olsen |
| *Media* (AK-83) | 10-19 Jun 45 | unknown |
| *Menkar* (AK-123) | 10-19 May 45 | Lt. Comdr. Niels P. Thompsen |
| *Mintaka* (AK-94) | 21-31 May 45 | Lt. Comdr. Michael J. Johnson |
| *Rotanin* (AK-108) | 21-31 May 45 | Lt. Comdr. George H. Lehleitner |
| *Sterope* (AK-96) | 10-31 May 45 | Lt. Comdr. John B. Krestensen |
| *Zaurak* (AK-117) | 13-19 Jun 45 | Comdr. John S. Kapuscinski |
| *Achernar* (AKA-53) | 1-19 Apr 45 | Lt. Comdr. John R. Lange |
| *Algol* (AKA-54) | 1-10 Apr 45 | Lt. Comdr. Axton T. Jones |
| *Algorab* (AKA-8) | 1-9 Apr 45 | Comdr. Thomas C. Green |
| *Allegan* (AKA-225) | 3 May-30 Jun 45 | Lt. Comdr. Joseph S. Hulings |
| *Alshain* (AKA-55) | 1-5 Apr 45 | Comdr. Benjamin W. Strickland |
| *Andromeda* (AKA-15) | 1 Apr-14 Jun 45 | Lt. Comdr. Duncan Ward |
| *Aquarius* (AKA-16) | 1-19 Apr 45 | Comdr. Ira Edwin Eskridge<br>Comdr. Edwin C. Whitfield |
| *Arcturus* (AKA-1) | 1-14 Apr 45 | Comdr. Charles H. K. Miller |
| *Arneb* (AKA-56) | 1-10 Apr 45 | Comdr. Edward Thomas Collins |
| *Athene* (AKA-22) | 15 Apr-11 Jun 45 | Comdr. Edward R. Nelson Jr. |
| *Aurelia* (AKA-23) | 1-14 Apr 45 | Comdr. Ernest G. MacMurdy |
| *Betelgeuse* (AKA-11) | 1-9 Apr 45 | Comdr. W. E. Webb |
| *Capricornus* (AKA-57) | 1-9 Apr 45 | Lt. Comdr. Benjamin F. McGuckin |
| *Caswell* (AKA-72) | 1-9 Apr 45 | Lt. Comdr. F. M. Diffley |
| *Centaurus* (AKA-17) | 1 Apr-14 Jun 45 | Comdr. John Paul Gray |
| *Cepheus* (AKA-18) | 1-15 Apr 45 | Comdr. Robert C. Sarratt Jr. |
| *Chara* (AKA-58) | 1-6 Apr 45 | Comdr. John Pascoe Clark |
| *Circe* (AKA-25) | 1-6 Apr 45 | Lt. Comdr. Victor James Barnhart |
| *Corvus* (AKA-26) | 1-10 Apr 45 | Lt. Comdr. Charles M. Gregson |
| *Devosa* (AKA-27) | 1-10 Apr 45 | Lt. Comdr. Glenn L. Overmyer Jr. |
| *Diphda* (AKA-59) | 1-10 Apr 45 | Lt. Comdr. Robert C. Willson |
| *Fomalhaut* (AKA-5) | 24 Mar-30 Jun 45 | Comdr. Roger V. Mullany |
| *Hydrus* (AKA-28) | 1-15 Apr 45 | Lt. Comdr. Richard J. Wissinger |
| *Lacerta* (AKA-29) | 1-9 Apr 45 | Lt. Comdr. Louis Funkenstein Jr. |
| *Lenoir* (AKA-74) | 17-22 Apr 45 | Lt. Comdr. Marcus. L. Whitford |
| *Leo* (AKA-60) | 1-14 Apr 45 | Comdr. Thomas E. Healy |
| *Lumen* (AKA-30) | 17-22 Apr 45 | Lt. Comdr. Thomas A. Marshall Jr. |
| *Merrick* (AKA-97) | 18 Jun 45 | Lt. Comdr. Walter E. Reed |
| *Muliphen* (AKA-61) | 1-11 Apr 45 | Lt. Comdr. Walter W. Williamson |

| | | |
|---|---|---|
| *New Hanover* (AKA-73) | 9-19 Apr 45 | Lt. Comdr. John R. Haines |
| *Oberon* (AKA-14) | 26 Mar-26 Apr 45 | Lt. Comdr. Harold T. Cameron |
| *Pamina* (AKA-34) | 16-19 May 45 | Lt. Comdr. Edwin P. Teague |
| *Prentis* (AKA-102) | 21-30 Jun 45 | Lt. Comdr. George P. Walker<br>Lt. Comdr. Richard H. Tibbets |
| *Procyon* (AKA-2) | 1-7 Apr 45 | Comdr. Campbell Harris Minckler |
| *Rankin* (AKA-103) | 11-28 Jun 45 | Lt. Comdr. Thomas D. Price |
| *Sheliak* (AKA-62) | 1-19 Apr 45 | Comdr. Searcy J. Lowery |
| *Shoshone* (AKA-65) | 1-11 Apr 45 | Lt. Comdr. Stanley E. Melville |
| *Southampton* (AKA-66) | 1-11 Apr 45 | Lt. Comdr. Lester V. Cooke |
| *Starr* (AKA-67) | 1-10 Apr 45 | Comdr. Frederick O. Goldsmith |
| *Stokes* (AKA-68) | 9-19 Apr 45 | Lt. Comdr. George W. Graber |
| *Suffolk* (AKA-69) | 26 Mar-30 Apr 45 | Lt. Comdr. E.C. Clusman |
| *Tate* (AKA-70) | 26 Mar-22 Apr 45 | Comdr. Rupert E. Lyon |
| *Theenim* (AKA-63) | 1-15 Apr 45 | Comdr. Gordon A. Littlefield |
| *Tolland* (AKA-64) | 9-15 Apr 45 | Comdr. Edward J. Kingsland |
| *Torrance* (AKA-76) | 26 Mar-30 Apr 45 | Lt. Comdr. George A. Euerle |
| *Trego* (AKA-78) | 3-11 Jun 45 | Lt. Comdr. James F. Hunnewell |
| *Trousdale* (AKA-79) | 11-22 Apr 45 | Lt. Comdr. William J. Lane |
| *Tyrell* (AKA-80) | 1-10 Apr 45 | Lt. Comdr. John L. McLean |
| *Union* (AKA-106) | 1-9 Apr 45 | Comdr. Hartwell T. Doughty |
| *Uvalde* (AKA-88) | 1-9 Apr 45 | Lt. Comdr. William M. McCloy |
| *Valencia* (AKA-81) | 17-22 Apr 45 | Lt. Comdr. Rodney A. Blake |
| *Venango* (AKA-82) | 12-22 Apr 45 | Lt. Comdr. Thurman A. Whitaker |
| *Virgo* (AKA-20) | 1-16 May 45 | Lt. Comdr. Herbert E. Randall |
| *Whiteside* (AKA-90) | 9-15 Apr 45 | Capt. Charles P. Woodson |
| *Wyandot* (AKA-92) | 26 Mar-29 Apr 45 | Lt. Comdr. Robert B. Alderman |
| *Yancey* (AKA-93) | 9-15 Apr 45 | Lt. Comdr. Edward R. Rice |
| *Antares* (AKS-3) | 9 May-18 Jun 45 | Lt. Comdr. Nicholas T. Gansa |
| *Castor* (AKS-1) | 1 Apr-30 Jun 45 | Comdr. Harold Burton Herty<br>Capt. Franklin C. Huntoon |
| *Kochab* (AKS-6) | 8-30 Jun 45 | Lt. Comdr. Rosewell E. King |

**Okinawa Operation: Raids in support of Okinawa Gunto operation**

| | | |
|---|---|---|
| *Bucyrus Victory* (AK-234) | 26 Mar-22 May 45 | Lt. Comdr. Felix A. Geissert |
| *Lakewood Victory* (AK-236) | 3-25 Apr 45 | Lt. Comdr. Eric H. Petrelius |
| *Las Vegas Victory* (AK-229) | 25 Mar-1 Jun 45 | Comdr. William F. Lally |
| *Manderson Victory* (AK-230) no BS | 12 Apr-15 Jun 45 | Lt. Comdr. John Larsen |
| *Mayfield Victory* (AK-232) | 13 Apr-18 May 45 | Lt. Comdr. Niels H. Olsen |
| *Mercury* (AK-42) | 23 Mar-13 May 45 | Lt. Comdr. Nelson D. Salmon |
| *Fomalhaut* (AKA-5) no BS | 11 May-1 Jun 45 | Comdr. Roger V. Mullany |

**Borneo Operations: Tarakan Island operation**

| | | |
|---|---|---|
| *Titania* (AKA-13) | 27 Apr-5 May 45 | Comdr. Malcolm W. Callahan |

**Borneo Operations: Balikipapan operation**

| | | |
|---|---|---|
| *Poinsett* (AK-205) | 28 Jun-7 Jul 45 | Lt. Comdr. Robert M. Baughman |
| *Titania* (AKA-13) no BS | 26 Jun-4 Jul 45 | Comdr. Malcolm W. Callahan |

| Third Fleet operations against Japan | | |
|---|---|---|
| *Alcyone* (AKA-7) | 17 Jul-3 Aug 45 | Comdr. Hermann P. Knickerbocker |
| *Thuban* (AKA-19) | 9-15 Aug 45 | Lt. Comdr. James H. Strudley |

# Appendix F: USN/RN Ships Sunk or Damaged at Okinawa (26 March-30 April 1945)

| Name | Type | Date | Cause | Extent |
|---|---|---|---|---|
| ACHERNAR | AKA 53 | 2 April | Suicide plane | Major |
| ADAMS | DM 27 | 26 March | Suicide plane | Major |
| AGENOR | ARL 3 | 5 April | Operational (collision) | Major |
| ALPINE | APA 92 | 3 April | Suicide plane | Major |
| AMMEN | DD 527 | 21 April | Air Attack | Minor |
| ARIKARA | ATF 98 | 11 April | Operational (grounded) | Minor |
| AUDRAIN | APA 59 | 6 April | AA fire | Minor |
| BARNETT | APA 5 | 26 April | AA fire | Minor |
| BENNETT | DD 473 | 7 April | Suicide plane | Major |
| BENNION | DD 662 | 28 April | Suicide plane | Minor |
| BERRIEN | APA 62 | 11 April | Operational (collision) | Minor |
| BLACK | DD 666 | 11 April | Strafing | Minor |
| BORIE | DD 704 | 2 April | Operational (collision) | Major |
| BOWERS | DE 637 | 16 April | Suicide plane | Major |
| BOZEMAN VICTORY | XAK | 27 April | Suicide boat | Minor |
| BROWN | DD 546 | 28 April | Air attack | Minor |
| BRUSH | DD 745 | 12 April | Air attack | Minor |
| BRYANT | DD 665 | 16 April | Suicide plane | Major |
| BULL | APD 78 | 20 April | Operational | Minor |
| BUSH | DD 529 | 6 April | Suicide plane | SUNK |
| BUTLER | DMS 29 | 28 April | Suicide near miss | Minor |
| CANADA VICTORY | AKE | 27 April | Suicide plane | SUNK |
| CASSIN YOUNG | DD 793 | 12 April | Suicide plane | Major |
| CHILTON | APA 38 | 3 April | Suicide plane | Minor |
| C. J. BADGER | DD 657 | 9 April | Suicide boat | Major |
| COLHOUN | DD 801 | 6 April | Suicide plane | SUNK |
| COLORADO | BB 45 | 19 April | Operational (powder expl.) | Minor |
| COMFORT | AH 6 | 7 April | Air attack | Minor |
| COMFORT | AH 6 | 28 April | Suicide plane | Major |
| CONKLIN | DE 439 | 12 April | Air attack | Minor |
| CORREGIDOR | CVE 58 | 20 April | Storm | Major |
| COWANESQUE | AO 79 | 4 April | Storm | Minor |
| DALY | DD 519 | 28 April | Suicide near miss | Minor |
| DAMON M. CUMMINGS | DE 643 | 12 April | Air attack | Minor |
| DEFENSE | AM 317 | 6 April | Suicide plane | Minor |
| DEVASTATOR | AM 318 | 6 April | Suicide plane | Minor |
| DICKERSON | APD 21 | 3 April | Suicide plane | SUNK |
| DORSEY | DMS 1 | 27 March | Suicide plane | Minor |
| EMMONS | DMS 22 | 6 April | Suicide plane | SUNK |
| ENGLAND | DE 635 | 27 April | Air attack | Minor |
| ENTERPRISE | CV 6 | 11 April | Suicide plane (near miss) | Major |
| ESSEX | CV 9 | 11 April | Near bomb miss | Minor |
| FIEBERLING | DE 640 | 6 April | Suicide plane | Minor |
| FOREMAN | DE 633 | 6 April | Air attack | Major |
| FOREMAN | DE 633 | 26 March | Suicide plane | Minor |
| FRANKS | DD 554 | 2 April | Operational (collision) | Major |

| Name | Type | Date | Cause | Extent |
|---|---|---|---|---|
| GLADIATOR | AM 319 | 12 April | Suicide plane | Minor |
| GOODHUE | APA 107 | 3 April | Suicide plane | Minor |
| GREGORY | DD 802 | 7 April | Air Attack | Major |
| GILMER | APD 11 | 26 March | Suicide plane | Major |
| HAGGARD | DD 555 | 29 April | Suicide plane | Major |
| HALE | DD 642 | 11 April | Suicide near miss | Minor |
| HALE | DD 642 | 27 April | Operational (collision) | Major |
| HALLIGAN | DD 584 | 26 March | Mine | SUNK |
| HAMBLETON | DMS 20 | 5 April | Suicide near miss | Minor |
| HANCOCK | CV 19 | 6 April | Suicide plane | Major |
| HANK | DD 702 | 11 April | Suicide near miss | Minor |
| HARDING | DMS 28 | 16 April | Suicide plane | Major |
| HARRISON | DD 573 | 6 April | Suicide near miss | Minor |
| HARRY F. BAUER | DM 26 | 5 April | Dud torpedo | Minor |
| HAYNSWORTH | DD 700 | 7 April | Suicide plane | Major |
| HAZELWOOD | DD 531 | 29 April | Suicide plane | Major |
| HENRICO | APA 45 | 2 April | Suicide plane | Major |
| HICKOX | DD 673 | 26 April | Air attack | Minor |
| HINSDALE | APA 120 | 1 April | Suicide plane | Major |
| HOBBS VICTORY | AKE | 6 April | Suicide plane | SUNK |
| HOBSON | DMS 26 | 16 April | Suicide plane | Major |
| HOPPING | APD 51 | 9 April | Shore battery | Major |
| HOWORTH | DD 592 | 6 April | Suicide plane | Major |
| HUDSON | DD 475 | 22 April | Air attack | Major |
| HUTCHINS | DD 476 | 6 April | Suicide plane | Minor |
| HUTCHINS | DD 476 | 27 April | Suicide boat | Major |
| HYMAN | DD 732 | 6 April | Suicide plane & torpedo | Major |
| IDAHO | BB 42 | 12 April | Suicide plane | Major |
| INDEFATIGABLE | CV | 1 April | Suicide plane | Minor |
| INDIANAPOLIS | CA 35 | 31 March | Suicide plane | Major |
| INTREPID | CV 11 | 16 April | Suicide plane | Major |
| ISHERWOOD | DD 520 | 22 April | Suicide plane | Major |
| JEFFERS | DMS 27 | 12 April | Suicide plane | Major |
| KIDD | DD 661 | 11 April | Suicide plane | Major |
| KIMBERLY | DD 521 | 26 March | Suicide plane | Major |
| KLINE | APD 120 | 20 April | Operational (collision) | Minor |
| KNUDSON | APD 101 | 26 March | Suicide plane | Minor |
| LAFFEY | DD 724 | 15 April | Suicide plane | Major |
| LCI | 82 | 4 April | Suicide boat | SUNK |
| LCI | 354 | 9 April | Operational | Minor |
| LCI | 407 | 16 April | Suicide plane | Minor |
| LCI | 462 | 2 April | Operational (grounded) | Minor |
| LCI | 558 | 9 April | Air attack | Minor |
| LCI(G) | 568 | 2 April | Suicide plane | Minor |
| LCI | 580 | 2 April | Operational | Minor |
| LCI | 580 | 8 April | Operational (grounded) | Minor |
| LCI | 580 | 28 April | Air attack | Major |

| Name | Type | Date | Cause | Extent |
|---|---|---|---|---|
| LCI(G) | 588 | 28 March | Suicide boat | Minor |
| LCI | 754 | 16 April | Operational | Minor |
| LCI | 765 | 5 April | Operational | Minor |
| LCI | 816 | 30 April | Operational (grounded) | Minor |
| LCI(M) | 807 | 1 April | Operational (mortar expl) | Major |
| LCS | 15 | 22 April | Suicide plane | SUNK |
| LCS | 33 | 12 April | Suicide plane | SUNK |
| LCS | 36 | 11 April | Suicide plane | Major |
| LCS | 37 | 28 April | Suicide boat | Major |
| LCS | 38 | 11 April | Operational | Minor |
| LCS | 51 | 15 April | Suicide plane | Minor |
| LCS | 57 | 12 April | Operational (grounded) | Minor |
| LCS | 57 | 12 April | Suicide plane (3) | Major |
| LCS | 88 | 11 April | Air attack | Major |
| LCS | 116 | 15 April | Suicide plane | Major |
| LCT | 876 | 9 April | Suicide plane | SUNK |
| LEUTZE | DD 481 | 6 April | Suicide plane | Major |
| LINDSEY | DM 32 | 12 April | Suicide plane | Major |
| LOGAN VICTORY | AKE | 6 April | Suicide plane | SUNK |
| LONGSHAW | DD 559 | 9 April | Air attack | Minor |
| LSM | 12 | 16 April | Operational (broached) | SUNK |
| LSM | 28 | 18 April | Air attack | Major |
| LSM | 188 | 28 March | Suicide plane | Major |
| LSM | 192 | 1 April | Operational | Major |
| LSM | 189 | 15 April | Air Attack | Minor |
| LSM | 193 | 17 April | Operational (grounded) | Major |
| LSM | 279 | 12 April | AA fire | Minor |
| LSM | 312 | 13 April | Operational | Minor |
| LST | 70 | 5 April | Operational (broaching) | Major |
| LST | 71 | 5 April | Operational (collision) | Major |
| LST | 78 | 11 April | Air attack | Major |
| LST | 166 | 5 April | Operational | Minor |
| LST | 267 | 19 April | Operational (collision) | Minor |
| LST | 343 | 9 April | Operational (broaching) | Major |
| LST | 347 | 6 April | Air attack | SUNK |
| LST | 447 | 9 April | Suicide plane | SUNK |
| LST | 449 | 16 April | Shore battery | Minor |
| LST | 554 | 5 April | Storm | Major |
| LST | 557 | 10 April | Shore battery | Minor |
| LST | 568 | 6 April | Operational | Minor |
| LST | 570 | 6 April | Operational | Minor |
| LST | 599 | 3 April | Suicide plane | Major |
| LST | 608 | 5 April | Operational (beaching) | Minor |
| LST | 609 | 5 April | Operational (beaching) | Minor |
| LST | 612 | 5 April | Operational (beaching) | Minor |
| LST | 624 | 5 April | Operational (grounded) | Major |
| LST | 625 | 5 April | Operational | Minor |
| LST | 658 | 6 April | Operational | Minor |
| LST | 675 | 5 April | Operational (grounded) | Major |
| LST | 698 | 7 April | Operational (beaching) | Major |
| LST | 723 | 5 April | Operational | Minor |
| LST | 756 | 5 April | Operational (beaching) | Major |

| Name | Type | Date | Cause | Extent |
|---|---|---|---|---|
| LST | 762 | 5 April | Operational | Minor |
| LST | 782 | 5 April | Operational | Minor |
| LST | 884 | 1 April | Suicide plane | Major |
| LST | 890 | 7 April | Operational (collision) | Minor |
| LST | 929 | 19 April | Operational (collision) | Minor |
| MANLOVE | DE 36 | 10 April | Strafing | Minor |
| MANNERT L. ABELE | DD 733 | 12 April | Suicide plane | SUNK |
| MARYLAND | BB 46 | 6 April | Bomb | Major |
| McDERMUT | DD 677 | 16 April | AA fire | Major |
| MINOT VICTORY | XAK | 14 April | Air attack | Minor |
| MOBILE | CL 63 | 19 April | Operational (ex.in turret) | Minor |
| MORRIS | DD 417 | 6 April | Suicide plane | Major |
| MULLANY | DD 528 | 6 April | Suicide planes (2) | Major |
| MURRAY | DD 576 | 27 March | Aerial torpedo | Major |
| NEVADA | BB 36 | 26 March | Suicide plane near miss (2) | Major |
| NEVADA | BB 36 | 5 April | Shore battery | Minor |
| NEWCOMB | DD 586 | 6 April | Suicide planes (2) | Major |
| NEW MEXICO | BB 40 | 12 April | Suicide plane | Major |
| NEW YORK | BB 34 | 18 April | Suicide plane | Minor |
| NORTH CAROLINA | BB 55 | 6 April | AA fire | Minor |
| NORMAN SCOTT | DD 690 | 12 April | Air attack | Minor |
| OAKLAND | CL 95 | 12 April | Air attack | Minor |
| O'BRIEN | DD 725 | 26 March | Suicide plane | Major |
| PC | 462 | 7 April | Operational | Minor |
| PC | 851 | 16 April | Operational | Minor |
| PGM | 11 | 14 April | Operational | Minor |
| PGM | 18 | 7 April | Mine | SUNK |
| PINKNEY | APH 2 | 28 April | Air attack | Major |
| PORTERFIELD | DD 682 | 10 April | AA fire | Minor |
| PORTERFIELD | DD 682 | 27 March | Suicide plane | Major |
| PRINGLE | DD 477 | 16 April | Suicide plane | SUNK |
| PRITCHETT | DD 561 | 3 April | Bomb | Major |
| PURDY | DD 734 | 12 April | Suicide plane | Major |
| RALL | DE 304 | 12 April | Suicide plane | Major |
| RALPH TALBOT | DD 390 | 27 April | Air attack | Major |
| RATHBURNE | APD 25 | 27 April | Suicide plane | Major |
| RECRUIT | AM 285 | 6 April | Suicide plane | Minor |
| REMEY | DD 688 | 12 April | Air attack | Minor |
| R. H. SMITH | DM 23 | 26 March | Air attack | Minor |
| RIDDLE | DE 185 | 12 April | Suicide plane | Major |
| RODMAN | DMS 21 | 6 April | Suicide plane | Major |
| ROOKS | DD 804 | 6 April | Air attack | Minor |
| SAMUEL S. MILES | DE 183 | 12 April | Suicide near miss | Minor |
| SAN JACINTO | CVL 30 | 6 April | Bomb near miss | Minor |
| SC | 667 | 11 April | Operational (grounded) | Major |

| Name | Type | Date | Cause | Extent |
|---|---|---|---|---|
| S. HALL YOUNG | XAK | 30 April | Air attack | Minor |
| SHANNON | DM 25 | 29 April | Suicide near miss | Minor |
| SHEA | DM 30 | 22 April | Air attack | Minor |
| SIGSBEE | DD 502 | 13 April | Suicide plane | Major |
| SKIRMISH | AM 303 | 26 March | Suicide plane | Minor |
| SKYLARK | AM 63 | 26 March | Mines | SUNK |
| SKIRMISH | AM 303 | 2 April | Air Attack | Minor |
| SPEAR | AM 322 | 19 April | Air attack | Minor |
| SPROSTON | DD 577 | 4 April | Bomb near miss | Minor |
| STANLY | DD 478 | 12 April | Suicide plane | Major |
| STARR | AKA 87 | 9 April | Suicide boat | Minor |
| STERETT | DD 407 | 9 April | Suicide plane | Major |
| STRINGHAM | APD 6 | 24 April | Operational | Major |
| SWALLOW | AM 65 | 22 April | Suicide plane | SUNK |
| TALUGA | AO 62 | 16 April | Suicide plane | Major |
| TAUSSIG | DD 746 | 6 April | Bomb near miss (2) | Minor |
| TELFAIR | APA 210 | 3 April | Suicide plane | Minor |
| TENNESSEE | BB 43 | 12 April | Air attack | Major |
| TERROR | CM 5 | 30 April | Air attack | Major |
| THORNTON | AVD 11 | 5 April | Operational (collision) | SUNK |
| TOLMAN | DM 28 | 18 April | Operational (grounded) | Major |
| TRATHEN | DD 530 | 14 April | Shore Battery | Major |
| TWIGGS | DD 591 | 28 April | Suicide plane | Major |
| TYRRELL | AKA 80 | 2 April | Suicide plane | Minor |
| ULSTER (Brit.) | DD | 1 April | Bomb | Major |
| VAMMEN | DE 644 | 2 April | Operational | Minor |
| WADSWORTH | DD 516 | 28 April | Suicide plane | Minor |
| WAKE ISLAND | CVE 65 | 3 April | Suicide near miss | Major |
| WALTER C. WANN | DE 412 | 12 April | Air attack | Minor |
| WESSON | DE 184 | 7 April | Suicide plane | Major |
| WEST VIRGINIA | BB 48 | 1 April | Suicide plane | Minor |
| WHITEHURST | DE 634 | 12 April | Suicide plane | Major |
| WICHITA | CA 45 | 28 April | Shore battery | Minor |
| WILSON | DD 408 | 16 April | Suicide near miss | Minor |
| WITTER | DE 636 | 6 April | Suicide planes (2) | Major |
| WILLIAM J. DITTER | DM 31 | 30 April | Air attack | Minor |
| WYANDOT | AKA 92 | 29 March | Bomb near misses | Minor |
| YMS | 92 | 8 April | Mine | Major |
| YMS | 96 | 10 April | Operational (collision) | Major |
| YMS | 103 | 7 April | Mine | SUNK |
| YMS | 311 | 6 April | Suicide plane | Minor |
| YMS | 321 | 6 April | Air attack | Minor |
| YMS | 427 | 7 April | Shore battery | Minor |
| ZELLARS | DD 777 | 12 April | Suicide plane | Major |

| | |
|---|---|
| *Ships with Major damage | 100 |
| *Ships with Minor damage | 112 |
| *Ships SUNK | 22 |
| Total................. | 234 |

Los and damaged according to cause:

| | Damaged | SUNK |
|---|---|---|
| Suicide planes | 90 | 14 |
| Miscellaneous air attack** | 47 | 1 |
| Operational | 52 | 2 |
| Mines | 1 | 4 |
| Shore battery fire | 7 | 0 |
| AA fire | 6 | 0 |
| Suicide boat | 6 | 1 |
| Storm | 3 | 0 |
| Totals................ | 212 | 22 |

Lost or damaged by types.***

| Type | Damaged | Sunk |
|---|---|---|
| BB | 10 | 0 |
| CV | 5 | 0 |
| CA | 2 | 0 |
| CL | 2 | 0 |
| CVL | 1 | 0 |
| CVE | 2 | 0 |
| DD | 57 | 5 |
| DE | 16 | 0 |
| CM | 1 | 0 |
| DM | 8 | 0 |
| DMS | 7 | 1 |
| AVD | 0 | 1 |
| APD | 7 | 1 |
| APA | 9 | 0 |
| AKA | 4 | 0 |
| AO | 2 | 0 |
| AM | 7 | 2 |
| AH | 2 | 0 |
| APH | 1 | 0 |
| ARL | 1 | 0 |
| ATF | 1 | 0 |
| XAK-AKE | 3 | 3 |
| PC | 2 | 0 |
| SC | 1 | 0 |
| PGM | 1 | 1 |
| YMS | 5 | 1 |
| LST | 27 | 2 |
| LSM | 7 | 1 |
| LCT | 0 | 1 |
| LCI | 12 | 1 |
| LCI(M) | 1 | 0 |
| LCS | 8 | 2 |
| Totals.... | 212 | 22 |

* NOTE: Some ships were hit more than once for each entry in the list above, but this summary only takes into account the number of entries.

** NOTE: Some of the miscellaneous air attacks may have been by suicide planes.

*** NOTE: Includes British Pacific Fleet ships.

Commander in Chief, U.S. Pacific Fleet and Pacific Ocean Areas, Operations in Pacific Ocean Areas – April 1945, 16 October 1945.

# Bibliography/Notes


Blunt, William. *Lolita and the Hollywood Fleet.* Self-published, 2020.

Bruhn, David D. *Battle Stars for the "Cactus Navy": America's Fishing Vessels and Yachts in World War II.* Berwyn Heights, MD: Heritage Books, 2014.

—*Eyes of the Fleet: The U.S. Navy's Seaplane Tenders and Patrol Aircraft in World War II.* Berwyn Heights, MD: Heritage Books, 2016.

—*Ingram's Fourth Fleet: U.S. and Royal Navy Operations Against German Runners, Raiders, and Submarines in the South Atlantic in World War II.* Berwyn Heights, MD: Heritage Books, 2017.

—*MacArthur and Halsey's "Pacific Island Hoppers": The Forgotten Fleet of World War II.* Berwyn Heights, MD: Heritage Books, 2014.

—*Wooden Ships and Iron Men: The U.S. Navy's Coastal and Motor Minesweepers, 1941-1953.* Westminster, MD: Heritage Books, 2009.

Bruhn, David D., Rob Hoole. *Nightraiders: U.S. Navy, Royal Navy, Royal Australian Navy, and Royal Netherlands Navy Mine Forces Battling the Japanese in the Pacific in World War II.* Berwyn Heights, MD: Heritage Books, 2018

Carter, Worrall Reed. *Beans, Bullets, and Black Oil.* Washington, DC: Department of the Navy, 1953.

Cooper, Anthony. *Darwin Spitfires: The Real Battle for Australia.* Barnsley, UK: Pen & Sword Aviation, 2013.

Cressman, Robert. *The Official Chronology of the US Navy in World War II.* Annapolis, MD: Naval Institute, 2016.

Dean, H. R. *The Royal New Zealand Air Force in South-East Asia 1941–42.* Wellington, NZ: Historical Publications Branch, 1952.

Dyer, George Carroll. *The Amphibians Came to Conquer: The Story of Admiral Richmond Kelly Turner.* Washington, DC: U.S. Department of the Navy, 1972.

Lott, Arnold S. *Most Dangerous Sea.* Annapolis, MD: U.S. Naval Institute, 1959.

Mansbridge, Francis. *Launching History: The Saga of the Burrard Dry Dock.* Madeira, BC: Harbour Publishing, 2002.

McLaren T.A., Vickie Jensen. *Ships of Steel – A British Columbia Shipbuilder's Story.* Madeira, BC: Harbour Publishing, 2000.

Morison, Samuel Eliot. *History of United States Naval Operation in World War II: Breaking the Bismarcks Barrier, 22 July 1942-1 May 1944.* Boston, Mass.: Little, Brown, 1984.

—*The Two Ocean War*. Boston, MA: Little, Brown, 1963.

Shaw Jr., Henry I., Bernard C. Nalty, and Edwin T. Turnbladh. *History of U.S. Marine Corps Operations in World War II Volume III: Central Pacific Drive*. Washington, DC: Headquarters, U.S. Marine Corps, 1966.

Turner, Mike, Hector Donohue. *Australian Minesweepers at War*. Canberra: Sea Power Centre – Australia, 2018.

U.S. Government. *Building the Navy's Bases in World War II: History of the Bureau of Yards and Docks and the Civil Engineer Corps 1940-1946, Vol II*. Washington, DC: U.S. Government Printing Office, 1947.

—*Combat Narratives, Miscellaneous Actions in the South Pacific*. Washington, DC: Office of Naval Intelligence, 1943.

—*Combat Narratives, Solomon Islands Campaign: I, The Landing in the Solomons 7-8 August 1942*. Washington DC: Office of Naval Intelligence, 1943.

—*Combat Narratives Solomon Islands Campaign, II The Battle of Savo Island 9 August 1942, III The Battle of the Eastern Solomons 23-25 August 1942*. Washington, DC: Office of Naval Intelligence, 1943.

—*Combat Narratives Solomon Islands Campaign: VI Battle of Guadalcanal 11-15 November 1942*. Washington, DC: Office of Naval Intelligence, 1944.

—*Combat Narratives Solomon Islands Campaign X, Operations in the New Georgia Area 21 June-5 August 1943* (Washington, DC: Office of Naval Intelligence 1944).

—*Combat Narratives Solomon Islands Campaign: XII The Bougainville Landing and the Battle of Empress Augusta Bay 27 October-1 November 1943*. Washington, DC: Office of Naval Intelligence, 1945.

—*Combat Narratives: The Aleutians Campaign June 1942-August 1943*. Washington, DC: Office of Naval Intelligence, 1945.

—*United States Naval Administration in World War II Office of Naval Operations History of the Naval Armed Guard Afloat*. Washington, DC: Director of Naval History, 1949.

—*Naval Aviation in the Pacific*. Washington, DC: Office of the Chief of Naval Operations, 1947.

Water, Sydney D., *The Royal New Zealand Navy*. Wellington: Historical Publications Branch, 1956.

## FOREWORD NOTES:

[1] Francis Mansbridge, *Launching History: The Saga of the Burrard Dry Dock*, 77-79; T. A. McLaren & Vickie Jensen, *Ships of Steel – A British Columbia Shipbuilder's Story*, 49-51.

[2] McLaren & Jensen, *Ships of Steel – A British Columbia Shipbuilder's Story*, 65.

[3] Ibid, 75.

## PREFACE NOTES:

[1] Worrall Reed Carter, *Beans, Bullets, and Black Oil*, 1, 5.
[2] Commander Service Force, U.S. Pacific Fleet, History of Service Force – forwarding of, 31 January 1946.
[3] Samuel Eliot Morison, *The Two Ocean War*, 423-424.
[4] *Combat Narratives, Miscellaneous Actions in the South Pacific* (Washington, DC: Office of Naval Intelligence, 1943).
[5] *Building the Navy's Bases in World War II: History of the Bureau of Yards and Docks and the Civil Engineer Corps 1940-1946, Vol II* (Washington, DC: U.S. Government Printing Office, 1947), 221.
[6] ONI *Combat Narratives, Miscellaneous Actions in the South Pacific.*
[7] Ibid.
[8] "The China Fleet – Small Ships WW2" (https://www.navyhistory.org.au/the-china-fleet-small-ships-ww2/3/: accessed 15 October 2020).
[9] "William Frederick Halsey, Jr. 30 October 1882 - 16 August 1959" (https://www.history.navy.mil/content/history/nhhc/research/library/research-guides/modern-biographical-files-ndl/modern-bios-h/halsey-william-f.html: accessed 14 October 2020).
[10] Ibid.
[11] Ibid.
[12] "Battle of Leyte Gulf (1944)" by Jozef Straczek (https://www.navy.gov.au/history/feature-histories/battle-leyte-gulf-1944); "The Role of the RAN in the Liberation of the Philippines" by Guy Royle (https://www.navyhistory.org.au/the-role-of-the-ran-in-the-liberation-of-the-philippines/2/): accessed 9 October 2020).
[13] "England's Shadow Fleet: White Ensign in the Pacific" by Neil M. Heckman, *Sea Classics*, May 2004 (https://web.archive.org/web/20071113144344/http://findarticles.com/p/articles/mi_qa4442/is_200405/ai_n16061305/pg_8: accessed 16 October 2020).
[14] "The British Pacific Fleet Task Force 57 Politics & Logistics: Sakishima Gunto, Okinawa Campaign, 1945" (http://www.armouredcarriers.com/task-force-57-british-pacific-fleet: accessed 13 January 2020); "Operation Meridian" (https://www.armouredcarriers.com/operation-meridian-the-palembang-strike: accessed 4 June 2020).
[15] "British Pacific Fleet 1944-46" (https://www.navyhistory.org.au/british-pacific-fleet-1944-46/2/: accessed 15 October 2020).
[16] Ibid.
[17] Ibid.
[18] "The British Pacific Fleet" (https://www.navy.gov.au/history/feature-histories/british-pacific-fleet); "British Pacific Fleet – 1945" (https://www.navyhistory.org.au/british-pacific-fleet-1945/); "Royal Navy Maintenance and Supply Ships"

(http://fortships.tripod.com/Royal%20Navy.htm); "WW2 Fort Class" (http://historicalrfa.org/ww2-fort-class): accessed 16 October 2020).
[19] "British Pacific Fleet – 1945."
[20] "Operation Meridian"; Mike Turner and Hector Donohue, *Australian Minesweepers at War*, 134-139.
[21] "England's Shadow Fleet: White Ensign in the Pacific."
[22] "Battle Honours of RN ships & Naval Air Squadrons" (http://www.royalnavyresearcharchive.org.uk/ESCORT/Battle_hons.htm: accessed 17 August 2020).
[23] Navy and Marine Corps Awards Manual, 1953 (https://www.ibiblio.org/hyperwar/USN/ref/Awards/Awards-IV-17.html#sec2-20: accessed 17 August 2020).
[24] "The China Fleet – Small Ships WW2" (https://www.navyhistory.org.au/the-china-fleet-small-ships-ww2/3/: accessed 15 October 2020).

## CHAPTER 1 NOTES:

[1] USS *Alchiba* (AKA6) Torpedo Damage Solomon Islands 28 November and 7 December 1942, undated, numbered 12170636-59.
[2] USS *Alchiba* (AKA6) Torpedo Damage Solomon Islands; *Combat Narratives, Miscellaneous Actions in the South Pacific.*
[3] *Combat Narratives, Miscellaneous Actions in the South Pacific.*
[4] USS *Alchiba* (AKA6) Torpedo Damage Solomon Islands.
[5] USS *Alchiba* (AKA6) Torpedo Damage Solomon Islands; *Combat Narratives, Miscellaneous Actions in the South Pacific.*
[6] USS *Alchiba* (AKA6) Torpedo Damage Solomon Islands.
[7] Ibid.
[8] Ibid.
[9] Ibid.
[10] Ibid.
[11] Ibid.
[12] Ibid.
[13] Ibid.
[14] "H-Gram 013: Night of the Long Lances 7 December 2017" (https://www.history.navy.mil/about-us/leadership/director/directors-corner/h-grams/h-gram-013.html#1: accessed 26 October 2020).
[15] Bureau of Naval Personnel Information Bulletin No. 315 (June 1943).

## CHAPTER 2 NOTES:

[1] Commander Service Force, U.S. Pacific Fleet, History of Service Force – forwarding of, 31 January 1946.
[2] Ibid.
[3] Ibid.
[4] Ibid.
[5] Commanding Officer, USS *Vega*, Action - Report of, 10 December 1941.

[6] Ibid.
[7] Ibid.
[8] Ibid.
[9] Commanding Officer, USS *Antares*, Air Raid on Oahu December 7, 1941; Report on, 10 December 1941; David D. Bruhn, *Wooden Ships and Iron Men: The U.S. Navy's Coastal and Motor Minesweepers, 1941-1953*, 12-13.
[10] Ut supra.
[11] Commanding Officer, USS *Antares*, Air Raid on Oahu December 7, 1941.
[12] Ibid.
[13] Commanding Officer, USS *Castor*, December 7th Raid – Report on, 11 December 1941.
[14] Ibid.
[15] Ibid.
[16] Commanding Officer, USS *Castor*, December 7th Raid – Report on, 11 December 1941; Commanding Officer, USS *Neosho*, Raid on Pearl Harbor, T.H., December 7, 1941 - Report on, 11 December 1941.
[17] Commanding Officer, USS *Castor*, December 7th Raid – Report on, 11 December 1941.
[18] Commander Service Force, U.S. Pacific Fleet, History of Service Force – forwarding of, 31 January 1946.
[19] Ibid.
[20] Ibid.
[21] Ibid.
[22] Ibid.
[23] Commander Service Force, U.S. Pacific Fleet, History of Service Force – forwarding of, 31 January 1946; U.S. Bureau of Yards and Docks, *Building the Navy's Bases in World War II: History of the Bureau of Yards and Docks and the Civil Engineer Corps 1940-1946, Vol II*, 191.
[24] Ut supra.
[25] Commander Service Force, U.S. Pacific Fleet, History of Service Force – forwarding of, 31 January 1946.

## CHAPTER 3 NOTES:

[1] *Combat Narratives, Solomon Islands Campaign: I, The Landing in the Solomons 7-8 August 1942*, 73.
[2] Ibid, 1.
[3] Ibid.
[4] Morison, *The Two-Ocean War*, 140.
[5] *Combat Narratives, Solomon Islands Campaign: I, The Landing in the Solomons 7-8 August 1942*, 1-2.
[6] Ibid, 2.
[7] *Combat Narratives, Solomon Islands Campaign: I, The Landing in the Solomons 7-8 August 1942*, 2; "Robert Lee Ghormley 15 October 1883 - 21 June 1958" (https://www.history.navy.mil/content/history/nhhc/our-collections/photography/numerical-list-of-images/nhhc-series/nh-series/NH-97000/NH-97742.html: accessed 3 November 2020).

[8] *Combat Narratives, Solomon Islands Campaign: I, The Landing in the Solomons 7-8 August 1942*, 5-6.
[9] Ibid, 6.
[10] Ibid, 7-8.
[11] Ibid, 24-25.
[12] "The Battle of Savo Island August 9th, 1942 Strategical and Tactical Analysis" (https://www.history.navy.mil/research/library/online-reading-room/title-list-alphabetically/b/battle-savo-island-strat-tact-analysis.html: accessed 9 November 2020).
[13] *Combat Narratives, Solomon Islands Campaign: I, The Landing in the Solomons 7-8 August 1942*, 9, 13, 24, 26.
[14] Ibid, 34.
[15] Ibid, 44.
[16] Ibid, 46.
[17] *Combat Narratives, Solomon Islands Campaign: I, The Landing in the Solomons 7-8 August 1942*, 46; USS *Alchiba* War Diary, August 1942; "The Battle of Savo Island August 9th, 1942 Strategical and Tactical Analysis."
[18] "The Battle of Savo Island August 9th, 1942 Strategical and Tactical Analysis."
[19] *Combat Narratives, Solomon Islands Campaign: I, The Landing in the Solomons 7-8 August 1942*, 46-47; *Mugford, DANFS.*
[20] "The Battle of Savo Island August 9th, 1942 Strategical and Tactical Analysis"; Commanding Officer, USS *Mugford*, Addendum to Deck Log, USS *Mugford*, Month of August, 1942, 28 November 1942.
[21] *Combat Narratives, Solomon Islands Campaign: I, The Landing in the Solomons 7-8 August 1942*, 58, 60-61; "The Battle of Savo Island August 9th, 1942 Strategical and Tactical Analysis."
[22] *Combat Narratives, Solomon Islands Campaign: I, The Landing in the Solomons 7-8 August 1942*, 58, 61; Commanding Officer, USS *Betelgeuse*, Report of action of Guadalcanal Island, Solomon Islands on 7, 8 and 9 August 1942; forwarding of, 15 August 1942; "The Battle of Savo Island August 9th, 1942 Strategical and Tactical Analysis."
[23] *Combat Narratives, Solomon Islands Campaign: I, The Landing in the Solomons 7-8 August 1942*, 58, 61; Commanding Officer, USS *Betelgeuse*, Report of action of Guadalcanal Island, Solomon Islands on 7, 8 and 9 August 1942; forwarding of, 15 August 1942.
[24] Commanding Officer, USS *Betelgeuse*, Report of action of Guadalcanal Island, Solomon Islands on 7, 8 and 9 August 1942; forwarding of, 15 August 1942.
[25] USS *George F. Elliott* War Diary, August 1942.
[26] Ibid.
[27] USS *George F. Elliott* War Diary, August 1942; "The Battle of Savo Island August 9th, 1942 Strategical and Tactical Analysis."
[28] "Battle of Savo Island, 9 August 1942" (https://www.history.navy.mil/browse-by-topic/wars-conflicts-and-

operations/world-war-ii/1942/guadalcanal/battle-of-savo-island.html: accessed 9 November 2020).
[29] "Battle of Savo Island, 9 August 1942."
[30] "The Battle of Savo Island August 9th, 1942 Strategical and Tactical Analysis."
[31] "Battle of Savo Island, 9 August 1942"; *Jarvis, DANFS.*

## CHAPTER 4 NOTES:

[1] Brochure *New Guinea: The U.S. Army Campaigns of World War II*, prepared by Edward J. Drea for the U.S. Army Center of Military History (https://history.army.mil/brochures/new-guinea/ng.htm: accessed 10 November 2020)
[2] David D. Bruhn, *MacArthur and Halsey's "Pacific Island Hoppers": The Forgotten Fleet of World War II*, 16.
[3] Ibid.
[4] Ibid.
[5] Ibid, 16-17.
[6] Ibid, 17.
[7] Ibid, 17-18.
[8] Ibid, 18.
[9] Ibid.
[10] Ibid, 18-19.
[11] Ibid, 19.
[12] "New Guinea campaign (January 1942-September 1945" (http://www.historyofwar.org/articles/campaign_new_guinea.html: accessed 10 November 2020).
[13] Ibid.
[14] Brochure *New Guinea: The U.S. Army Campaigns of World War II.*
[15] Bruhn, *MacArthur and Halsey's "Pacific Island Hoppers,"* 20.
[16] Ibid.
[17] Ibid, 21.

## CHAPTER 5 NOTES:

[1] "The RAN's Chinese Coastal Steamers" by Petar Djokovic (https://www.navy.gov.au/history/feature-histories/ran%E2%80%99s-chinese-coastal-steamers: accessed 10 October 2020).
[2] Ibid.
[3] "HMAS *Ping Wo* FL150 Stores Ship in RAN Service in Australia during WWII" (https://www.ozatwar.com/ran/hmaspingwo.htm: accessed 8 October 2020); "The RAN's Chinese Coastal Steamers" by Petar Djokovic.
[4] "HMAS *Ping Wo* FL150 Stores Ship in RAN Service in Australia during WWII."
[5] Ibid.
[6] Ibid
[7] Ibid.

[8] *The Western Australian* (Perth, WA 1879-1954) 5 April 1946.
[9] "Niagara mined off Northland coast 19 June 1940" (https://nzhistory.govt.nz/niagara-mined-off-northland-coast: accessed 8 October 2020).
[10] Ibid.
[11] David D. Bruhn, *Ingram's Fourth Fleet: U.S. and Royal Navy Operations Against German Runners, Raiders, and Submarines in the South Atlantic in World War II*, 62-63.
[12] Ibid, 62.
[13] "KMS Hilfskreuzer Orion (HSK-1)" (https://www.asisbiz.com/il2/Ar-196/BoFlGr-Orion.html: accessed 8 October 2020); Bruhn, *Ingram's Fourth Fleet*, 63-64.
[14] Sydney D. Waters, *The Royal New Zealand Navy* (Wellington: Historical Publications Branch, 1956), 120.
[15] Ibid.
[16] "The RAN's Chinese Coastal Steamers" by Petar Djokovic.
[17] Ibid.
[18] "Battle of Leyte Gulf (1944)" by Jozef Straczek (https://www.navy.gov.au/history/feature-histories/battle-leyte-gulf-1944); "The Role of the RAN in the Liberation of the Philippines" by Guy Royle (https://www.navyhistory.org.au/the-role-of-the-ran-in-the-liberation-of-the-philippines/2/): accessed 9 October 2020).
[19] "Circa 1942-43: Stores carrier HMAS BARALABA - coaler maybe too hot for the tropics" (https://www.flickr.com/photos/41311545@N05/6317961706: accessed 9 October 2020).
[20] "John Serjeant" (http://johnoxley.org.au/history/historical-articles/recollections-john-serjeant-1943/: accessed 9 October 2020).
[21] "Circa 1942-43: Stores carrier HMAS BARALABA - coaler maybe too hot for the tropics"; Bruhn, *MacArthur and Halsey's "Pacific Island Hoppers,"* 20; William Blunt, *Lolita and the Hollywood Fleet* (Self-published, 2020). (https://lolitaandthehollywoodfleet.com/LolitaAndTheHollywoodFleet.html#Contents: accessed 9 October 2020).
[22] Bruhn, *MacArthur and Halsey's "Pacific Island Hoppers,"* 21.
[23] Ibid.
[24] "Circa 1942-43: Stores carrier HMAS *BARALABA* - coaler maybe too hot for the tropics."

## CHAPTER 6 NOTES:

[1] USS *Fomalhaut* War Diary, August 1942.
[2] *Combat Narratives Solomon Islands Campaign, II The Battle of Savo Island 9 August 1942, III The Battle of the Eastern Solomons 23-25 August 1942* (Washington, DC: Office of Naval Intelligence, 1943); "Battle of the Eastern Solomons August 24, 1942" (http://www.cv6.org/1942/solomons/solomons.htm): both accessed 13 November 2020).

[3] "IJN Submarine RO-34: Tabular Record of Movement" (http://www.combinedfleet.com/RO-34.htm; "H-010-1: "Operation Shoestring" (https://www.history.navy.mil/content/history/nhhc/about-us/leadership/director/directors-corner/h-grams/h-gram-010/h-010-1.html): both accessed 13 November 2020.
[4] "IJN Submarine RO-34: Tabular Record of Movement."
[5] *Combat Narratives Solomon Islands Campaign, II The Battle of Savo Island 9 August 1942, III The Battle of the Eastern Solomons 23-25 August 1942*; "Battle of the Eastern Solomons August 24, 1942."
[6] *Combat Narratives Solomon Islands Campaign, II The Battle of Savo Island 9 August 1942, III The Battle of the Eastern Solomons 23-25 August 1942*
[7] Morison, *The Two-Ocean War*, 179.
[8] Ibid.
[9] Ibid.
[10] Ibid, 179-180.
[11] "Battle of the Eastern Solomons, 23–25 August 1942" (https://www.history.navy.mil/browse-by-topic/wars-conflicts-and-operations/world-war-ii/1942/guadalcanal/battle-of-the-eastern-solomons.html: accessed 13 November 2020).
[12] "H-010-2: The Battle of the Eastern Solomons, 23–25 August 1942" (https://www.history.navy.mil/about-us/leadership/director/directors-corner/h-grams/h-gram-010/h-010-2.html: accessed 13 November 2020).
[13] Commander in Chief, U.S. Pacific Fleet, Solomon Islands Campaign – Torpedoing of *Saratoga*, *Wasp*, and *North Carolina*, 31 October 1942.
[14] "IJN Submarine *I-26*: Tabular Record of Movement" (http://www.combinedfleet.com/I-26.htm: accessed 14 November 2020).
[15] "IJN Submarine *I-26*: Tabular Record of Movement"; *Saratoga*, *DANFS*.
[16] "IJN Submarine I-11: Tabular Record of Movement" (http://www.combinedfleet.com/I-11.htm: accessed 14 November 2020).
[17] Ibid.
[18] Commander Task Force Seventeen, Submarine attack on September 6, 1942 – report of, 17 September 1942.
[19] Ibid.
[20] General Orders: Commander 2d Carrier Task Force Pacific: Serial 01680 (September 1, 1945).
[21] "IJN Submarine I-19: Tabular Record of Movement" (http://www.combinedfleet.com/I-19.htm: accessed 14 November 2020).
[22] Ibid.
[23] "IJN Submarine I-19: Tabular Record of Movement"; Commanding Officer, USS *San Francisco*, Action Report – torpedoing of USS *Wasp*, *North Carolina*, and *O'Brien*, September 15, 1942; forwarding of, 30 September 1942.
[24] Commanding Officer, USS *San Francisco*, Action Report – torpedoing of USS *Wasp*, *North Carolina*, and *O'Brien*, September 15, 1942; forwarding of, 30 September 1942.
[25] Ibid.

[26] Ibid.
[27] Ibid.
[28] Commanding Officer, USS *San Francisco*, Action Report – torpedoing of USS *Wasp*, *North Carolina*, and *O'Brien*, September 15, 1942; forwarding of, 30 September 1942; *Wasp*, *DANFS*.
[29] Commander in Chief, U.S. Pacific Fleet, Solomon Islands Campaign – Torpedoing of *Saratoga*, *Wasp*, and *North Carolina*, 31 October 1942.
[30] "IJN Submarine I-4: Tabular Record of Movement" (http://www.combinedfleet.com/I-4.htm: accessed 14 November 2020).
[31] "IJN Submarine I-4: Tabular Record of Movement."
[32] "IJN Submarine I-4: Tabular Record of Movement"; "Tryon Departs for White Poppy (Part 4) – USS *Monssen*" (https://usstryon.wordpress.com/2013/02/: accessed 15 November 2020).
[33] Commanding Officer, USS *Alhena*, USS *Alhena* – torpedoing of, 7 October 1942, "Tryon Departs for White Poppy (Part 4) – USS *Monssen*"; Short History of the USS *Alhena*, AKA-9, Period Covered June 15, 1941 to September 15, 1945 (https://upload.wikimedia.org/wikipedia/commons/d/db/USS_ALHENA_-_War_History.pdf: accessed 15 November 2020).
[34] Commanding Officer, USS *Alhena*, USS *Alhena* – torpedoing of, 7 October 1942; USS *Alhena* (AKA9) Torpedo Damage South Pacific Sept. 29, 1942, War Damage Report No. 17, Bureau of Ships, Navy Department, 24 May 1943.
[35] Short History of the USS *Alhena*, AKA-9, Period Covered June 15, 1941 to September 15, 1945; USS *Alhena* (AKA9) Torpedo Damage South Pacific Sept. 29, 1942, War Damage Report No. 17, Bureau of Ships, Navy Department, 24 May 1943; "IJN Submarine I-4: Tabular Record of Movement."
[36] Short History of the USS *Alhena*, AKA-9, Period Covered June 15, 1941 to September 15, 1945; USS *Alhena* (AKA9) Torpedo Damage South Pacific Sept. 29, 1942, War Damage Report No. 17, Bureau of Ships, Navy Department, 24 May 1943; "Tryon Departs for White Poppy (Part 4) – USS *Monssen*."
[37] Commanding Officer, USS *Alhena*, USS *Alhena* – torpedoing of, 7 October 1942; USS *Alhena* (AKA9) Torpedo Damage South Pacific Sept. 29, 1942, War Damage Report No. 17, Bureau of Ships, Navy Department, 24 May 1943.
[38] Short History of the USS *Alhena*, AKA-9, Period Covered June 15, 1941 to September 15, 1945.
[39] USS *Libra* War Diary, November 1942.
[40] "Naval Battle of Guadalcanal, 12-15 November 1942 Order of Battle" (https://www.history.navy.mil/browse-by-topic/wars-conflicts-and-operations/world-war-ii/1942/guadalcanal/naval-battle-of-guadalcanal/testtest.html: accessed 16 November 2020).
[39] USS *Libra* War Diary, November 1942.
[40] *Combat Narratives Solomon Islands Campaign: VI Battle of Guadalcanal 11-15 November 1942* (Washington, DC: Office of Naval Intelligence, 1944).

[41] *Combat Narratives Solomon Islands Campaign: VI Battle of Guadalcanal 11-15 November 1942*; USS *Zeilin* and USS *Libra* War Diary, November 1942.
[42] *Combat Narratives Solomon Islands Campaign: VI Battle of Guadalcanal 11-15 November 1942.*
[43] Ibid.
[44] Ibid.
[45] Ibid.
[46] USS *Zeilin* War Diary, November 1942.
[47] USS *Zeilin* War Diary, November 1942; *Combat Narratives Solomon Islands Campaign: VI Battle of Guadalcanal 11-15 November 1942.*
[48] USS *Betelgeuse* War Diary, November 1943.
[49] Ibid.
[50] Ibid.
[51] USS *Libra* War Diary, November 1943.
[52] USS *Libra* War Diary, November 1943; *Combat Narratives Solomon Islands Campaign: VI Battle of Guadalcanal 11-15 November 1942.*
[53] Commanding Officer, USS *Betelgeuse*, Report of Actions off Guadalcanal Island, Solomon Islands on 11 and 12 November 1942, 15 November 1942; *Combat Narratives Solomon Islands Campaign: VI Battle of Guadalcanal 11-15 November 1942.*
[54] Ut supra.
[55] *Combat Narratives Solomon Islands Campaign: VI Battle of Guadalcanal 11-15 November 1942*; Commander in Chief, U.S. Pacific Fleet, Action Report, Air attack 12 November 1942, Cincpac File No. Pac-90-1h A16-3/SOL Serial 03772.
[56] Commanding Officer, USS *Betelgeuse*, Report of Actions off Guadalcanal Island, Solomon Islands on 11 and 12 November 1942, 15 November 1942.
[57] Ibid.
[58] Ibid.
[59] Ibid.
[60] Ibid.
[61] Ibid.
[62] Ibid.
[63] Ibid.
[64] Ibid.
[65] Ibid.
[66] Commander in Chief, U.S. Pacific Fleet, Action Report, Air attack 12 November 1942, Cincpac File No. Pac-90-1h A16-3/SOL Serial 03772.
[67] USS *Libra* War Diary, November 1942.
[68] USS *Libra* War Diary, November 1942; Commander Amphibious Force, South Pacific, Report of Operations of Task Force Sixty-Seven and Task Group 62.4 – Reinforcement of Guadalcanal November 8-15, 1942, and Summary of Third Battle of Savo, 3 December 1942.
[69] Commanding Officer, USS *Alchiba*, USS *Alchiba* – History of, 17 December 1945.
[70] Ibid.

[71] Ibid.
[72] Ibid.
[73] "Japanese Withdrawal, February 1943" (https://www.history.navy.mil/content/history/nhhc/browse-by-topic/wars-conflicts-and-operations/world-war-ii/1942/guadalcanal/japanese-withdrawal.html: accessed 20 November 2020).
[74] Ibid.
[75] USS *Carina* War Diary, February 1943.

## CHAPTER 7 NOTES:

[1] "HMAS *Patricia Cam*" (https://www.navy.gov.au/hmas-patricia-cam); "Japanese Float Plane bombs HMAS *Patricia Cam* near the Wessel Islands, NT, on 22 January 1943" (https://www.ozatwar.com/ran/ran.htm: accessed 6 October 2020).
[2] Ut supra.
[3] "HMAS *Patricia Cam*"; "Japanese Float Plane bombs HMAS *Patricia Cam* near the Wessel Islands, NT, on 22 January 1943"; "The Pat Cam Story in Brief" (https://www.pastmasters.net/hmas-patricia-cam.html: accessed 11 October 2020).
[4] "HMAS *Patricia Cam*."
[5] "Japanese Float Plane bombs HMAS *Patricia Cam* near the Wessel Islands, NT, on 22 January 1943.
[6] "HMAS *Patricia Cam*."
[7] "HMAS *Patricia Cam*"; "The Pat Cam Story in Brief."
[8] Ut supra.
[9] "HMAS *Patricia Cam*"; "Japanese Float Plane bombs HMAS *Patricia Cam* near the Wessel Islands, NT, on 22 January 1943."
[10] "Japanese Float Plane bombs HMAS *Patricia Cam* near the Wessel Islands, NT, on 22 January 1943; "The Pat Cam Story in Brief."
[11] Ut supra.
[12] Ut supra.
[13] "Japanese Float Plane bombs HMAS *Patricia Cam* near the Wessel Islands, NT, on 22 January 1943.
[14] "Japanese Float Plane bombs HMAS *Patricia Cam* near the Wessel Islands, NT, on 22 January 1943"; "An Australian taken Prisoner by the Japanese in Australian Waters off the Northern Territory on 22 January 1943 'The Kentish Affair'" (https://www.ozatwar.com/japsbomb/kentishaffair.htm); "Proceedings of a Military Court Held at Hong Kong" http://www.pastmasters.net/uploads/2/6/7/5/26751978/military_trial_proceedings_hk_1948.pdf): both accessed 6 October 2020).

## CHAPTER 8 NOTES:

[1] David D. Bruhn, *Battle Stars for the "Cactus Navy": America's Fishing Vessels and Yachts in World War II* (Berwyn Heights, MD: Heritage Books, 2014), 207.
[2] *Carina, DANFS.*

[3] *Carina, DANFS*; Commanding Officer, USS *Carina*, Damage to USS *Carina* by bombs, 4 March 1943.
[4] Commanding Officer, USS *Carina*, Damage to USS *Carina* by bombs, 4 March 1943.
[5] *Carina, DANFS*.
[6] Commanding Officer, USS *Arided*, Ship's History – Submission of, 18 November 1945.
[7] Ibid.
[8] Commanding Officer, USS *Arided*, Ship's History – Submission of, 18 November 1945; "United States Third Fleet History" (https://www.c3f.navy.mil/About-Us/History/: accessed 23 November 2020).
[9] Carter, *Beans, Bullets, and Black Oil: The Story of Fleet Logistics Afloat in the Pacific During World War II*, 48-49.
[10] Ibid.
[11] "Biography of the USS *Cassiopeia* AK 75" (https://www.seabees93.net/GI%20Cassiopeia%20-%20Friederich%20intro.htm: accessed 22 November 2020).
[12] USS *Sterett* War Diary, March 1943.
[13] "Class: Crater (AK-70)" (http://www.shipscribe.com/usnaux/AK/AK70.html: accessed 21 November 2020).
[14] Ibid.
[15] USS *Aaron Ward* War Diary, March 1943.
[16] Commanding Officer, USS *Adhara*, War Diary and Action Report, 1 May 1943.
[17] Ibid.
[18] Ibid.
[19] Commander Transport Group, South Pacific, Action Report – 7 April 1943, 22 April 1943.
[20] Commander Transport Group, South Pacific, Action Report – 7 April 1943, 22 April 1943; Commanding Officer, USS *Aludra*, War Diary, and Action Report, 1 May 1943.
[21] Commanding Officer, USS *Aludra*, War Diary, and Action Report, 1 May 1943.
[22] Ibid.
[23] Commander Transport Group, South Pacific, Action Report – 7 April 1943, 22 April 1943; Commanding Officer, USS *Aludra*, War Diary, and Action Report, 1 May 1943.
[24] U.S. Government, *United States Naval Administration in World War II Office of Naval Operations History of the Naval Armed Guard Afloat* (Washington, DC: Director of Naval History, 1949), 1-15. (https://www.ibiblio.org/hyperwar/USN/Admin-Hist/173-ArmedGuards/index.html#index: accessed 22 November 2020).
[25] Ibid.
[26] Ibid.

[27] USS *Titania* War Diary, May 1953.
[28] USS *Fomalhaut* War Diary, May 1943.
[29] Ibid.
[30] Ibid.
[31] USS *O'Bannon*, USS *Waters* War Diary, June 1943.
[32] USS *Skylark*, USS *O'Bannon*, War Diary, June 1943; Commanding Officer, USS *Celeno*, Ship's History – Forwarding of, 9 November 1945.
[33] Commanding Officer, USS *Celeno*, Ship's History – forwarding of, 9 November 1945; Commanding Officer, USS *Celeno*, Action Report of Air Attack on this Vessel on June 16, 1943, between Lunga and Koli Point, One and One-Half Miles Off-Shore, 20 June 1943.
[34] Ut supra.
[35] Commanding Officer, USS *Celeno*, Ship's History – forwarding of, 9 November 1945; Commanding Officer, USS *Celeno*, Action Report of Air Attack on this Vessel on June 16, 1943, between Lunga and Koli Point, One and One-Half Miles Off-Shore, 20 June 1943; *Celeno*, *DANFS*.
[36] Commanding Officer, USS *Celeno*, Action Report of Air Attack on this Vessel on June 16, 1943, between Lunga and Koli Point, One and One-Half Miles Off-Shore, 20 June 1943.
[37] Commanding Officer, USS *Celeno*, Ship's History – forwarding of, 9 November 1945.
[38] Ibid.
[39] Ibid.
[40] Commanding Officer, USS *Ward*, Submarine Attack on Task Unit 32.4.4, 4 August 1943; Commanding Officer, USS *Skylark*, Submarine Attack on Task Unit 32.4.4, 18 July 1943; Commanding Officer, USS *O'Bannon*, Report of Sinking of USS *Aludra* and USS *Deimos* and The Rescue Operations, during early morning of 23 June 1943, 1 July 1943.
[41] USS *O'Bannon* War Diary, June 1943.
[42] Commander, Transport Group, South Pacific, Report of Submarine Attack on Task Unit 32.4.4 on 23 June 1943, resulting in the loss of the USS *Aludra* and USS *Deimos*, 31 July 1943; Commanding Officer, USS *Skylark*, Submarine Attack on Task Unit 32.4.4, 18 July 1943.
[43] "*Skylark* (AM 63)" (http://www.navsource.org/archives/11/02063.htm: accessed 23 November 2020).
[44] "IJN Submarine *RO-103*: Tabular Record of Movement" (http://www.combinedfleet.com/RO-103.htm: accessed 21 November 2020); Commander, Transport Group, South Pacific, Report of Submarine Attack on Task Unit 32.4.4 on 23 June 1943, resulting in the loss of the USS *Aludra* and USS *Deimos*, 31 July 1943.
[45] Commander, Transport Group, South Pacific, Report of Submarine Attack on Task Unit 32.4.4 on 23 June 1943, resulting in the loss of the USS *Aludra* and USS *Deimos*, 31 July 1943.
[46] Commanding Officer, USS *Deimos*, Torpedoing of the USS *Deimos* (AK78), 27 June 1943; *Aludra*, *DANFS*.

[47] Commanding Officer, USS *Deimos*, Torpedoing of the USS *Deimos* (AK78), 27 June 1943.
[48] Ibid.

## CHAPTER 9 NOTES:

[1] "Milingimbi" (https://www.pastmasters.net/milingimbi.html: accessed 10 October 2020).
[2] Anthony Cooper, *Darwin Spitfires: The Real Battle for Australia* (Barnsley, UK: Pen & Sword Aviation, 2013).
[3] Ibid.
[4] Ibid.
[5] "Maroubra HMAS p.73" (https://www.navy.gov.au/sites/default/files/documents/MANUNDA_TO_MAYO_BROTHERS.pdf): accessed 10 October 2020).
[6] "Australian Naval History on 21 September 1942" (https://www.navyhistory.org.au/21-september-1942/);
[7] Cooper, *Darwin Spitfires: The Real Battle for Australia.*
[8] Ibid.

## CHAPTER 10 NOTES:

[1] "H-016-2: The Aleutians Campaign, 1942–43" (https://www.history.navy.mil/content/history/nhhc/about-us/leadership/director/directors-corner/h-grams/h-gram-016/h-016-2.html: accessed 24 November 2020).
[2] *ICE CAP NEWS*, Vol. 56 No. 3 July 2011; *Building the Navy's Bases in World War II: History of the Bureau of Yards and Docks and the Civil Engineer Corps 1940-1946 Volume II*, 184.
[3] *Building the Navy's Bases in World War II: History of the Bureau of Yards and Docks and the Civil Engineer Corps 1940-1946 Volume II*, 187.
[4] Ibid.
[5] *Building the Navy's Bases in World War II: History of the Bureau of Yards and Docks and the Civil Engineer Corps 1940-1946 Volume II*, 187; "Battle for the Aleutians: WWII's Forgotten Alaskan Campaign" (https://www.history.com/news/battle-for-the-aleutians-wwiis-forgotten-alaskan-campaign: accessed 26 November 2020).
[6] *Spica*, *DANFS*.
[7] *Spica*, *DANFS*; "Interview with Charles Donovan Aleutian World War II National Historic Area Oral History Program January 4, 2010" (https://www.nps.gov/aleu/learn/photosmultimedia/upload/Donovan-transcript-no-images-or-audio-508.pdf: 24 November 2020).
[8] USS *Spica* War Diary, May 1943.
[9] Ibid.
[10] *Building the Navy's Bases in World War II: History of the Bureau of Yards and Docks and the Civil Engineer Corps 1940-1946 Volume II*, 187.
[11] USS *Spica* War Diary, May 1943.
[12] Ibid.

[13] USS *Charleston*, USS *Phelps* War Diary, May 1943.
[14] Commander North Pacific Force, Report of Anti-Aircraft Action, 26 June 1943.
[15] Ibid.
[16] "H-016-2: The Aleutians Campaign, 1942–43"; "Battle for the Aleutians: WWII's Forgotten Alaskan Campaign"; "The Battle of Attu: 60 Years Later" (https://www.nps.gov/articles/battle-of-attu.htm: accessed 26 November 2020).
[17] *Spica, DANFS.*
[18] *Building the Navy's Bases in World War II: History of the Bureau of Yards and Docks and the Civil Engineer Corps 1940-1946 Volume II*, 187.
[19] *Combat Narratives The Aleutians Campaign June 1942-August 1943*, 100.
[20] "Hong Kong" (http://www.canadiansoldiers.com/history/battlehonours/waragainstjapan/Hong%20Kong.htm: accessed 30 November 2020).
[21] "Operation Cottage" (http://www.canadiansoldiers.com/history/operations/operationcottage.htm : accessed 30 November 2020).
[22] "Operation Cottage"; "History of the First Special Service Force" (http://www.firstspecialserviceforce.net/history.html: accessed 20 November 2020).
[23] "History of the First Special Service Force."
[24] *Combat Narratives The Aleutians Campaign June 1942-August 1943*, 102.
[25] Ibid, 102-103.
[26] "Operation Cottage."
[27] *Combat Narratives The Aleutians Campaign June 1942-August 1943*, 105.
[28] "The Kiska Air Battle" (https://www.junobeach.org/canada-in-wwii/articles/home-defence/the-kiska-air-battle/); "Timeline" (http://www.kadiak.org/timeline.html); "Squadron Leader Kenneth Arthur Boomer" (https://www.veterans.gc.ca/eng/remembrance/memorials/canadian-virtual-war-memorial/detail/2319633) accessed 2 December 2020..
[29] "Liberators' Legacy: Memorials to the Aleutians Air War" (https://www.historynet.com/liberators-legacy-memorials-to-the-aleutians-air-war.htm: accessed 2 December 2020).
[30] "The War Against Japan" (http://www.canadiansoldiers.com/history/campaigns/waragainstjapan/waragainstjapan.htm: accessed 30 November 2020).
[31] "No. 1 Special Wireless Group Royal Canadian Corps of Signals (RCCS) in Australia during WW2" (https://www.ozatwar.com/sigint/1swg.htm); "Chinese Canadians of Force 136" (https://www.thecanadianencyclopedia.ca/en/article/chinese-canadians-of-force-136): both accessed 30 November 2020).
[32] George Duddy correspondence of 8 December 2020.

## CHAPTER 11 NOTES:

[1] "H-020-2: Central Solomon Islands Campaign: Kula Gulf, Kolombangara, Vella Gulf, PT-109, and Battles with No Names (Not High-Velocity Learning)" (https://www.history.navy.mil/content/history/nhhc/about-us/leadership/director/directors-corner/h-grams/h-gram-020/h-020-2-central-solomons-campaign.html: accessed 27 November 2020); Samuel Eliot Morison, *The Two Ocean War*, 248.
[2] Ibid.
[3] *Combat Narratives Solomon Islands Campaign X, Operations in the New Georgia Area 21 June-5 August 1943*, 1.
[4] Ibid, 1-2.
[5] Bruhn, *MacArthur and Halsey's "Pacific Island Hoppers,"* 50.
[6] *Combat Narratives Solomon Islands Campaign X, Operations in the New Georgia Area 21 June-5 August 1943*, 4-5.
[7] Ibid, 5-6.
[8] Ibid, 9.
[9] Ibid, 9-10.
[10] Ibid, 11-12.
[11] Ibid, 11.
[12] Ibid.
[13] Ibid, 11-12.
[14] Ibid, 12.
[15[ Ibid.
[16] Ibid.
[17] Ibid.
[18] Commanding Officer, USS *McCawley*, Action report; seizure and occupation of Rendova, torpedo plane, dive bombing, and submarine attack on USS *McCawley*, 4 July 1953.
[19] Commanding Officer, USS *McCawley*, Action report; seizure and occupation of Rendova, torpedo plane, dive bombing, and submarine attack on USS *McCawley*, 4 July 1953; USS *Libra* War Diary, June 1943.
[20] Commanding Officer, USS *McCawley*, Action report; seizure and occupation of Rendova, torpedo plane, dive bombing, and submarine attack on USS *McCawley*, 4 July 1953; *Combat Narratives Solomon Islands Campaign X, Operations in the New Georgia Area 21 June-5 August 1943*, 13.
[21] George Carroll Dyer, *The Amphibians Came to Conquer: The Story of Admiral Richmond Kelly Turner* (Washington, DC: U.S. Department of the Navy, 1972), 561-562; Commanding Officer, USS *McCawley*, Action report; seizure and occupation of Rendova, torpedo plane, dive bombing, and submarine attack on USS *McCawley*, 4 July 1953; *Combat Narratives Solomon Islands Campaign X, Operations in the New Georgia Area 21 June-5 August 1943*, 13.
[22] "U.S. Pacific Fleet Third Amphibious Force Command History, October 1943" (https://apps.dtic.mil/dtic/tr/fulltext/u2/a637630.pdf: accessed 29 November 2020).
[23] Ibid.

## CHAPTER 12 NOTES:

[1] Bruhn, *Battle Stars for the "Cactus Navy,"* 219.
[2] Ibid, 219-220.
[3] Ibid, 220.
[4] "Battle of Empress Augusta Bay 1-2 November 1943" (https://www.history.navy.mil/browse-by-topic/wars-conflicts-and-operations/world-war-ii/1943/beyond-guadalcanal/empress-augusta-bay.html: accessed 3 December 2020).
[5] Ibid.
[6] Ibid.
[7] *Combat Narratives Solomon Islands Campaign: XII The Bougainville Landing and the Battle of Empress Augusta Bay 27 October- 1 November 1943.*
[8] *Alchiba, DANFS.*
[9] *Alhena, DANFS.*
[10] *Combat Narratives Solomon Islands Campaign: XII The Bougainville Landing and the Battle of Empress Augusta Bay 27 October- 1 November 1943.*
[11] Ibid.
[12] Ibid.
[13] Ibid.
[14] Commander Transport Group, Third Amphibious Force, Report of Landing Operations in Northern Empress Augusta Bay Area, Bougainville Island, 1-2 November 1943, 22 December 1943.
[15] Ibid.
[16] Ibid.
[17] Commanding Officer, USS *Titania*, Report of Operation [name redacted], 2 November 1943.
[18] Commander Transport Group, Third Amphibious Force, Report of Landing Operations in Northern Empress Augusta Bay Area, Bougainville Island, 1-2 November 1943, 22 December 1943.
[19] Ibid.
[20] Ibid.
[21] Ibid.
[22] Ibid.

## CHAPTER 13 NOTES:

[1] Commander in Chief, U.S. Pacific Fleet, Ship's history – Forwarding of, 27 November 1945.
[2] David D. Bruhn, *Eyes of the Fleet: The U.S. Navy's Seaplane Tenders and Patrol Aircraft in World War II*, 235.
[3] Ibid, 236.
[4] Ibid, 237.
[5] Commander in Chief, U.S. Pacific Fleet and Pacific Ocean Areas, Operations in Pacific Ocean Areas – November 1943, 18 February 1944.
[6] Ibid.
[7] Ibid.
[8] Ibid.
[9] Ibid.

[10] Commander in Chief, U.S. Pacific Fleet, Ship's history – forwarding of, 27 November 1945.
[11] Ibid.
[12] Ibid.
[13] Commander in Chief, U.S. Pacific Fleet and Pacific Ocean Areas, Operations in Pacific Ocean Areas – November 1943, 18 February 1944; Commander in Chief, U.S. Pacific Fleet, Ship's history – Forwarding of, 27 November 1945; Dyer, *The Amphibians Came to Conquer: The Story of Admiral Richmond Kelly Turner*, 692.
[14] Commander in Chief, U.S. Pacific Fleet and Pacific Ocean Areas, Operations in Pacific Ocean Areas – November 1943, 18 February 1944.
[15] Dyer, *The Amphibians Came to Conquer: The Story of Admiral Richmond Kelly Turner*, 692-693.
[16] Commander Transports Division Four, Action Report – Tarawa Atoll, Gilbert Islands, 4 December 1943; Commanding Officer, USS *Thuban*, Action; report of – Tarawa Atoll, Gilbert Islands, 6 December 1943.
[17] Commander in Chief, U.S. Pacific Fleet and Pacific Ocean Areas, Operations in Pacific Ocean Areas – November 1943, 18 February 1944.
[18] Ibid.
[19] Commander in Chief, U.S. Pacific Fleet, Ship's history – forwarding of, 27 November 1945.
[20] Ibid.
[21] Commander in Chief, U.S. Pacific Fleet and Pacific Ocean Areas, Operations in Pacific Ocean Areas – November 1943, 18 February 1944.
[22] Ibid.
[23] USS *Alcyone* War Diary, November 1943.
[24] Bruhn, *"Battle Stars for the "Cactus Navy,"* 244.

## CHAPTER 14 NOTES:

[1] Commander Southern Transport Group, Report on Operations of the Southern Transport Group during Kwajalein Capture and early occupation, 15 February 1944.
[2] Commander in Chief, U.S. Pacific Fleet and Pacific Ocean Areas, Operations in the Pacific Ocean Areas – February 1944, 1 June 1944; "Marshall Islands" (https://www.britannica.com/place/Marshall-Islands: accessed 22 December 2020).
[3] Commander in Chief, U.S. Pacific Fleet and Pacific Ocean Areas, Operations in the Pacific Ocean Areas – February 1944, 1 June 1944.
[4] Henry I. Shaw, Jr., Bernard C. Nalty, and Edwin T. Turnbladh, *History of U.S. Marine Corps Operations in World War II Volume III: Central Pacific Drive* (Washington, DC: Headquarters, U.S. Marine Corps, 1966), 125-127; Burton Wright III, *Eastern Mandates*, U.S. Army Center of Military History brochure.
[5] Commander in Chief, U.S. Pacific Fleet and Pacific Ocean Areas, Operations in the Pacific Ocean Areas – February 1944, 1 June 1944.
[6] Commander in Chief, U.S. Pacific Fleet and Pacific Ocean Areas, Operations in the Pacific Ocean Areas – February 1944, 1 June 1944;

Commanding Officer, USS *Electra*, Ship's History – forwarding of, 14 October 1945.
[7] Commander in Chief, U.S. Pacific Fleet and Pacific Ocean Areas, Operations in the Pacific Ocean Areas – February 1944, 1 June 1944.
[8] "H-Gram 026: Operations Flintlock, Catchpole, and Hailstone" (https://www.history.navy.mil/content/history/nhhc/about-us/leadership/director/directors-corner/h-grams/h-gram-026.html: accessed 17 December 2020).
[9] Bruhn, *Eyes of the Fleet*, 247-248.
[10] "United States Maritime Commission C2 Type Ships" (http://www.usmm.org/c2ships.html); "Ships of the U.S. Navy, 1940-1945 Maritime Commission Types" (https://www.ibiblio.org/hyperwar/USN/ships/ships-mc.html): both accessed 17 December 2020).
[11] Bruhn, *Eyes of the Fleet*, 249.
[12] Ibid, 250.
[13] Dyer, *The Amphibians Came to Conquer: The Story of Admiral Richmond Kelly Turner*, 829.
[14] Secret Information Bulletin No. 17, Battle Experience: Supporting Operations for the Occupation of the Marshall Islands including the Westernmost Atoll, Eniwetok, February, 1944, Headquarters of the Commander in Chief, United States Fleet.
[15] Dyer, *The Amphibians Came to Conquer: The Story of Admiral Richmond Kelly Turner*, 828-829.
[16] Bruhn, *Eyes of the Fleet*, 250.
[17] Ibid.
[18] Bruhn, *Eyes of the Fleet*, 250; "H-Gram 026: Operations Flintlock, Catchpole, and Hailstone."
[19] Ibid, 250.

## CHAPTER 15 NOTES:

[1] Commander Attack Group, Admiralty Islands Operation, 29 February 1944 – Report on, dated 16 March 1944; Samuel Eliot Morison, *History of United States Naval Operation in World War II: Breaking the Bismarcks Barrier, 22 July 1942-1 May 1944*, 435.
[2] Bruhn, *MacArthur and Halsey's "Pacific Island Hoppers,"* 145-146.
[3] Bruhn, *MacArthur and Halsey's "Pacific Island Hoppers,"* 146-147; Morison, *The Two-Ocean War*, 294.
[4] Commander Task Force 77 Operation Plan No. 3-44, 3 April 1944.
[5] "RAN Beach Commandos" by Petar Djokovic (https://www.navy.gov.au/history/feature-histories/ran-beach-commandos: accessed 19 December 2020).
[6] Ibid.
[7] "Occasional Paper 77: HMAS Assault. WWII Combined Operations Directorate Establishment – Port Stephens NSW" by Dennis J Weatherall (navyhistory.orgau: accessed 19 December 2020).

[8] "Captain Allan Paterson Cousin" (https://www.navy.gov.au/biography/captain-allan-paterson-cousin#:~:text=Allan%20Paterson%20Cousin%20(1900%2D1976,born%20wife%20Jane%2C%20n%C3%A9e%20McLean: accessed 19 December 2020).
[9] Bruhn, *MacArthur and Halsey's "Pacific Island Hoppers,"* 149-150.
[10] Ibid, p. 150-151.

## CHAPTER 16 NOTES:

[1] Commander in Chief, U.S. Pacific Fleet and Pacific Ocean Areas, Operations in the Pacific Ocean Area – June 1944, 7 November 1944; "United States Navy at War, Second Official Report to the Secretary of the Navy Covering Combat Operation March 1, 1944, to March 1, 1945" by Fleet Admiral Ernest J. King (https://www.history.navy.mil/content/history/nhhc/research/library/online-reading-room/title-list-alphabetically/u/us-navy-at-war-second-official-report.html: accessed 21 December 2020).
[2] Ut supra.
[3] "United States Navy at War, Second Official Report to the Secretary of the Navy Covering Combat Operation March 1, 1944, to March 1, 1945."
[4] Commander in Chief, U.S. Pacific Fleet and Pacific Ocean Areas, Operations in the Pacific Ocean Area – June 1944, 7 November 1944.
[5] Commander in Chief, U.S. Pacific Fleet and Pacific Ocean Areas, Operations in the Pacific Ocean Area – June 1944, 7 November 1944; Commander Transports, Group Three, Fifth Amphibious Force, Pacific Fleet, Report of Guam Operation, 1 September 1944; Commander Task Group Fifty-Three Point Two, Report of operations of Task Group 53.2 covering the periods preparatory to, during and subsequent to the assault and capture of Guam, 29 August 1944.
[6] Commander Transports, Group Three, Fifth Amphibious Force, Pacific Fleet, Report of Guam Operation, 1 September 1944.
[7] USS *Mercury* War Diary, June 1944; Commanding Officer, USS *Mercury*, Ship's History, 17 April 1946.
[8] Ut supra.
[9] USS *Mercury* War Diary, June 1944.
[10] Ibid.
[11] Ibid.
[12] Commander in Chief, U.S. Pacific Fleet and Pacific Ocean Areas, Operations in the Pacific Ocean Area – July 1944, 22 December 1944.
[13] Ibid.
[14] Ibid.
[15] Ibid.
[16] Ibid.
[17] Ibid.

## CHAPTER 17 NOTES:

[1] Commander Angaur Assault Unit Baker, Action Report – Assault Landing on Angaur, 1 October 1944.
[2] Commander in Chief, U.S. Pacific Fleet and Pacific Ocean Areas, Operations in the Pacific Ocean Areas – September 1944, 7 March 1945.
[3] Ibid.
[4] Ibid.
[5] Morison, *The Two-Ocean War*, 326; Robert Cressman, *The Official Chronology of the US Navy in World War II* (Annapolis, MD: Naval Institute, 2016).
[6] Commander Task Force 32, Amphibious Operation to Capture Peleliu and Angaur – Report on, 16 October 1944; Commander in Chief, U.S. Pacific Fleet and Pacific Ocean Areas, Operations in the Pacific Ocean Areas – September 1944, 7 March 1945.
[7] Commander Task Force 32, Amphibious Operation to Capture Peleliu and Angaur – Report on, 16 October 1944.
[8] Ibid.
[9] Ibid.
[10] Commander Task Force 32, Amphibious Operation to Capture Peleliu and Angaur – Report on, 16 October 1944; "The Battle of Angaur" (https://www.stamfordhistory.org/ww2_angaur.htm#:~:text=Angaur%20was%20the%20base%20of,incapacitated%20through%20accident%20or%20sickness: accessed 23 December 2020).
[11] Commander Task Force 32, Amphibious Operation to Capture Peleliu and Angaur – Report on, 16 October 1944; Morrison, *The Two-Ocean War*, 427; Arnold S. Lott, *Most Dangerous Sea*, 178-179.

## CHAPTER 18 NOTES:

[1] Bruhn, *Eyes of the Fleet*, 329.
[2] Ibid.
[3] Ibid, 330.
[4] Commander Task Group 79.2, Leyte Operation – Report of, 4 November 1944.
[5] Commander in Chief, U.S. Pacific Fleet and Pacific Ocean Areas, Operations in the Pacific Ocean Areas – October 1944, 31 May 1945.
[6] Ibid.
[7] Bruhn, *Eyes of the Fleet*, 330.
[8] Ibid.
[9] Commander Task Force 78 Operation Plan No. 101-44.
[10] Commander Task Group 79.1, Report of Task Group 79.1 participation in the Amphibious Operations for the capture of Leyte, P.I., 26 October 1944.
[11] Commander Task Force 79.2, Leyte Operation – Report of, 4 November 1944.
[12] Commander Task Group 78.3. Task Group 78.3 Actions Reports of Panaon Operation – forwarding of, 3 December 1944.
[13] Ibid.
[14] HMAS *Manoora* Action Report, SC192/2/0699.

[15] Ibid.
[16] Ibid.
[17] Commander in Chief, U.S. Pacific Fleet and Pacific Ocean Areas, Operations in the Pacific Ocean Areas – October 1944, 31 May 1945; *Honolulu*, *Sonoma*, and *Fremont*, *DANFS*.
[18] "HMAS *Australia* (II)" (https://www.navy.gov.au/hmas-australia-ii: accessed 29 December 2020).
[19] USS *Auriga* War Diary, October 1944.
[20] Ibid.
[21] Ibid.
[22] Ibid.
[23] Ibid.
[24] Ibid.
[25] Ibid.
[26] Ibid.
[27] Ibid.
[28] USS *Auriga* War Diary, October 1944; Register of the Commissioned Officers, Cadets, Midshipmen, and Warrant Officers of the United States Naval Reserve July 1, 1941 (https://www.ibiblio.org/hyperwar/USN/ref/BuPers/USNR-1941.pdf: accessed 30 December 2020).
[29] USS *Auriga* War Diary, October 1944.
[30] Ibid.
[31] USS *Auriga* War Diary, October 1944; Senior Medical Officer, USS *Mount Olympus*, Casualty and Casualty Evacuation Report, 31 October 1944.
[32] USS *Auriga* War Diary, October 1944; Register of the Commissioned Officers, Cadets, Midshipmen, and Warrant Officers of the United States Naval Reserve July 1, 1941.
[33] Public Information Officer, News Story, USS *Hyperion* (AK-107), 6 October 1945.

## CHAPTER 19 NOTES:

[1] Commander in Chief, U.S. Pacific Fleet and Pacific Ocean Areas, Operations in Pacific Ocean Areas – January 1945, 31 July 1945.
[2] Ibid.
[3] Commander Luzon Attack Force, Action Report – Luzon Attack Force, Lingayen Gulf – Musketeer Mike One Operation, 15 May 1945; Commander Task Unit 79.3.2, Operation and Action Report of Amphibious Assault on Lingayen Area, Luzon, Philippines, 12 January 1945.
[4] USS *Aquarius* War Diary, January 1945.
[5] USS *Aquarius* War Diary, January 1945; *DuPage*, *DANFS*.
[6] Ship's History USS *Almaack* (AKA-10).
[7] Commander Task Unit 79.3.2, Operation and Action Report of Amphibious Assault on Lingayen Area, Luzon, Philippines, 12 January 1945; Ship's History USS *Almaack* (AKA-10).

[8] Commander Task Unit 79.3.2, Operation and Action Report of Amphibious Assault on Lingayen Area, Luzon, Philippines, 12 January 1945; Ship's History USS *Almaack* (AKA-10); Commanding Officer, History of the USS *Alshain*, Submission of, 26 October 1945.
[9] Commander Task Unit 79.3.2, Operation and Action Report of Amphibious Assault on Lingayen Area, Luzon, Philippines, 12 January 1945.
[10] Commander Task Unit 79.3.2, Operation and Action Report of Amphibious Assault on Lingayen Area, Luzon, Philippines, 12 January 1945; Commanding Officer, History of the USS *Alshain*, Submission of, 26 October 1945.
[11] Factual History of the USS *Chara* (AKA-58), January 1944 - December 1945.
[12] Factual History of the USS *Chara* (AKA-58), January 1944 - December 1945; Ship's History USS *Almaack* (AKA-10); *Columbia*, *DANFS*.
[13] Factual History of the USS *Chara* (AKA-58), January 1944 - December 1945; Commanding Officer, History of the USS *Alshain*, Submission of, 26 October 1945; Ship's History USS *Almaack* (AKA-10).
[14] Ship's History USS *Almaack* (AKA-10).
[15] Ship's History USS *Almaack* (AKA-10); *War Hawk*, *DANFS*.
[16] Commanding Officer, History of the USS *Alshain*, Submission of, 26 October 1945.
[17] Commanding Officer, USS *Alcyone*, History of the USS *Alycone* (AKA-7), submission of, 21 October 1945.
[18] Commander Luzon Attack Force, Action Report – Luzon Attack Force, Lingayen Gulf – Musketeer Mike One Operation, 15 May 1945; *DuPage*, *Zeilin*, *DANFS*.

## CHAPTER 20 NOTES:

[1] "The Long Blue Line: USS *Serpens*—the Coast Guard's greatest loss" by William H. Thiesen (https://coastguard.dodlive.mil/2018/01/the-long-blue-line-uss-serpens-the-coast-guards-greatest-loss/: accessed 1 January 2021).
[2] "H-039-5: The Explosion of Ammunition Ship USS *Mount Hood* (AE-11), 10 November 1944" (https://www.history.navy.mil/content/history/nhhc/about-us/leadership/director/directors-corner/h-grams/h-gram-039/h-039-5.html: accessed 1 January 2021); *Serpens*, *DANFS*.
[3] "The Coast Guard During World War II" (https://www.nps.gov/parkhistory/online_books/npswapa/extContent/wapa/coast_guard/cg1.htm: accessed 2 January 2021).
[4] "The Sea-Going Moving Van" by John C. Cole (http://www.ak97.org/documents/ak_story.pdf: accessed 2 January 2021).
[5] Ibid.
[6] Ibid.
[7] Ibid.
[8] Ibid.
[9] Ibid.

[10] USS *Serpens* War Diary, December 1944.
[11] USS *Serpens* (AK-97) Ship's History, Public Information Division, U.S. Coast Guard Headquarters, Washington, DC; ComSoPac Dispatch, 30 January 1945; Commanding Officer, USS *Serpens* (AK-97), Official Report of Loss of Ship, 31 January 1945 (http://www.ak97.org/official.html: accessed 2 January 2021); Commander Naval Bases, South Solomons Sub-Area, USS *Serpens* Loss of – Additional Bomb Data, 21 February 1945.
[12] ComSoPac Dispatch, 30 January 1945; "H-039-5: The Explosion of Ammunition Ship USS Mount Hood (AE-11), 10 November 1944" (https://www.history.navy.mil/content/history/nhhc/about-us/leadership/director/directors-corner/h-grams/h-gram-039/h-039-5.html: accessed 5 December 2020).
[13] Commanding Officer, USS Serpens (AK-97), Official Report of Loss of Ship, 31 January 1945.
[14] Ibid.
[15] Ibid.
[16] "Alone on a wide, wide sea" by Donna Alvis-Banks, *The Roanoke Times*, Sunday, May 29, 2005 (http://www.ak97.org/survivor.html: accessed 1 January 2021.)
[17] Ibid.
[18] "The Long Blue Line: USS *Serpens*—the Coast Guard's greatest loss" by William H. Thiesen.
[19] "USS *Serpens* AK 97 Commemorative Website - Personal Documents Page" (http://www.ak97.org/personal.html: accessed 2 January 2021).

## CHAPTER 21 NOTES:

[1] Bruhn, *Eyes of the Fleet*, 353.
[2] Ibid.
[3] *U.S. Naval Aviation in the Pacific* (Washington, DC: Office of the Chief of Naval Operations, 1947), 39.
[4] David D. Bruhn, Rob Hoole, *Nightraiders: U.S. Navy, Royal Navy, Royal Australian Navy, and Royal Netherlands Navy Mine Forces Battling the Japanese in the Pacific in World War II* , 241.
[5] Ibid, 242.
[6] Commander in Chief, U.S. Pacific Fleet and Pacific Fleet Areas, Operations in the Pacific Fleet Areas – February 1945, 27 August 1945.
[7] Commander Task Group 53.2, Iwo Jima Operation – Report on, 13 April 1945; Commander Transport Squadron Sixteen, Action Report – Iwo Jima; Period 11 January to 28 February 1945, 15 April 1945.
[8] Commander Transport Squadron Sixteen, Action Report – Iwo Jima; Period 11 January to 28 February 1945, 15 April 1945.
[9] Commander Task Group 53.2, Iwo Jima Operation – Report on, 13 April 1945.
[10] Commander in Chief, U.S. Pacific Fleet and Pacific Ocean Areas, Operations in Pacific Ocean Areas – February 1945, 27 August 1945.

## CHAPTER 22 NOTES:

[1] Commander in Chief, U.S. Pacific Fleet and Pacific Ocean Areas, Operations in the Pacific Ocean Areas – April 1945, 16 October 1945.
[2] Ibid.
[3] Ibid.
[4] Commander Amphibious Forces, U.S. Pacific Fleet, General Action Report, Capture of Okinawa Gunto, Phases I and II, 17 February 1945 to 17 May 1945 – Submission of, 25 July 1945.
[5] Commander Amphibious Group Seven, Action Report – Capture of Okinawa Gunto, Phases 1 and 2, 26 May 1945.
[6] Ibid.
[7] Ibid.
[8] Commander Transport Division Fifty-One, Action Report, Capture of Okinawa Gunto – Phase One and Two, 19 May 1945.
[9] Ibid.
[10] Commander Amphibious Group Seven, Action Report – Capture of Okinawa Gunto, Phases 1 and 2, 26 May 1945; Commander Transport Division Fifty-One, Action Report, Capture of Okinawa Gunto – Phase One and Two, 19 May 1945.
[11] Ut supra.
[12] Commander in Chief, U.S. Pacific Fleet and Pacific Ocean Areas, Operations in the Pacific Ocean Areas – April 1945, 16 October 1945.
[13] Ibid.
[14] Ibid.
[15] Ibid.
[16] Ibid.
[17] Ibid.
[18] Morison, *The Two Ocean-War*, 531-532.
[19] Commander Fifth Fleet, Action Report, Ryukyus Operation through 27 May 1945, 21 June 1945.
[20] Commander Amphibious Forces, U.S. Pacific Fleet, General Action Report, Capture of Okinawa Gunto, Phases I and II, 17 February 1945 to 17 May 1945 – Submission of, 25 July 1945.
[21] Commanding Officer, USS *Terrell*, History of the USS *Terrell*, 24 October 1945.
[22] USS *Achernar* War Diary, April 1945.
[23] Ibid.
[24] Ibid.
[25] Commanding Officer, USS *Starr*, Material, Public Information, 7 December 1945.
[26] Ibid.
[27] Ibid.
[28] Ibid.
[29] Commander Fifth Fleet, Action Report, Ryukyus Operation through 27 May 1945, 21 June 1945.

[30] "The British Pacific Fleet Task Force 57 Politics & Logistics: Sakishima Gunto, Okinawa Campaign, 1945" (http://www.armouredcarriers.com/task-force-57-british-pacific-fleet: accessed 13 January 2020); "Operation Meridian" (https://www.armouredcarriers.com/operation-meridian-the-palembang-strike: accessed 4 June 2020).
[31] Commander Fifth Fleet, Action Report, Ryukyus Operation through 27 May 1945, 21 June 1945.
[32] Ibid.
[33] Ibid.
[34] "Battle of Okinawa 1 April–22 June 1945" (https://www.history.navy.mil/content/history/nhhc/browse-by-topic/wars-conflicts-and-operations/world-war-ii/1945/battle-of-okinawa.html: accessed 6 January 2021); Morison, *The Two-Ocean War*, 556.
[35] Commander Amphibious Group Five, Action Report – The Occupation of Okinawa, 16 March to 14 April 1945, 15 April 1945.
[36] Ibid.
[37] Commander Amphibious Forces, U.S. Pacific Fleet, General Action Report, Capture of Okinawa Gunto, Phases I and II, 17 February 1945 to 17 May 1945 – Submission of, 25 July 1945.
[38] Commander Transport Squadron Sixteen, Action Report – Okinawa, Phase I; Period 9 April to 16 April 1945, 30 April 1945.

## CHAPTER 23 NOTES:

[1] "The Short but Brilliant Life of the British Pacific Fleet" by Nicholas E. Sarantakes (http://www.sarantakes.com/PDF%20Other%20Writings/B7-JFQ.pdf: accessed 20 October 2020).
[2] Waters, *The Royal New Zealand Navy*, 368.
[3] Ibid, 368-369.
[4] "Fact File: Second Quebec Conference 12-16 September 1944" (http://www.bbc.co.uk/history/ww2peopleswar/timeline/factfiles/nonflash/a1144838.shtml); "4-506 To Henry L. Stimson, September 13, 1944" (https://www.marshallfoundation.org/library/digital-archive/to-henry-l-stimson-4/): accessed 20 October 2020.
[5] "4-506 To Henry L. Stimson, September 13, 1944."
[6] Waters, *The Royal New Zealand Navy*, 369.
[7] Ibid, 371.
[8] "The British Pacific and East Indies Fleets: The forgotten fleets that fought the Japanese in the Pacific and Indian Oceans" (http://www.royalnavyresearcharchive.org.uk/BPF-EIF/Ships/PEPYS.htm#.X4zQie1lCM8: accessed 19 October 2020).
[9] Ibid.
[10] "Fleet Train (British Pacific Fleet)" (https://www.navalhistoryarchive.org/index.php/Fleet_Train_(British_Pacific_Fleet: accessed 19 October 2020).
[11] Benjamin Jones, "Ashore, Afloat and Airborne: The Logistics of British Naval Airpower, 1914-1945," Thesis submitted for the Degree of Doctor of

Philosophy, Kings College, London, 2007, 231 (https://kclpure.kcl.ac.uk/portal/files/2935516/498772.pdf: accessed: 18 October 2020).
[12] "Fleet Train (British Pacific Fleet)."
[13] Jones, "Ashore, Afloat and Airborne: The Logistics of British Naval Airpower, 1914-1945,"230-232.
[14] Ibid, 232.
[15] Ibid, 237-238.
[16] "An interesting life at sea" by Gerry Baxter (http://www.historicalrfa.org/archived-stories68/1285-an-interesting-life-at-sea22: accessed 17 October 2020).
[17] "RFA Serbol" (http://www.historicalrfa.org/rfa-serbol-ships-details: accessed 21 October 2020).
[18] "An interesting life at sea" by Gerry Baxter.
[19] Ibid.
[20] Ibid.
[21] "RFA Serbol."
[22] Third Supplement to *The London Gazette* of Friday, the 31st of January, 1947, of 4 February, 1947.
[23] H. R. Dean, *The Royal New Zealand Air Force in South-East Asia 1941–42* (Wellington, NZ: Historical Publications Branch, 1952), 19 (http://nzetc.victoria.ac.nz/tm/scholarly/tei-WH2-2Epi-c8-WH2-2Epi-h.html: accessed 21 October 2020).
[24] Ibid, 23.
[25] Ibid.
[26] Ibid, 23-24.
[27] Ibid, 24.
[28] Ibid.
[29] Ibid.
[30] Ibid.
[31] Ibid.
[32] Ibid.
[33] Ibid, 25.
[34] Ibid.
[35] Ibid.
[36] Third Supplement to *The London Gazette* of Friday, the 31st of January, 1947, of 4 February, 1947.
[37] Ibid.
[38] Ibid.
[39] Ibid.
[40] Waters, *The Royal New Zealand Navy*, 401-402.
[41] "The British Pacific and East Indies Fleets: The forgotten fleets that fought the Japanese in the Pacific and Indian Oceans"; "Fleet Train (British Pacific Fleet)."
[42] Ibid.
[43] Ibid.

## CHAPTER 24 NOTES:

[1] Bruhn, *MacArthur and Halsey's 'Pacific Island Hoppers,'* 242.
[2] USS *Titania* War Diary, April 1945; "HMAS *Manoora* (I)" (https://www.navy.gov.au/hmas-manoora-i: accessed 24 December 2020).
[3] USS *Titania* War Diary, April 1945; Commanding Officer, HMAS *Manoora*, Action Report on Assault Landings at Tarakan Island on 1st May 1945, 4 May 1945.
[4] USS *Titania* War Diary, April 1945; "HMAS *Manoora* (I)"; Bruhn, *MacArthur and Halsey's 'Pacific Island Hoppers,'* 243.
[5] USS *Titania* War Diary, May 1945; Bruhn, *MacArthur and Halsey's 'Pacific Island Hoppers,'* 244.
[6] Bruhn, *MacArthur and Halsey's 'Pacific Island Hoppers,'* 244.
[7] USS *Titania* War Diary, May 1945.
[8] Bruhn, *MacArthur and Halsey's 'Pacific Island Hoppers,'* 245.
[9] Ibid.
[10] Ibid, 246.
[11] Commander Task Group 78.1, Action Report, CTG 78.1 (ComGruPhibSix) – Brunei Bay, Borneo Operations (10-17 June 1945), 19 June 1945.
[12] Ibid.
[13] Commander in Chief, U.S. Pacific Fleet and Pacific Ocean Areas, Operations in Pacific Ocean Areas – July 1945, 1 December 1945.
[14] Commander in Chief, U.S. Pacific Fleet and Pacific Ocean Areas, Operations in Pacific Ocean Areas – June 1945, 21 November 1945.
[15] Lott, *Most Dangerous Sea*, 158-161.
[16] Commander Task Group 78.2, Action Report – Balikpapan-Manggar-Borneo June 15-July 6, 1945, 14 August 1945.
[17] History of the USS *Poinsett* (AK-205).
[18] USS *Poinsett* War Diary, June-July 1945.
[19] Ibid.
[20] Ibid.

## APPENDIX A. NOTES:

[1] "Ships of the U.S. Navy, 1940-1945, Maritime Commission Types" (https://www.ibiblio.org/hyperwar/USN/ships/ships-mc.html); "Vela (AK-89)" (http://www.navsource.org/archives/09/13/130089.htm): both accessed 8 January 2021).

## APPENDIX B. NOTES:

[1] "Ships of the U.S. Navy, 1940-1945, AK – Cargo Ships" (https://www.ibiblio.org/hyperwar/USN/ships/ships-ak.html: accessed 8 January 2021).
[2] Ibid.
[3] Ibid.

[4] "Ships of the U.S. Navy, 1940-1945, AKA – Attack Cargo Ships" (https://www.ibiblio.org/hyperwar/USN/ships/ships-aka.html: accessed 8 January 2021).
[5] Ibid.
[6] Ibid.
[7] Ibid.

## APPENDIX C. NOTES

[1] "The 'Park' and 'Fort' ships" (https://www.pc.gc.ca/en/culture/historique-historic/mer-port/navires-vessels: accessed 1 December 2020).
[2] "Mariners" (http://www.mariners-l.co.uk/Park1.html: accessed 1 December 2020).
[3] "Introduction" (http://fortships.tripod.com/introduction.htm); "Fort Class Ships from WWII" (http://www.historicalrfa.org/ww2-fort-class): both accessed 1 December 2020).
[4] "Ocean Ship" (https://www.wikiwand.com/en/Ocean_ship: accessed 1 December 2020).
[5] "Ocean Ship; "Liberty Ships and Victory Ships, America's Lifeline in War (Teaching with Historic Places)" (https://www.nps.gov/articles/liberty-ships-and-victory-ships-america-s-lifeline-in-war-teaching-with-historic-places.htm#:~:text=The%20Victory%20ships%20were%20different,approximately%2018.5%20miles%20per%20hour): both accessed 1 December 2020).
[6] "Canadian Shipbuilders and Boatbuilders" (http://shipbuildinghistory.com/canada.htm#Active%20Shipbuilders: accessed 3 December 2020).
[7] "10,000 ton Victory-class ships [technical drawings]"; "The 'Park' and 'Fort' ships."
[8] "Introduction."
[9] "10,000 ton Victory-class ships [technical drawings]."
[10] Ibid.
[11] Ibid.

## APPENDIX D. NOTES:

[1] "The China Fleet – Small Ships WW2" (https://www.navyhistory.org.au/the-china-fleet-small-ships-ww2/3/: accessed 15 October 2020).

# Index

## About the Author

Commander David D. Bruhn, US Navy (Retired) served twenty-two years on active duty and two in the Naval Reserve, as both an enlisted man and as an officer, between 1977 and 2001.

After completion of basic training, he served as a sonar technician aboard USS *Miller* (FF-1091) and USS *Leftwich* (DD-984). He was commissioned in 1983 following graduation from California State University at Chico. His initial assignment was to USS *Excel* (MSO-439), serving as supply officer, damage control assistant, and chief engineer. He then served in USS *Thach* (FFG-43) as chief engineer and Destroyer Squadron Thirteen as material officer.

After graduation from the Naval Postgraduate School, Commander Bruhn was assigned to Secretary of the Navy and Chief of Naval Operations staffs as a budget analyst and resources planner before attending the Naval War College in 1996, following which he commanded the mine countermeasures ships USS *Gladiator* (MCM-11) and USS *Dextrous* (MCM-13) in the Persian Gulf.

Commander Bruhn's final assignment was executive assistant to a senior (SES 4) government service executive at the Ballistic Missile Defense Organization in Washington, D.C.

Following military service, he was a high school teacher and track coach for ten years, and is now a USA Track & Field official. He lives in northern California with his wife Nancy and has two grown sons, David and Michael.

www.ingramcontent.com/pod-product-compliance
Lightning Source LLC
LaVergne TN
LVHW020523100826
845148LV00010B/1324

* 9 7 8 1 5 5 6 1 3 8 2 0 1 *